# Buildings of Tomorrow: Goals and Challenges for Design and Operation of High-Performance Buildings

# Buildings of Tomorrow: Goals and Challenges for Design and Operation of High-Performance Buildings

Editors

**Mitja Košir**
**Manoj Kumar Singh**

MDPI • Basel • Beijing • Wuhan • Barcelona • Belgrade • Manchester • Tokyo • Cluj • Tianjin

*Editors*
Mitja Košir
University of Ljubljana
Ljubljana
Slovenia

Manoj Kumar Singh
Shiv Nadar University
Uttar Pradesh
India

*Editorial Office*
MDPI
St. Alban-Anlage 66
4052 Basel, Switzerland

This is a reprint of articles from the Special Issue published online in the open access journal *Sustainability* (ISSN 2071-1050) (available at: https://www.mdpi.com/journal/sustainability/special_issues/Buildingstomorrow_High-PerformanceBuildings).

For citation purposes, cite each article independently as indicated on the article page online and as indicated below:

LastName, A.A.; LastName, B.B.; LastName, C.C. Article Title. *Journal Name* **Year**, *Volume Number*, Page Range.

ISBN 978-3-0365-4881-4 (Hbk)
ISBN 978-3-0365-4882-1 (PDF)

Cover image courtesy of Manoj Kumar Singh

# Contents

# About the Editors

**Mitja Košir**

Dr. Mitja Košir is working as an associate professor at the University of Ljubljana, Faculty of Civil and Geodetic Engineering, Slovenia, where he is Head of the Chair of Buildings and Constructional Complexes. His primary research interest is building physics, focusing on building energy performance and daylighting in connection with the design of high-performance buildings. Over the last decade, he has been mainly involved in researching climate change impacts on buildings' past, present, and future energy performance and how to utilize bioclimatic adaptations to counter climate change impacts on building performance. He has authored over 80 publications in scientific journals, conferences, and books. He has actively taught at the Faculty of Civil and Geodetic Engineering and Faculty of Health Sciences, University of Ljubljana, Slovenia, where he lectures about bioclimatic design, building energy efficiency, daylighting, and smart buildings.

**Manoj Kumar Singh**

Dr. Manoj Kumar Singh is working as an assistant professor at Shiv Nadar University, India. He did his M.Tech. in Energy Technology from Tezpur University and PhD in Bio-Climatic Building Design from the Indian Institute of Technology, Delhi. His research interests are in the area of thermal comfort, bioclimatic building design and sustainability, building energy simulation, and energy performance of building envelopes. His research on thermal comfort is also included in ASHRAE Standard-55 2017. He is the recipient of the Ministry of New and Renewable Energy Fellowship, Belgium Government Postdoctoral Fellowship, and JSPS Postdoctoral Fellowship, Japan. He has more than 70 scientific publications in reputed international journals and conferences, with an h index of 26.

# Preface to "Buildings of Tomorrow: Goals and Challenges for Design and Operation of High-Performance Buildings"

In the last few decades, building design has been shifting toward higher energy efficiency and better performance. Although the main focus is usually on the reduction in energy use for the operation and construction of buildings, awareness of the benefits of higher occupant comfort and health has shifted the focus toward a more holistic treatment of building design. This notion was further emphasized during the last two and a half years. Firstly, this was due to the COVID-19 pandemic and the realization that the indoor environment is directly related to occupants' health, and secondly, the energy insecurity fueled by the Ukrainian war. Therefore, we have to realize that contemporary high-performance buildings will not only have to be energy-efficient but will also have to address synergetic interconnectedness between the indoor environment and user health and comfort, while at the same time being sustainable and resilient—a task that is not easily achieved, and which is further complicated by the issues of anthropogenically induced global warming that already necessitates the adaptation of buildings to the future climate during the design phase. With the exposed complexity and interconnectedness of parameters influencing the design of high-performance buildings, a crucial research question emerges: "how can we accomplish appropriate optimization among opposing and contrasting demands of different fields governing the design of high-performance buildings?" This question, of course, is not answered in the present reprint book of a Special Issue of the *Sustainability* journal. Nevertheless, the papers published in it represent essential contributions that broaden the knowledge in the field of architectural engineering and, as such, provide a small but valuable contribution to creating a sustainable and resilient built environment.

The content of the Special Issue and the present reprint book can be roughly divided into two parts. The first one includes papers primarily concerned with the functioning of the building and its components concerning energy use. In contrast, the second part addresses the occupant's comfort concerning the building. The book's first part consists of chapters 1 to 5 and covers some interesting aspects related to building design. **Chapter 1** deals with building envelope optimization, and the integration of passive-cooling measures in building design by adopting a building simulation approach. **Chapter 2** highlights the risks associated with buildings designed with the bioclimatic approach in the context of uncertain future climates. This chapter especially talks about the overheating problem in Central Europe's residential buildings. **Chapter 3** mentions retrofitting buildings with phase-change materials and aerogel to adapt the building to extreme heatwave conditions. It also reports that using the above materials significantly reduces energy use, peak cooling load, $CO_2$ emissions, and operational energy cost for a typical Australian house in the Melbourne climate. **Chapter 4** highlights the impact of the building shape factor on energy demand and $CO_2$ emission in the cold Oceanic climate of southern Chile. Through case studies, the authors concluded that a shape factor below 0.767 leads to a decrease in energy demand under the studied climate. **Chapter 5** addresses the issue of the urban heat island effect (UHI) and associated energy consumption in buildings. Through the paper, the authors conducted a systematic literature review of white roofing materials in emerging economies in the context of parameters such as energy performance cost–benefit, maintenance, and consumer indifference.

The second part of the book consists of chapters 6 to 9. An adaptive thermal comfort study in university hostel dormitories is presented in **Chapter 6** of the book. This chapter puts forth the characteristics of the subject's seasonal thermal perception and adaptive actions to restore comfort in the hostel dormitories of the composite climate of India. **Chapter 7** reflects the impact of the high-albedo materials used in the tall buildings on pedestrian streets in an urban environment. Authors in their study found that diffusely reflective façades did not increase the incident radiation at the pedestrian level by more than 30%. However, in the case of a specular reflective façade, the situation worsened due to an increase in incident radiation by 100% to 300% and should therefore be avoided. A student spends a considerable amount of time in education buildings during their education, starting from kindergarten to the university level. It is also evident from the published research that adequate thermal comfort impacts students' learning curve. **Chapter 8** of the book highlights the recent advancement in thermal comfort in educational buildings and the associated issues. Lastly, **Chapter 9**, through a literature review, addresses the parameters that affect thermal comfort and the instruments used in field surveys to record thermal-comfort parameters. This chapter emphasizes understanding occupants' behavior and individualized approaches.

Ultimately, we must acknowledge that the Special Issue and this reprint book would not exist without the authors' contributions. Therefore, we thank everyone for their valuable and interesting contributions that will undoubtedly increase our knowledge in the field of high-performance buildings. Of course, the Special Issue would never have materialized without the opportunity to edit it given to us by MDPI and the editorial board of the *Sustainability* journal, for which we are grateful. Lastly, we would like to extend our appreciation to our families, loved ones, and our current and past colleagues for their support and for their contributions, in some form, to the creation of the reprint book and Special Issue.

**Mitja Košir and Manoj Kumar Singh**
*Editors*

 *sustainability*

*Article*

# Multi-Objective Techno-Economic Optimization of Design Parameters for Residential Buildings in Different Climate Zones

**Muhammad Usman * and Georg Frey**

Automation and Energy Systems, Saarland University, 66123 Saarbrücken, Germany;
georg.frey@aut.uni-saarland.de
* Correspondence: muhammad.usman@aut.uni-saarland.de

**Abstract:** The comprehensive approach for a building envelope design involves building performance simulations, which are time-consuming and require knowledge of complicated processes. In addition, climate variation makes the selection of these parameters more complex. The paper aims to establish guidelines for determining a single-family household's unique optimal passive design in various climate zones worldwide. For this purpose, a bi-objective optimization is performed for twenty-four locations in twenty climates by coupling TRNSYS and a non-dominated sorting genetic algorithm (NSGA-III) using the Python program. The optimization process generates Pareto fronts of thermal load and investment cost to identify the optimum design options for the insulation level of the envelope, window aperture for passive cooling, window-to-wall ratio (WWR), shading fraction, radiation-based shading control, and building orientation. The goal is to find a feasible trade-off between thermal energy demand and the cost of thermal insulation. This is achieved using multi-criteria decision making (MCDM) through criteria importance using intercriteria correlation (CRITIC) and the technique for order preference by similarity to ideal solution (TOPSIS). The results demonstrate that an optimal envelope design remarkably improves the thermal load compared to the base case of previous envelope design practices. However, the weather conditions strongly influence the design parameters. The research findings set a benchmark for energy-efficient household envelopes in the investigated climates. The optimal solution sets also provide a criterion for selecting the ranges of envelope design parameters according to the space heating and cooling demands of the climate zone.

**Keywords:** residential building; building envelope; multi-objective genetic algorithm; TRNSYS; climate zone; multi-criteria decision making; CRITIC; TOPSIS

**Citation:** Usman, M.; Frey, G. Multi-Objective Techno-Economic Optimization of Design Parameters for Residential Buildings in Different Climate Zones. *Sustainability* **2022**, *14*, 65. https://doi.org/10.3390/su14010065

Academic Editors: Mitja Košir and Manoj Kumar Singh

Received: 25 November 2021
Accepted: 20 December 2021
Published: 22 December 2021

**Publisher's Note:** MDPI stays neutral with regard to jurisdictional claims in published maps and institutional affiliations.



## 1. Introduction

Energy consumption in buildings accounts for a major part of the worldwide final energy use. The building sector consumes 30% of global energy use and produces 28% of global $CO_2$ emissions. Residential buildings alone account for 22% of total energy use [1,2]. The household sector in Asia has the maximum share of 35% in global building energy consumption. The European housing sector comes second and is responsible for approximately 28% of global energy use in residential buildings. The projected growth in households' energy consumption is 1.4% per year from 2018 to 2050 [3]. The worldwide energy usage of residential buildings increased by 30% from 1990 to 2014. In emerging economies, this increment was more than 50% during that period [4]. This growing energy demand and associated $CO_2$ emissions have led to new design approaches for energy conservation in buildings. Subsequently, energy-efficient buildings are recognized as a sustainable solution in reducing energy consumption and $CO_2$ emissions.

The thermal performance of the building envelope determines the energy consumption for thermal comfort, which has an explicit impact on the overall energy demand of the buildings [5]. Therefore, employing an integrative approach in the architecture of building envelope at the preliminary design stage plays a critical role in improving the energy performance of buildings. Designers need comprehensive information of building performance and make decisions among a large number of design possibilities. It is asserted that operational cost, energy consumption, and overall performance of buildings are dependent on the early design approach [6]. In the preliminary design stage, the designer needs to consider building orientation, ventilation rate, air leakage, solar gain, window-to-wall ratio (WWR), window shading, and thermal mass, which impact the building performance collectively or independently [7,8]. Climate is another decisive element in the architectural design of buildings. The variation in climate conditions of different regions makes the design phase more diverse and complex. The design parameters' solution sets differ for each climate, and the energy-saving potential also changes likewise [9]. The heat gain from electrical appliances is also a critical parameter to evaluate thermal load in a simulated environment, particularly the cooling load. A realistic electric heat gain profile can help calculate the amount of heat to be removed from the building in summer [10] and avoid overheating in winter [11].

Although building envelope design is an imperative prospect in building performance, it is not an easy task due to the wide variety of passive measures. It is difficult to quantitatively analyze the building optimization problem using a traditional design approach. Researchers have coupled building simulation tools such as Ecotect, EnergyPlus, Doe-2, and TRNSYS (Transient Systems Simulation Program) with optimization algorithms to determine energy-efficient solutions [7,12]. Optimization algorithms explore design alternatives with desired outcomes, and make it possible to trade-off between objective functions concurrently with the provided constraints of the design variables [13]. Genetic algorithms are widely applied in building energy optimization problems [14–21]. A population-based metaheuristic genetic algorithm transforms the population under the explicit rule of survival of the fittest to reach a desired state of the objective functions. Genetic algorithms can deal with the non-linearity in the optimization of building performance, and they also explore the global optimum solution and do not limit to local optimal points [22].

Penna et al. [23] performed a three-objective optimization analysis on a single-family house in two different climate locations of Italy by using the Transient Systems Simulation Program (TRNSYS) and non-dominated sorting genetic algorithm (NSGA-II). The goal was to minimize the energy cost and discomfort hours, and maximize energy performance by insulation of the envelope, high performance window glazing, and replacement of HVAC equipment. An optimal cost solution could save more than 57% energy consumption for both locations. Rabani et al. [24] proposed an optimization scheme to automate the identification of the best-suited window configuration, envelope, shading system, and energy supply system on an office building in Norway. Ascione et al. [25] coupled Energy Plus with NSGA-II to optimize the architecture design of residential buildings in four cities of Spain (Mediterranean climate). Window dimensions, window shading, type of window glazing, and the type of walls and roofs were determined to trade-off between the heating and cooling demands of the building. Ferrara et al. [26] coupled the energy model of a nearly zero-energy French household with the acoustic model in MATLAB, and optimized the building through the particle swarm optimization (PSO) algorithm in GenOpt for energy, cost, and acoustic performance. A building simulation tool IDA-ICE was coupled with NSGA-II in GenOpt, and the findings of the study showed a decease up to 77% in energy use. The best performance was achieved with a shading device control based on solar radiation and indoor temperature. Chang et al. [27] developed a flexible multi-objective optimization framework to improve energy performance, thermal comfort, and reduce emissions and building costs while optimizing various envelope parameters. The framework was tested on four residential buildings in Tokyo to find out the retrofit area of the envelope for optimal performance.

Since multi-objective optimization produces a series of Pareto solutions, selecting the best one(s) is challenging. Thus, some studies implemented multi-criteria decision-making (MCDM) techniques to rank the Pareto solutions and choose the best one(s). Delgarm et al. [28] optimized the building energy performance and indoor comfort using a multi-objective optimization technique and used the technique for order preference by similarity to ideal solution (TOPSIS) to choose the optimal solution from Pareto fronts. In a study [29], three MCDM methods, namely, Analytic Hierarchy Process (AHP), TOPSIS, and Choquet, were used to determine the ranking of different façade design alternatives. MCDM requires the assignment of appropriate weights to the performance criteria involved. The weighing process includes a pairwise comparison of attributes by the experts in the subjective method. On the other hand, objective weighting methods such as mean weight, entropy method, standard deviation, and criteria importance through intercriteria correlation (CRITIC) determine the weights based on the variation in the objectives in Pareto solutions [30]. There are studies in the literature that used CRITIC for weighing the attributes and TOPSIS for ranking of the energy system alternatives in combination with multi-objective optimization. C. Xu et al. [31] used an MCDM framework to optimize the capacity of a hybrid (wind/PV/hydrogen) energy system. The Pareto fronts for the three-objective optimization problem were generated using NSGA-II. The CRITIC technique was employed to assign weights to the objectives, followed by the TOPSIS method for ranking the hybrid energy system alternatives. M. Babatunde and D. Ighravwe [32] also used CRITIC and TOPSIS simultaneously to evaluate the techno-economic performance of six PV/wind/battery/diesel generator energy system alternatives. The results also identify the most and least important technical and economic criteria of a hybrid energy system. T. Salameh et al. [33] employed the MSCDM technique to optimize hybrid energy system alternatives. The authors used three weighing methods—no priority, entropy, and CRITIC— and the TOPSIS technique was used to decide the best solution from nine alternatives. Additionally, the reliability of TOPSIS was also asserted by other ranking methods, such as WASPA, MOORA, and EDAS.

Building performance is strongly influenced by the climate conditions and needs to be considered at the primary design stage. For this purpose, Zhao and Du [34] optimized the orientation, configuration of windows, and shading system of an office building in four different climates, from severe cold to hot, of China. It was concluded that the design parameters in those climates varied to achieve the desired energy and thermal comfort performances. The window materials and depth of the overhangs were different for the optimal case in hot and cold climates. The installation angle decreased from $110°$ in severe cold weather to $80°$ in hot weather. A study investigated the optimal passive design of a residential building in twenty-five different climates [35]. The authors investigated the effect of window blinds during daytime and the natural ventilation rates, air changes per hour (ACH) = 1 and 1.5, for passive cooling. They considered the envelope thermal transmittance, WWR of the facade, and windows glazing for efficient passive design, but limited the optimization process to only five values for each design parameter. The optimization results showed that the recommended thermal transmittance of the walls and roof are 0.2 W/m$^2$ K and 0.6 W/m$^2$ K in severe cold and hot climates, respectively. The WWR in cold climates reached 80% with the aim of reducing the heating load only. Natural ventilation significantly decreased the cooling load in hot climates and reduced the overheating hours in cold climates. Naji et al. [36] conducted performance evaluation of a double-story detached house of 214 m$^2$ area using TRNSYS, EnergyPlus, and an evolutionary algorithm (NSGA-II). The optimal values of envelope insulation thickness and area, glazing, and shading of windows were determined in eight different locations corresponding to tropical, temperate, and continental climates in Australia. The small window area characterized the optimal solutions for tropical, hot desert, and humid subtropical, while the larger window area was optimal in oceanic climates. Low insulation of the envelope was optimal in tropical, while a high insulation envelope was suitable for oceanic climates. A high level of south shading was required for tropical and hot deserts, whereas cold climates required a lower shading. Harkouss et al. [37] investigated the performance of multi-story apartment buildings in different cities of Lebanon and

France with the purpose to minimize the electricity demand of the building. A multi-objective building optimization tool (MOBO) was coupled with Energy Plus to optimize the insulation thickness of the envelope, window area and shading, type of window glazing, and heating/cooling set points. The performance analysis indicated that it is essential to minimize the building thermal load using passive strategies and a high-performance thermal envelope. The results showed that the thermal load was decreased in a range of 6.7–33.1% in different cities. Delgarm et al. [38] used Energy Plus and NSGA-II to optimize the building orientation, window area, and window shading of a single thermal zone in four climates of Iran, which resulted in a 23.8–42.2% decrease in total energy consumption. The optimal WWR was 0.26 in the city with higher HDD and 0.08 in the city with higher CDD. The orientation and overhand depth were almost the same in all cities.

The building design is traditionally based on expert opinion and lacks consideration of various design possibilities and multidisciplinary performance analysis. In most regions, building energy standards are established but contain limited information on building envelope components. Usually, those standards provide the allowable thermal resistance or transmittance of the envelope components and WWR in some cases [39–55]. The comprehensive design approach involves building performance simulations, which are time-consuming and require knowledge of complicated processes. In addition, climate variation makes the selection of these parameters more complex.

Previous studies have reported the optimization methods and solution sets for different building design parameters. Those studies mainly focus on optimizing individual or combinations of design parameters, such as envelope insulation, window area, window glazing, and window shading. On the other hand, design optimization is performed for a single climate or different climates of a particular region [24–27,29,38]. Some authors have conducted multi-climate optimization but with limited design options [23,34–36]. Further, very few studies consider multi-objective and multi-climate optimization followed by the MCDM for a single-family house by combining the envelope design parameters from the literature [28,37]. Many researchers have optimized building designs in different locations and reported energy savings in space heating and cooling loads. However, it is hard to find a study that demonstrates the variation in the optimal envelope parameters and the respective improvement in thermal demand in all major climates.

The present work aims to establish guidelines for determining the unique optimal passive design of a single-family household in various climate zones worldwide. This is accomplished by coupling a dynamic energy simulation tool (TRNSYS) and a Python-based multi-objective optimization algorithm (NSGA III). NSGA III produces the Pareto fronts between the cost of thermal insulation and annual thermal load. In the second stage, CRITIC is employed to determine the objectives' weights based on the objective data from the optimization process. Then the TOPSIS method is applied to identify the optimal solution from the Pareto front.

In this study, a wide range of architecture parameters, including insulation of envelope, passive cooling through windows, WWR, window shading fraction, radiation-based shading control, and orientation, are optimized to minimize the thermal energy demand in twenty climates from the Köppen–Geiger classification. This research contributes to scientific originality by identifying the household's optimal design parameters to efficiently achieve thermal comfort in various climates. A comparison of the annual thermal load is provided between the optimal envelope design from this study and the base-case building with previous design practices. Therefore, the novelty of this work is to provide the benchmarking of envelope design with climate adaptability, which overcomes the limitations of the case building performance model. Moreover, since the climate is changing for the last decades globally, the developed guidelines provide a criterion to modify the building design in the future. Finally, the findings of this work explain the variation in design parameters and thermal loads (solar gains, infiltration load, and transmission load) in the investigated climates.

## 2. Materials and Methods

The optimal design parameters were evaluated by a two-stage optimization method in this study. In the first stage, genetic algorithm NSGA-III was coupled with the building energy simulation tool TRNSYS to generate Pareto fronts by minimizing the thermal insulation expenses and annual thermal load of a single-family household. In the second stage, the CRITIC method was used to determine the weight of the objectives, while the TOPSIS approach was applied to select the unique optimal solution from Pareto solutions.

A comprehensive investigation of energy-efficient design options of the building envelope was carried out for the preliminary design stage. The ideal thermal energy need of the building was evaluated instead of the heating or cooling energy of a specific HVAC system. The thermal loads were calculated for achieving the room temperature ($T_{room}$), 20 °C and 25 °C in winter and summer, respectively, and relative humidity of 50% throughout the year. The predicted mean vote (PMV) value was calculated for these conditions as a measure of thermal comfort. The scheme of the optimization process is presented in Figure 1, and described in Sections 2.1–2.4.

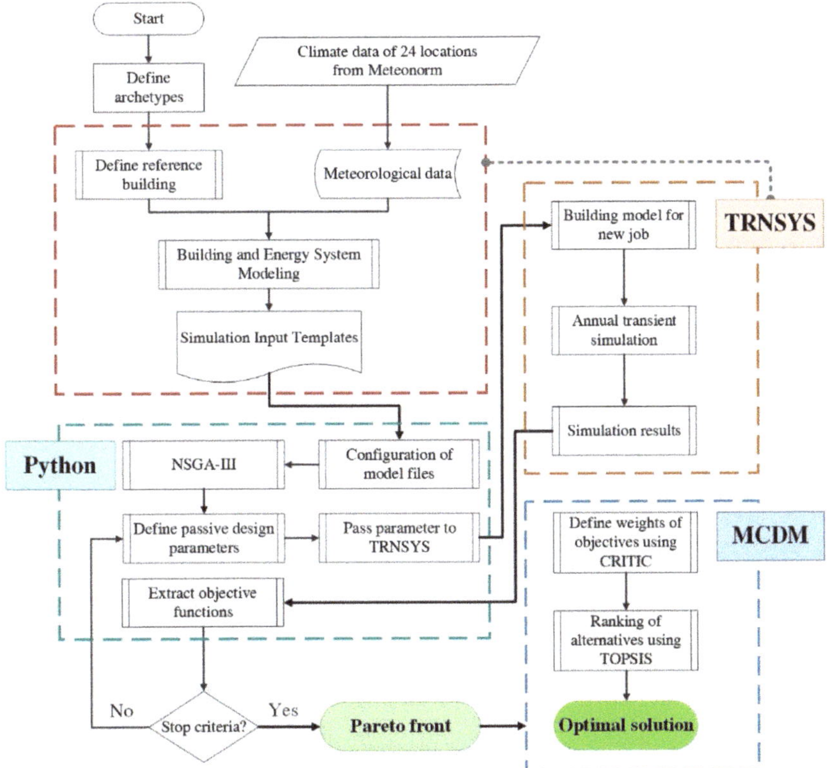


**Figure 1.** Python-based optimization scheme.

### 2.1. Base-Case Building

The base-case building model is a two-story residential building from International Energy Agency (IEA), Solar Heating and Cooling (SHC) Task 44/Annex 38 [56]. Building performance simulation was conducted using TRNSYS (Transient System Simulation Program) [57]. The building has a total net floor area of 140 m², an outside wall surface of 216 m², and an outer roof area of 81 m². Figure 2 shows the geometry and orientation of the building. The building was simulated as a single thermal zone. The internal walls (200 m²) and floor area of second story (70 m$^2$) were added in the TRNBuild tool to include the internal capacities of the building structure. The thermal properties of the envelope were defined based on the local building energy standards of each investigated location [39–55]. These standards provide the limits of thermal transmittance (U-value) for different components of the envelope. Tables 1 and 2 describe the construction and thermal properties of the building envelope. Expanded polystyrene (EPS) insulation and rockwool insulation layers were used for the external walls and roof, respectively. The thickness of the envelope insulation was defined such that the U-values of the external walls and roof met the criteria of allowed thermal transmittance in the local building energy standards. Double-glazed windows, having 4/16/4 geometry (4 mm inner pane, 16 mm space bar, and 4 mm outer pane), U-value of 1.4 W/m$^2$ K, and g-value of 0.622, were used for all facades.

**Figure 2.** Geometry of the base-case building.

**Table 1.** Construction and thermal properties of the opaque elements.

| Element | Layer | Thickness (m) | Density (kg/m$^3$) | Conductivity (W/mK) | U-Value (W/m$^2$ K) |
|---|---|---|---|---|---|
| External wall | plaster inside | 0.015 | 1200 | 0.60 | 0.18 [1] |
| | brick | 0.210 | 1380 | 0.70 | 0.16 [2] |
| | plaster outside | 0.003 | 1800 | 0.70 | 0.20 [3] |
| | | 0.200 [1] | | | 0.26 [4] |
| | | 0.230 [2] | | | 0.30 [5] |
| | EPS (expanded polystyrene) | 0.180 [3] | 17 | 0.04 | |
| | | 0.135 [4] | | | |
| | | 0.120 [5] | | | |
| Floor | wood | 0.015 | 600 | 0.15 | |
| | plaster floor | 0.080 | 2000 | 1.40 | |
| | sound insulation | 0.040 | 80 | 0.04 | 0.649 |
| | concrete | 0.150 | 2000 | 1.33 | |
| Roof ceiling | gypsum board | 0.025 | 900 | 0.21 | 0.13 [1] |
| | plywood | 0.015 | 300 | 0.08 | 0.17 [2] |
| | plywood | 0.015 | 300 | 0.08 | 0.15 [3] |
| | | 0.250 [1] | | | 0.22 [4] |
| | | 0.190 [2] | | | 0.20 [5] |
| | rockwool | 0.215 [3] | 60 | 0.03 | |
| | | 0.140 [4] | | | |
| | | 0.160 [5] | | | |
| Internal wall | clinker | 0.200 | 650 | 0.230 | 0.885 |

[1] Ostersund and Stockholm; [2] Saarbrücken; [3] Strasbourg; [4] Milan; [5] All investigated locations other than mentioned earlier (c.f. Table 3).

**Table 2.** Construction and thermal properties of the windows.

| Windows | Construction (mm) | Height (m) | Width (m) | Windows Area (m$^2$) | U-Value (W/m$^2$ K) | g-Value |
|---|---|---|---|---|---|---|
| North | | | | 3.0 | | |
| South | (4,16,4) | 1.0 | 1.0 | 12.0 | 1.4 | 0.622 |
| East | | | | 4.0 | | |
| West | | | | 4.0 | | |

### 2.1.1. Ventilation Load

The energy performance analysis in this research takes account of two ventilation loads due to air exchange through the leakage area of the building and air exchange through windows opening. Based on the Sherman Grimsrud model from ASHRAE fundamentals 1997 [58], a simple single-zone approach calculates the air infiltration rate into the building through the leakage area (Equation (1)).

$$Q_{inf} = (A_L/1000) \cdot \sqrt{C_s \Delta T + C_w V^2} \tag{1}$$

where, $Q_{inf}$ is the airflow rate (m$^3$/s), $A_L$ is the effective air leakage area (cm$^2$) of the house, $C_s$ is stack coefficient ((L/s)$^2$/(cm$^4$ K)), $\Delta T$ is the difference between the average indoor and outdoor temperature for the time interval of calculation (K), $C_w$ is wind coefficient ((L/s)$^2$/(cm$^4$ (m/s)$^2$)), and $V$ is average wind speed (m/s). The value of $C_s$ depends on the number of stories of the building. For instance, it has a value of 0.000145 for a single-story, 0.00029 for a two-story, and 0.000435 for three-story buildings. The wind coefficient $C_w$ depends on the height of the building and local shielding from surrounding objects, and the value is accordingly assigned as provided in ASHRAE fundamentals 1997.

The natural air exchange through six windows with aperture angle ($\alpha$) is the passive cooling rate. Passive ventilation through windows was activated based on the indoor temperature of the zone and the ambient temperature (T$_{amb}$) during the night. Control strategies were implemented such that passive cooling through windows occurred during the night (9 p.m. to 8 p.m.) along with active cooling. The 24 h average temperature (T$_{avg}$24) was used to indicate the seasonal variation. It was asserted from the climate data set that the average temperature below 12 °C occurs in the winter season, and there is no requirement for cooling in the building. Passive cooling starts when the indoor temperature rises above 24 °C, and the outdoor temperature is at least 2 °C below the indoor temperature. In comparison, active cooling starts when the indoor temperature rises above 25 °C. The margin of 1 °C before the start of active cooling was provided as an energy-saving measure. When the room temperature drops to 23 °C, the windows are closed. Thus, the passive cooling is operable between 23 °C and 25 °C indoor temperatures. If all the conditions described in Figure 3 were met, the windows tilt to an angle $\alpha$ for passive cooling. Thus, the ventilation load of the building is a sum of the heat gains/losses from two air-exchange rates.

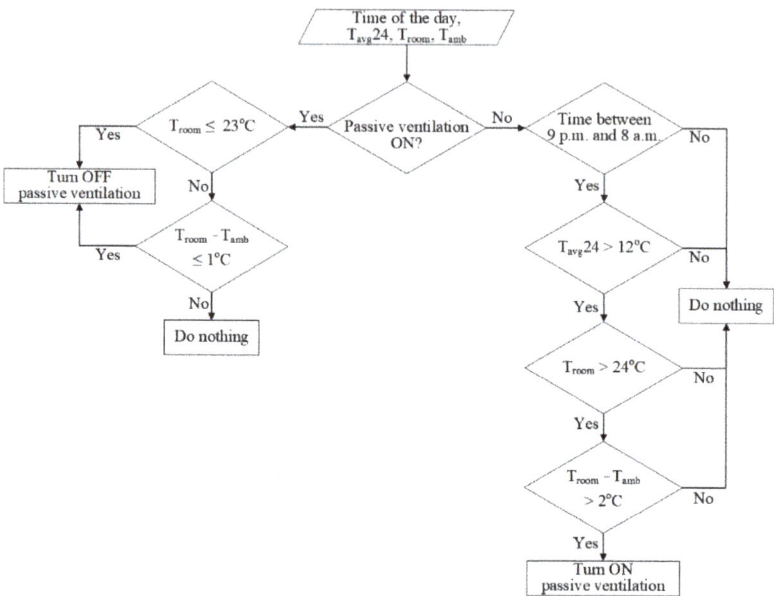



**Figure 3.** Temperature-based control of passive ventilation through windows.

### 2.1.2. Internal Gains

The investigated building was occupied by a family of four members. The occupancy schedule was adopted from ASHRAE [59], as shown in Figure 4, and is identical for all days of the week. The heat generated by the occupants is 115 W per person, which includes a sensible gain of 70 W and a latent gain of 45 W [60]. The sensible heat gain is divided into the radiative (42 W) and convective (28 W) parts. The latent heat is incorporated directly by the humidity with a mass flow of 0.059 kg/h.

Electrical appliances produce waste heat causing thermal gains in the zone. Since the electricity consumption varies with the diversity in the climate, annual hourly load profiles were generated based on the literature [61–75] for all the investigated countries to evaluate thermal gains. The load profiles do not include the electricity use for space heating, space cooling, and water heating. For realistic electric heat gains, the load profiles were modified, if required, to a household of 140 m² floor area. Kuusela et al. estimated the electricity consumption of common home appliances as a function of floor area. The relation between electricity consumption relative to a 140 m² household and floor area is shown in Figure 5 and mathematically represented in Equation (2) [76]. The electricity consumption of households in different locations was adjusted to a 140 m² floor area using Equation (2) because the area and geometry of the reference building were assumed to be identical for thermal load analysis in all climates.

$$y = -0.44 \, ln(x) + 3.19 \tag{2}$$

It was assumed that 58% of the electric energy is retained in the building as thermal gain. Table 3 provides the electric gains in the selected climates. The occupancy and electric thermal gains were added as external hourly profiles to the building model in TRNSYS.

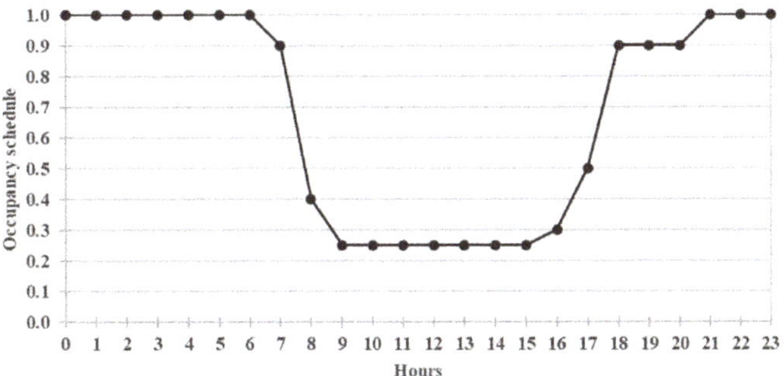

**Figure 4.** Occupancy schedule of a day from ASHRAE.

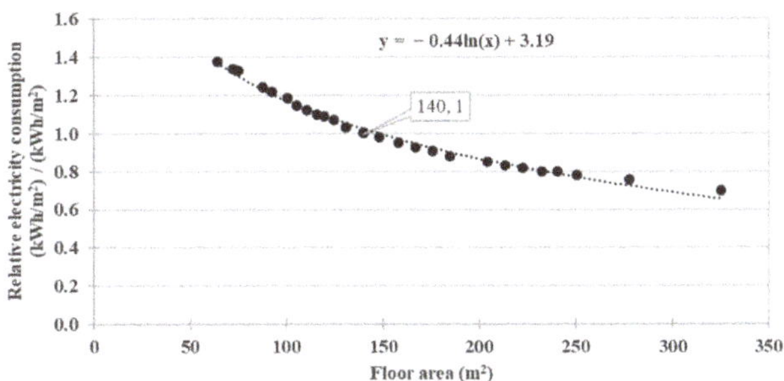

**Figure 5.** Relative electricity consumption as a function of area.

### 2.1.3. Solar Gain

All windows were equipped with shading blinds to avoid overheating in winter and higher cooling load in summer by limiting the solar gain. In this research work, a monthly schedule determined the shading fraction in the building model, and it was adjusted according to the month of the year, having the minimum shading fraction. According to a study, the average value of optimal shading during winter is approximately 23%, and the minimum value during a day is 10% [77]. In the summer season, the window shading can be varied between 25% and 100% for the optimization process [78]. Therefore, the minimum shading fraction is set to 0.11 in December for the base-case analysis. Figure 6 shows the variation in the shading for other months of the year relative to December. Furthermore, a windows-shading control was designed as a function of ambient temperature, the temperature of the zone, and global horizontal solar radiations. Figure 7 explains the control strategy to switch the shading device on and off. The temperature-based control of window shading restricts the solar heat gain into the building through windows. The window shade is turned on at 23.8 °C [56] to maintain the room temperature below 25 °C. This margin of 1.2 °C was used to delay the activation of the active cooling system. Although this value can be set to a lower level, it could result in higher lighting loads and heat gain from the lighting equipment. The window shading was turned off at 22.8, i.e., 1 °C lower than shading on temperature, to allow daylight into the building and avoid a further decrease in the temperature during summer. The WWR is the second factor that controls the solar gain through windows. WWR varies for all facades (c.f. Table 2), and the value of WWR is 20% for the south facade in the base case. For the optimization scheme, the upper limit of WWR is 40% according to the guideline of the ASHRAE standard 90.1–2019 [79], which states that the higher value of WWR is 40% for residential and non-residential buildings.

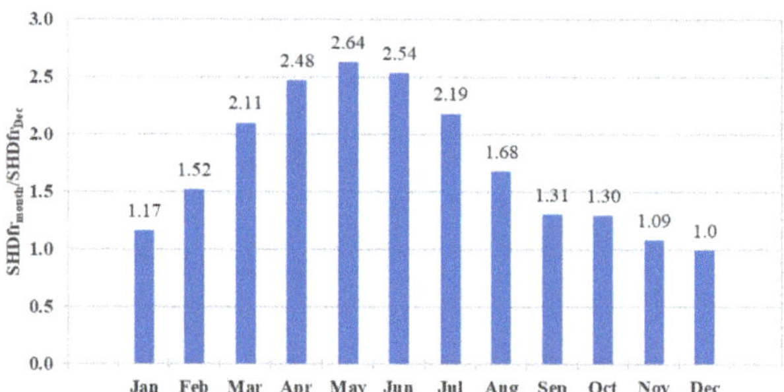

**Figure 6.** Relative shading fraction during the year.

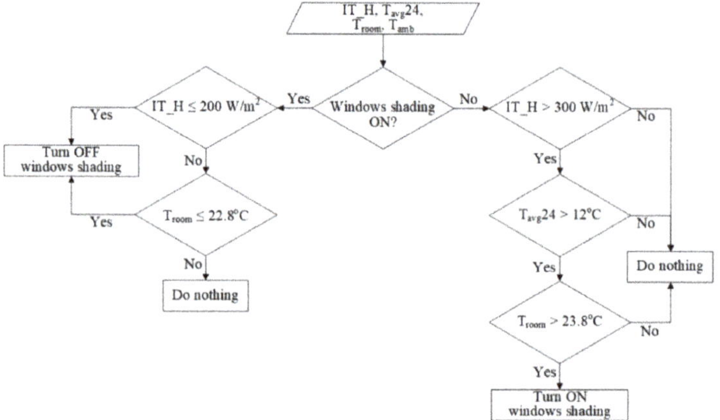


**Figure 7.** Solar radiation-based control of window shading.

### 2.2. Investigated Climates

This work investigated four major climate zones: A: Tropical; B: Dry; C: Temperate; and D: Continental, according to Köppen–Geiger climates classification [80] (first letter). These climate zones are subdivided based on the annual variation in ambient temperature and precipitation. The precipitation level (second letter) is defined as f (no dry season), m (Monsoon), s (dry summer), w (dry winter), S (semi-arid), and W (desert). Similarly, the temperature level (third letter) is categorized as a (hot summer), b (warm summer), c (cold summer), d (very cold winter), h (hot), and k (cold). Table 3 describes the average temperature ($T_{avg}$), cooling degree days ($CDD$), and heating degree days ($HDD$) for twenty selected climates of twenty-four locations. The Meteonorm tool generates the meteorological data for these locations. The $CDD_{10}$ and $HDD_{18}$ are defined as follows:

$$HDD_{18} = \sum_{t=1}^{365} (T_{base} - T_a) \tag{3}$$

$$CDD_{10} = \sum_{t=1}^{365} (T_a - T_{base}) \tag{4}$$

where $T_{base}$ is 18 °C and 10 °C for $HDD$ and $CDD$, respectively, $T_a$ is the average temperature of the day, and degree days are the yearly sum of the daily temperature differences.

**Table 3.** Climate characteristics and electric gains of the investigated locations.

| SN | Country | Location | Köppen Climate | IECC Climate | $T_{avg}$ (°C) | $HDD_{18}$ | $CDD_{10}$ | Electricity Consumption (kWh/m$^2$ a) | Electric Gains (kW$_{th}$/m$^2$ a) |
|---|---|---|---|---|---|---|---|---|---|
| 1 | Sweden | Ostersund | Dfc | 7 A | 3.9 | 5468 | 429 | 30.20 [61] | 17.52 |
| 2 | Sweden | Stockholm | Dfb | 5 A | 7.4 | 3922 | 841 | 30.20 [61] | 17.52 |
| 3 | Austria | Bischofshofen | Dfb | 5 A | 8.3 | 3660 | 994 | 23.87 [62] | 13.85 |
| 4 | China | Daocheng | Dwb | 6 A | 5.9 | 4434 | 378 | 11.67 [68] | 6.77 |
| 5 | Iran | Sarab | Dsb | 5 C | 9.1 | 3496 | 1305 | 32.51 [69] | 18.85 |
| 6 | Japan | Sapporo | Dfa | 5 A | 9.3 | 3523 | 1430 | 28.80 [70] | 16.71 |
| 7 | China | Beijing | Dwa | 4 B | 12.8 | 2875 | 2470 | 11.67 [68] | 6.77 |
| 8 | Iran | Arak | Dsa | 4 B | 14.4 | 2320 | 2523 | 32.51 [69] | 18.85 |
| 9 | Denmark | Odense | Cfb | 5 C | 8.9 | 3364 | 835 | 26.77 [71] | 15.53 |
| 10 | Germany | Saarbrücken | Cfb | 5 A | 9.8 | 3119 | 1074 | 34.36 [62] | 19.93 |
| 11 | UK | Birmingham | Cfb | 5 C | 10.8 | 3679 | 930 | 37.34 [72] | 21.66 |
| 12 | France | Strasbourg | Cfb | 4 A | 12.1 | 2470 | 1533 | 30.00 [62] | 17.40 |
| 13 | China | Kunming | Cwb | 3 C | 15.7 | 1137 | 2204 | 11.67 [68] | 6.77 |
| 14 | Spain | Vigo | Csb | 3 A | 15.4 | 1282 | 2042 | 19.92 [73] | 11.55 |
| 15 | Italy | Milan | Cfa | 4 A | 13.9 | 2099 | 2115 | 21.81 [74] | 12.65 |
| 16 | China | Hanzhong | Cwa | 3 A | 15.4 | 1853 | 2589 | 11.67 [68] | 6.77 |
| 17 | Portugal | Evora | Csa | 3 A | 16.1 | 1404 | 2397 | 27.15 [75] | 15.75 |
| 18 | Iran | Birjand | BWk | 3 B | 17.0 | 1693 | 3052 | 32.51 [69] | 18.85 |
| 19 | Pakistan | Quetta | BSk | 3 A | 17.9 | 1182 | 3312 | 22.19 [63] | 12.87 |
| 20 | Pakistan | Lahore | Bsh | 1 B | 24.7 | 348 | 5382 | 22.19 [63] | 12.87 |
| 21 | UAE | Dubai | Bwh | 0 B | 28.9 | 0 | 6910 | 39.93 [64] | 23.16 |
| 22 | Singapore | Singapore | Af | 0 A | 28.6 | 0 | 6782 | 28.04 [65] | 16.26 |
| 23 | India | Mumbai | Aw | 0 A | 28.1 | 0 | 6594 | 22.92 [66] | 13.30 |
| 24 | Indonesia | Jakarta | Am | 1 A | 26.6 | 0 | 6045 | 18.40 [67] | 10.67 |

### 2.3. Multi-Objective Optimization

TRNSYS is a stand-alone simulation program that calculates the thermal loads of the building and analyzes the performance of transient systems. TRNSYS uses text files as input to run the simulations and generates outputs also in text files. Python code devises an interface between TRNSYS and genetic algorithms to implement the multi-objective optimization process. This work used jMetalPy, an object-oriented Python-based framework, to solve the bi-objective optimization problem using a non-dominated sorting genetic algorithm (NSGA–III) [81]. The Python script reads the design variables and objective functions from the input and output files of TRNSYS, respectively, and formulates the optimization problem. jMetalPy generates a new set of design parameters for each simulation based on the values of objective functions from the previous run. Afterward, Python substitutes these values in the input files, and TRNSYS simulates the building energy behavior.

Multi-objective optimization results in a non-dominated solution set, called the Pareto front (PF), such that no other feasible solution exists that improves one objective without compromising the second objective.

### 2.3.1. Non-Dominated Sorting Genetic Algorithm (NSGA-III)

NSGA-III is an extension of NSGA-II and established as a baseline evolutionary multi-objective optimization algorithm. It uses several well-distributed reference points to select nondominated solutions for the next generation [82]. Thus, NSGA-III eliminates the drawback of non-diversity in NSGA-II during the generation of subsequent populations.

In the current optimization problem, NSGA-III carried out 5000 evaluations with a population size of 100 for each generation. Table 4 reports the input parameters for the bi-objective optimization process. The simulation-based optimization produced PFs of 100 non-dominated solutions for each location.

**Table 4.** Inputs of the genetic algorithm.

| NSGA-III Attributes | Value |
|---|---|
| Population size | 100 |
| No of variables | 7 |
| No of objectives | 2 |
| Maximum evaluations | 5000 |
| Mutation method | Polynomial |
| Mutation probability | 0.15 |
| Crossover method | Simulated binary crossover |
| Crossover probability | 0.8 |
| Termination criteria | Max evaluations |

### 2.3.2. Design Variables

The deciding factors of building thermal load are solar heat gains through windows, infiltration gains/losses through leakage area or the windows, and transmission gains/losses through opaque elements of the envelope. Therefore, the passive design parameters were selected based on their influence on heat gains and losses. The design variables for this study are building orientation, WWR, windows shading fraction, minimum solar radiation to turn on window shading, and the insulation thickness of the external walls and roof. Table 5 describes the detailed information of these variables.

**Table 5.** Design variables and their optimization bounds.

| Building Element | Variable | Lower Bound | Upper Bound |
|---|---|---|---|
| External wall insulation | EPS thickness ($EPS_{Thk}$), m | 0.10 | 0.25 |
| Roof insulation | rockwool thickness ($Rockwool_{Thk}$), m | 0.10 | 0.25 |
| Window aperture | $\alpha$ (degrees) | 5 | 20 |
| South faced window | Window-to-Wall ratio (WWR) | 0.2 | 0.4 |
| Windows shading | Minimum horizontal solar radiation (IT_H) for shading on | 250 | 500 |
| Windows shading | Shading fraction in December ($Shd_{Dec}$) | 0.10 | 0.33 |
| Building orientation | Orientation (N/S/E/W) | NA | NA |

### 2.3.3. Objective Functions

The goal is to reduce the energy demand for heating and cooling by adopting passive design measures while keeping the investment cost to a minimum. Since this work deals with the passive design of the building, the optimization process only considered the additional cost for the envelope's insulation. TRNSYS computed the annual thermal load of the household using a multi-zone building component (Type 56). The building geometry, envelope materials, and the schemes of building gains/losses were defined in the TRNBuild tool and imported as a text file in the Type 56 component to run the simulation. TRNSYS also calculated the investment cost as a function of insulation thickness. The market survey of the envelope insulation material in different investigated locations revealed that the cost is approximately the same for the insulation materials, having similar thermal properties. The cost of EPS and rockwool materials were 139 €/m$^3$ and 230 €/m$^3$, respectively [83], and the same are used in all locations. Regarding the cost associated with the window area, it was assumed that the cost-saving from constructing a smaller size insulated wall compensates for the extra cost for large WWR. Finally, TRNSYS produced the values of the objective function in text files. The following two objectives come up for bi-objective optimization.

- Minimize annual thermal load ($kW_{th}$): The annual thermal load is the sum of sensible and latent heating and cooling demands to maintain the comfort level in the building. All the design parameters influence this objective function.

- Minimize investment cost (€): This objective function only depends on the thickness of insulation materials and is calculated accordingly.

### 2.4. Multi-Criteria Decision Making

The CRITIC method uses standard deviation to measure the diversity of the attributes. The method assigns weights such that the criterion with a higher diversity gets a higher weight. This process normalizes the data and creates a correlation matrix to measure the information and importance of each criterion (Equations (5)–(8)) [31].

$$r_{ij} = \frac{x_{ij} - x_j^{max}}{x_j^{max} - x_j^{min}} \tag{5}$$

$$r_{jk} = \frac{\sum_{i=1}^{m} \left( r_{ij} - \overline{r_j} \right) \cdot \left( r_{ik} - \overline{r_k} \right)}{\sqrt{\sum_{i=1}^{m} \left( r_{ij} - \overline{r_j} \right)^2 \cdot \sum (r_{ik} - \overline{r_k})^2}} \tag{6}$$

$$c_j = \sigma_j \sum_{k=1}^{k} 1 - r_{jk} \tag{7}$$

$$w_j = \frac{c_j}{\sum_{j=1}^{n} c_j} \tag{8}$$

where $r_{ij}$ is the normalized performance value of the $i$th alternative on the $j$th criterion, $r_{jk}$ is the correlation coefficient between the $j \times k$ criteria matrix, $\sigma_j$ is the standard deviation of the $j$th criterion, $c_j$ represent the quantity of information contained in the $j$th criterion, and $w_j$ is the weight of the $j$th criterion.

The TOPSIS method makes the ranking decision of the criteria based on the shortest and farthest geometrical distances of criteria from ideal and non-ideal solutions (Equations (9)–(13)) [31]. TOPSIS procedure was carried out by first determining the normalized values of the criteria in a decision matrix (Equation (9)). The normalization process was followed by the design of a weighted normalized matrix (Equation (10)). The weights obtained from the CRITIC methods were assigned here. The next step was to evaluate the distance of alternatives from ideal and non-ideal solutions (Equations (11) and (12)). Based on the outcomes of these equations, the relative closeness of the alternatives was calculated using Equation (13). The alternative with the highest score was considered the best solution.

$$n_{ij} = \frac{x_{ij}}{\sqrt{\sum_{j=1}^{n} x_{ij}^2}} \tag{9}$$

$$v_{ij} = \overline{n_{ij}} \cdot w_j \tag{10}$$

$$D_i^+ = \sqrt{\sum_{j=1}^{n} \left( v_{ij} - v_j^+ \right)^2} \tag{11}$$

$$D_i^- = \sqrt{\sum_{j=1}^{n} \left( v_{ij} - v_j^- \right)^2} \tag{12}$$

$$D_i = \frac{D_i^-}{D_i^+ + D_i^-} \tag{13}$$

where $n_{ij}$ represents the normalized value of the $j$th criterion for the $i$th alternative, $w_j$ refers to the weight of the $j$th criterion, $v_{ij}$ represents the weighted normalized value of the $j$th criterion for $i$th alternative, $v_j^+$ and $v_j^-$ are the maximum and minimum values of the $j$th criterion, respectively, $D_i^+$ and $D_i^-$ represent the ideal and non-ideal distances for the $i$th alternative, respectively, and $D_i$ is the relative closeness of the $i$th alternative to the ideal solution.

## 3. Results

Firstly, the PFs are presented with trade-off solution points for all the investigated climates. The values of the design variables are provided along with the objective functions of the optimal solution. Secondly, thermal loads of the base-case and energy-optimal household models were extracted. The energy savings from the optimal passive design were also analyzed for heating and cooling loads. Finally, the variation in each design parameter with the changing climate was investigated.

### 3.1. Optimization Results

A multi-objective optimization does not produce a solution that minimizes or maximizes the objective functions simultaneously. Instead, it provides Pareto optimal solutions that are not affected by other solutions. Furthermore, no objective function can be improved without comprising at least another. The bi-objective optimization in this work generates the PFs for all the investigated climates with around 100 points. Figures 8–11 present the PFs of the continental, temperate, dry, and tropical climate zones, respectively. The 2D PFs plots between the cost of insulation and the annual thermal load were also drawn. The scatter plots are generally parallel in all climates. However, they are dispersed horizontally due to the diversity in climate conditions.

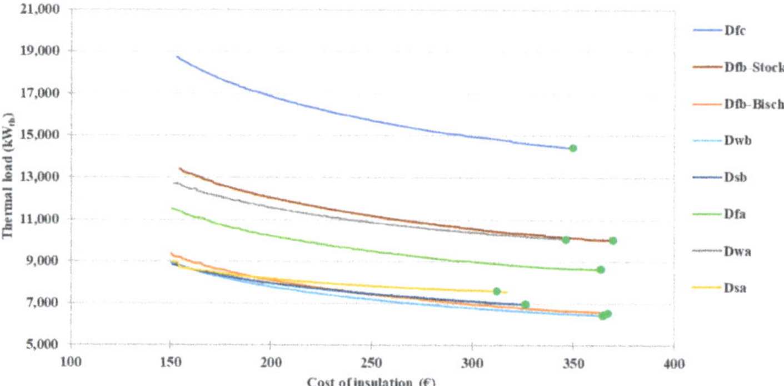

**Figure 8.** Pareto fronts of locations in a continental climate.

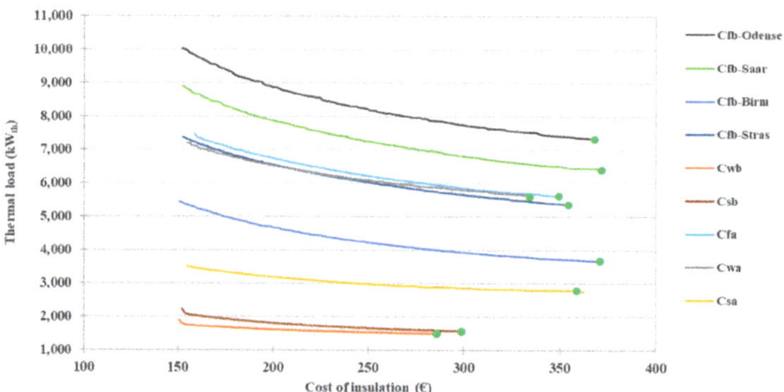

**Figure 9.** Pareto fronts of locations in a temperate climate.

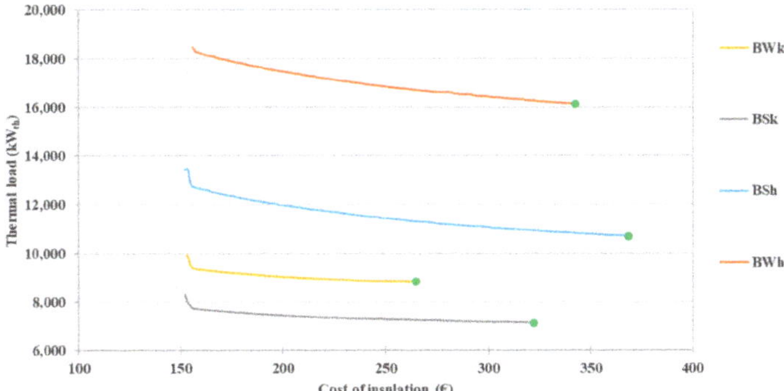

**Figure 10.** Pareto fronts of locations in a dry climate.

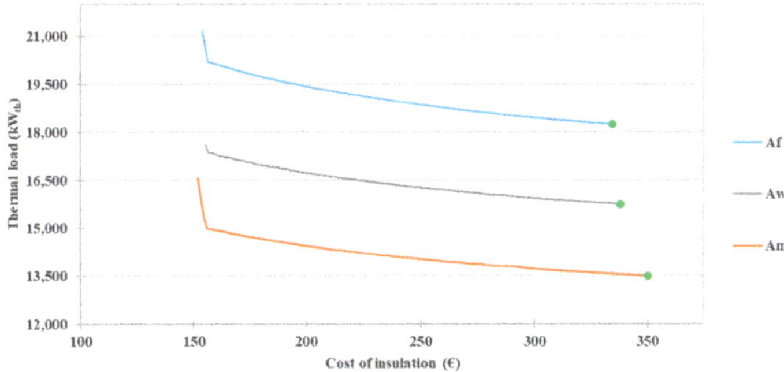

**Figure 11.** Pareto fronts of locations in a tropical climate.

The optimization results show that, initially, a minor increase in insulation cost results in a high reduction in the thermal load. This trend is more prominent in hot regions, as in the case of tropical and dry climate zones. In those climates, PFs are steeper at the start, but the slope gradually decreases as the cost of insulation increases. In this study, the optimum solution on a PF is determined by MCDM using the CRITIC and TOPSIS methods. The optimal solution is characterized by a low energy demand and high insulation thickness in all climate zones. The optimal trade-off solution is highlighted in green on PFs. It is interesting to note that the optimal solution is the one having the minimum thermal load or near to it in all cases. The phenomenon is justified by the higher weight obtained by the CRITIC method for the annual thermal load. In heating-dominant climates, the weight of thermal load ranges between 0.63 and 0.68, whereas this weight has a range of 0.60 to 0.72 in other climate zones.

Table 6 provides the values of design parameters, thermal transmittance of the external wall ($U_w$) and roof ($U_r$), and objective functions for the optimal solution in each location. In most cases, the optimal solutions require a highly insulated envelope and large window shading fraction. The WWR varies between the lower and upper bounds, i.e., 0.2 to 0.4. The solar radiation value remains near 250 W in most locations. Similarly, the average window aperture angle is 11.88 degrees but varies between 5 and 20 degrees in different climates. The U-value of the external wall and roof occurs around 0.15 W/m$^2$ K in climates with high space heating or cooling demand. However, it is relatively higher in moderate climates. In all cases, the U-value is decreased as compared to the base-case building after the optimization process other than Ostersund and Stockholm, where the U-value of the roof is increased from 0.13 W/m$^2$ K to 0.15 W/m$^2$ K.

**Table 6.** Trade-off solution set in the investigated climates.

| Climate | $EPS_{Thk}$ (m) | $Rockwool_{Thk}$ (m) | $\alpha$ (Degree) | WWR | IT_H (W) | $Shd_{Dec}$ | Orientation | $U_w$ (W/m$^2$ K) | $U_r$ (W/m$^2$ K) | Thermal Load (kW$_{th}$/a) | Cost of Insulation (€) |
|---|---|---|---|---|---|---|---|---|---|---|---|
| Dfc (Ostersund) | 0.247 | 0.211 | 9.8 | 0.37 | 279 | 0.330 | North | 0.150 | 0.154 | 14,317 | 350 |
| Dfb (Stockholm) | 0.237 | 0.196 | 10.9 | 0.31 | 289 | 0.330 | North | 0.156 | 0.164 | 10426 | 335 |
| Dfb (Bischofshofen) | 0.247 | 0.240 | 12.7 | 0.39 | 256 | 0.328 | North | 0.150 | 0.137 | 6729 | 367 |
| Dwb (Daocheng) | 0.245 | 0.241 | 5.2 | 0.40 | 287 | 0.253 | North | 0.151 | 0.136 | 6546 | 365 |
| Dsb (Sarab) | 0.234 | 0.188 | 11.0 | 0.33 | 255 | 0.329 | North | 0.157 | 0.170 | 7159 | 326 |
| Dfa (Sapporo) | 0.246 | 0.237 | 15.4 | 0.40 | 277 | 0.330 | North | 0.150 | 0.138 | 8845 | 364 |
| Dwa (Beijing) | 0.240 | 0.218 | 16.1 | 0.40 | 251 | 0.330 | North | 0.154 | 0.149 | 10,333 | 346 |
| Dsa (Arak) | 0.217 | 0.179 | 5.2 | 0.22 | 250 | 0.330 | North | 0.169 | 0.178 | 7757 | 317 |
| Cfb (Odense) | 0.246 | 0.237 | 10.0 | 0.34 | 304 | 0.328 | North | 0.150 | 0.138 | 7506 | 368 |
| Cfb (Saarbrucken) | 0.245 | 0.235 | 10.4 | 0.24 | 252 | 0.327 | North | 0.151 | 0.139 | 6595 | 372 |
| Cfb (Birmingham) | 0.240 | 0.249 | 13.1 | 0.32 | 279 | 0.329 | North | 0.154 | 0.132 | 3862 | 371 |
| Cfb (Strasbourg) | 0.244 | 0.214 | 12.3 | 0.28 | 260 | 0.330 | North | 0.151 | 0.152 | 5573 | 354 |
| Cwb (Kunming) | 0.160 | 0.219 | 7.3 | 0.22 | 250 | 0.330 | North | 0.222 | 0.148 | 1576 | 287 |
| Csb (Vigo) | 0.196 | 0.187 | 13.2 | 0.20 | 264 | 0.330 | North | 0.185 | 0.171 | 1674 | 298 |
| Cfa (Milan) | 0.231 | 0.223 | 16.1 | 0.28 | 251 | 0.330 | North | 0.159 | 0.146 | 5868 | 349 |
| Cwa (Hanzhong) | 0.232 | 0.203 | 8.4 | 0.32 | 255 | 0.328 | North | 0.159 | 0.159 | 5844 | 334 |
| Csa (Evora) | 0.235 | 0.224 | 19.2 | 0.20 | 261 | 0.330 | North | 0.157 | 0.145 | 2900 | 363 |
| BWk (Birjand) | 0.197 | 0.133 | 5.0 | 0.20 | 251 | 0.330 | North | 0.184 | 0.230 | 9000 | 265 |
| BSk (Quetta) | 0.223 | 0.185 | 12.0 | 0.20 | 250 | 0.330 | North | 0.165 | 0.173 | 7308 | 322 |
| Bsh (Lahore) | 0.249 | 0.219 | 16.3 | 0.20 | 256 | 0.330 | South | 0.149 | 0.148 | 10,826 | 368 |
| Bwh (Dubai) | 0.232 | 0.203 | 20.0 | 0.20 | 250 | 0.330 | South | 0.159 | 0.159 | 16,195 | 342 |
| Af (Singapore) | 0.250 | 0.167 | 13.4 | 0.20 | 251 | 0.330 | South | 0.148 | 0.189 | 17,933 | 334 |
| Aw (Mumbai) | 0.247 | 0.176 | 19.2 | 0.20 | 253 | 0.330 | South | 0.150 | 0.180 | 15,757 | 338 |
| Am (Jakarta) | 0.232 | 0.215 | 5.0 | 0.20 | 250 | 0.330 | North | 0.159 | 0.151 | 13,503 | 350 |

The variation in design parameters in the investigated locations can be characterized according to the degree days. The optimal solution sets are quite consistent in a specific range of degree days. Table 7 provides the mean, standard deviation (STD), and ranges in design parameters for climate zones having different degree days. The thermal transmittance should be low in heating-dominant locations. The mean value of thermal transmittance is the least in locations having $HDD_{18} > 3500$. The locations with $CDD_{10} > 3500$ have a mean thermal transmittance of 0.153 and 0.165 for the wall and roof, respectively. The WWR has a maximum mean value of 0.36 for $HDD_{18} > 3500$, decreasing to 0.2 for $CDD_{10} > 3500$. The window aperture angle has a mean value around 11 degrees for $HDD_{18} > 2000$. The maximum aperture angle occurs in locations having $CDD_{10} > 3500$, i.e., 14.77. For the horizontal solar radiation (IT_H), the mean value is maximum in the locations with $HDD_{18} > 3500$ and minimum in the locations with $CDD_{10} > 3500$. The STD is used to measure the spread of the solution set in different locations having similar climate conditions. The consistency of the solution sets is validated by very low STD in each category for all the design parameters, except for the aperture angle, which has a wide range and larger STD. A detailed analysis of variation in the design parameters for each climate zone is provided in Section 3.3.

**Table 7.** Comparison of the design parameters based on degree days.

| Category | | $U_w$ (W/m$^2$ K) | $U_r$ (W/m$^2$ K) | WWR | $\alpha$ (Degree) | IT_H (W) |
|---|---|---|---|---|---|---|
| $HDD_{18} > 3500$ | Range | 0.15–0.156 | 0.136–0.164 | 0.31–0.4 | 5.25–15.45 | 255–289 |
| | Mean | 0.152 | 0.144 | 0.362 | 11.192 | 278 |
| | STD | 0.003 | 0.013 | 0.040 | 3.505 | 11.90 |
| $3500 > HDD_{18} > 2000$ | Range | 0.15–0.169 | 0.138–0.178 | 0.22–0.4 | 5.23–16.15 | 250–304 |
| | Mean | 0.156 | 0.153 | 0.297 | 11.599 | 261 |
| | STD | 0.007 | 0.015 | 0.061 | 3.789 | 19.62 |
| $3500 > CDD_{10} > 2000$ | Range | 0.159–0.185 | 0.145–0.173 | 0.2–0.32 | 5.01–19.21 | 25–264 |
| | Mean | 0.179 | 0.171 | 0.224 | 10.859 | 255 |
| | STD | 0.024 | 0.031 | 0.048 | 5.083 | 6.04 |
| $CDD_{10} > 3500$ | Range | 0.149–0.159 | 0.148–0.189 | 0.2 | 5–20 | 250–256 |
| | Mean | 0.153 | 0.165 | 0.201 | 14.772 | 252 |
| | STD | 0.006 | 0.018 | 0.001 | 6.042 | 2.65 |

*3.2. Effect of Design Optimization on Thermal Loads*

The optimization results reveal that energy-optimal solutions significantly reduce the space heating and cooling demand in all climates. The PMV index falls between −0.5 and 0.5 throughout the year for each case, which complies with the recommended thermal limit in ASHRAE standard 55 [84]. The design parameters change simultaneously during optimization until they reach an optimum solution. Therefore, it could be hard to understand precisely how an individual parameter influences the objective functions. Nevertheless, the overall effect of architectural design is evident by the improvement in the space heating and cooling demands. Building thermal loads includes infiltration losses/gains, transmission losses/gains, equipment gains, occupancy gain, solar gains, heating gains, and cooling losses. The passive design of the building is associated only with the energy transfer by infiltration, transmission, and solar radiation. Therefore, this work evaluates the variation in those thermal loads in the investigated climates, as shown in Figure 12. The clustered columns show the heat exchange for the base-case and optimal passive designs of the household, whereas the scatter plots represent the change in those thermal loads after optimization on the secondary vertical axis. The cumulative change in heat loss during winter and heat gain during summer is presented for transmission and infiltration loads. The optimization results show that envelope insulation, windows shading, and WWR are the most influential design factors for energy efficiency in households. In contrast, horizontal solar radiation for window shading control and window aperture ($\alpha$) for passive ventilation do not significantly impact the thermal load.

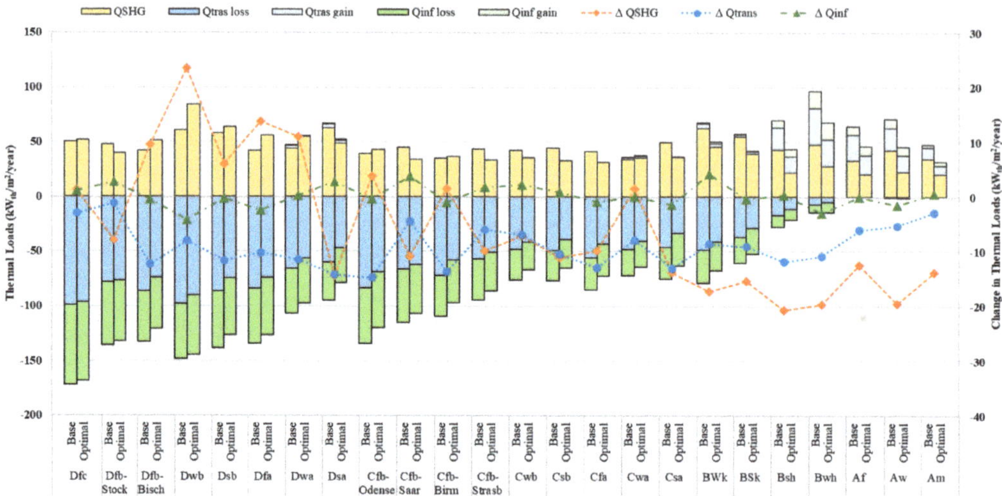

**Figure 12.** Households' thermal loads for the base-case and energy-optimal scenarios.

The optimal design increases the solar heat gains through windows (QSHG) in continental climates besides Dsa (Arak) and Dfb (Stockolm), whereas in temperate climates, solar gains slightly increase or decrease depending upon the space cooling demand of the location. In the case of tropical and dry climates, the QSHG is always reduced after optimization. Furthermore, the optimal passive design can decrease transmission losses during winter and transmission gains in summer. Thus, the net effect is lowering the transmission heat exchange (Qtrans) after optimization in all locations. The change in heat exchange due to infiltration (Qinf) is not very significant in cold and hot climates. Yet, it avoids overheating in winter or decreases cooling demand in summer to some extent through passive cooling. The following sections describe the improvement in energy demand of the household in four major climate zones of the Köppen–Geiger classification.

### 3.2.1. Continental Climate

The continental climate has the temperature of the coldest month below 0 °C and the temperature of the hottest month greater than 10 °C. Therefore, the locations in this zone are heating-dominant besides the hot and dry summer continental climate (Dsa). Even though the cooling energy demand is very low, the optimization further reduces it substantially. Figure 13 compares the heating and cooling energy demands of the base-case and optimal household and illustrates the energy saving after optimization. On average, the heating load decreases by 33.47%. The maximum reduction in heating demand is 56.93% in Dwb (Daocheng). In fact, the energy saving for space heating depends on the current practices of building energy standards. For example, after optimization, there is no significant change in the space heating loads in Dfc (Oestersund) and Dfb (Stockholm). This is due to the energy-efficient envelope standards in those locations.

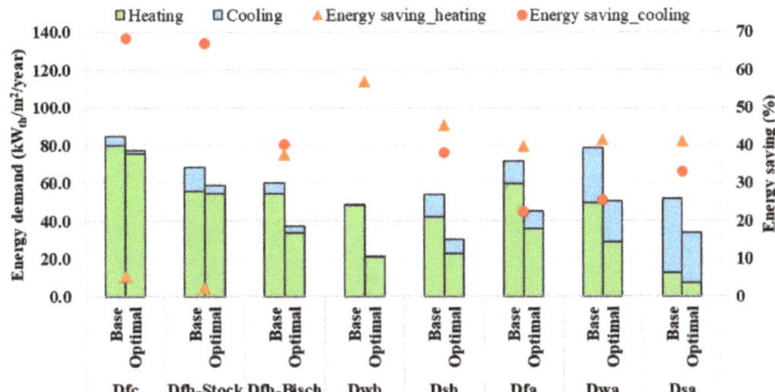

**Figure 13.** Heating and cooling loads and respective energy savings in a continental climate.

### 3.2.2. Temperate Climate

The temperate climate is characterized by the coldest month's average temperature between 0 °C and 18 °C and at least one month averaging above 10 °C. Figure 14 shows the energy demand of the base-case building and the energy demand and energy saving after optimization. The investigated locations show a mixed trend for space heating and cooling dominance. In contrast to the continental climate, the space cooling demand is relatively higher in temperate climates. The average energy savings for space heating and cooling are 33.36% and 50.98%, respectively. The energy saving is higher for lower energy demand and vice versa in both cases of heating and cooling. Therefore, a higher energy-saving potential in cooling load means these locations have lower cooling loads. Maximum heating demand and cooling demand, after optimization, are 42.59 $kW_{th}/m^2$ a in Cfb (Odense) and 18.54 $kW_{th}/m^2$ a in Cwa (Hanzhong), respectively.

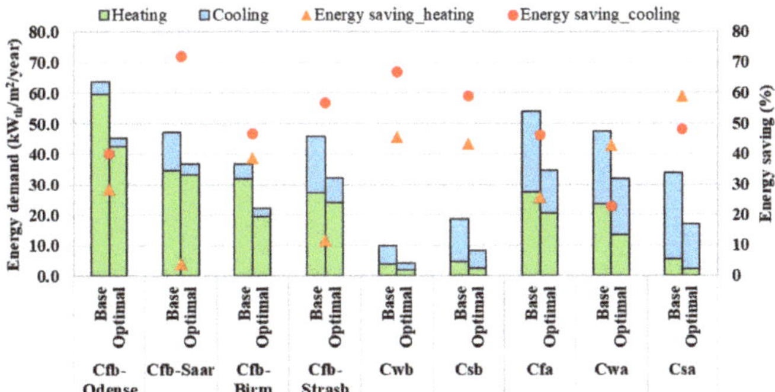

**Figure 14.** Heating and cooling loads and respective energy savings in a temperate climate.

### 3.2.3. Dry Climate

The dry climate is defined by very little precipitation during the year. Moreover, it has two subgroups based on the mean average temperature: hot, MAT $\geq$ 18 °C; and cold, MAT < 18 °C. Consequently, the locations in this zone have a long summer season and shortened winter season. The design optimization is equally effective in cooling-dominant climates, as shown in Figure 15. The energy saving in dry climates averages 33.3% for space cooling. Maximum cooling demand is 142.34 $kW_{th}/m^2$ a in the hot desert climate (BWh) of Dubai, which reduces to 99.45 $kW_{th}/m^2$ a after optimization.

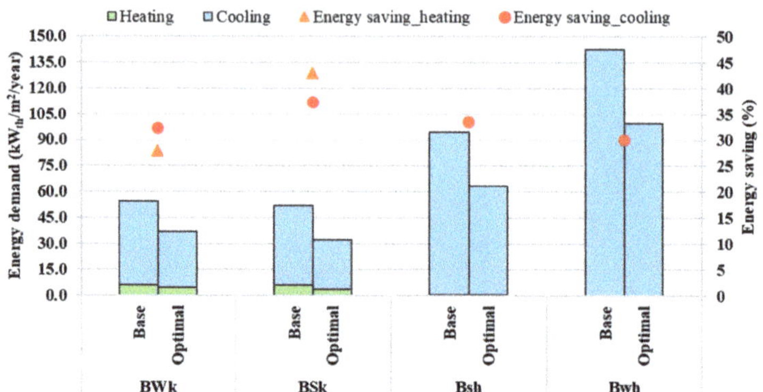

**Figure 15.** Heating and cooling loads and respective energy savings in a dry climate.

### 3.2.4. Tropical Climate

In a tropical climate, the average temperature of every month is 18 °C or higher, with significant year-round precipitation. The high humidity level throughout the year is another prominent feature of this climate. Therefore, there is no space-heating load in this climate, or it is so low as to be considered negligible. The space-cooling loads for three representative tropical climates and energy saving through design optimization are presented in Figure 16. The space cooling demand in Jakarta's tropical-monsoon climate (Am) is the lowest due to the higher precipitation. The highest energy saving is 29.4% in Mumbai, a tropical savanna climate (Aw). The average energy saving amounts to 25.95%, much lower than the heating-dominant continental and temperate climates.

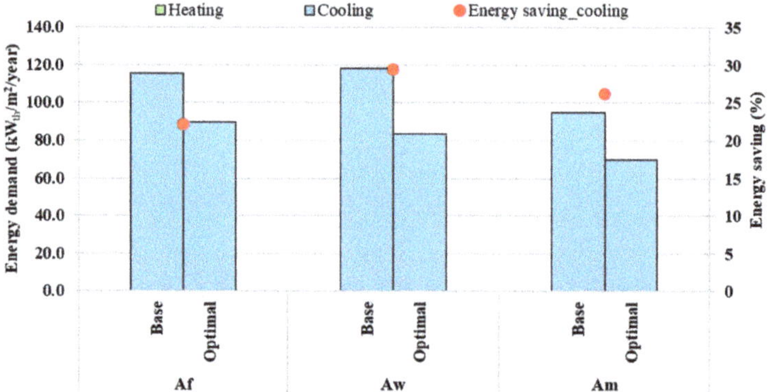

**Figure 16.** Heating and cooling loads and respective energy savings in a tropical climate.

### 3.3. Climatic Variation of Design Parameters

The optimization results show that the climate conditions significantly influence design parameters. Although each investigated climate requires a specific set of design parameters, the locations with similar climate conditions can be grouped to devise a climate adaptability pattern. Thus, the investigated climates were further categorized, as described in Table 8. The variation in each design parameter in the major climates zones is described in Sections 3.3.1–3.3.7.

**Table 8.** Categorization of the investigated climates.

| Category | Climates |
| --- | --- |
| Continental—cold | Dfc |
| Continental—warm summer | Dfb, Dwb, Dsb |
| Continental—hot summer | Dfa, Dwa, Dsa |
| Temperate—warm summer | Cfb, Cwb, Csb |
| Temperate—hot summer | Cfa, Cwa, Csa |
| Dry—cold | BWk, BSk |
| Dry—hot | Bwh, Bsh |
| Tropical | Af, Am, Aw |

### 3.3.1. External Wall Insulation

In general, the locations with extreme weather conditions have very insulated envelope to minimize transmission gains or losses. Figure 17 shows box plots of the change in the EPS thickness in different climates. The EPS thickness of the external wall is maximum, 0.247 m, in the continental—cold climate. It decreases to the mean value of 0.222 m in the temperate—warm climate because of the decreasing heating loads. In a temperate—hot climate, it is necessary to apply a higher level of insulation due to the significant cooling load. The dry—hot and tropical climates are cooling-dominant climates and require a larger insulation thickness of the external walls. Dry—cold climate has the lowest mean EPS thickness of 0.21 m, due to its lower space-heating and cooling loads than other climates.

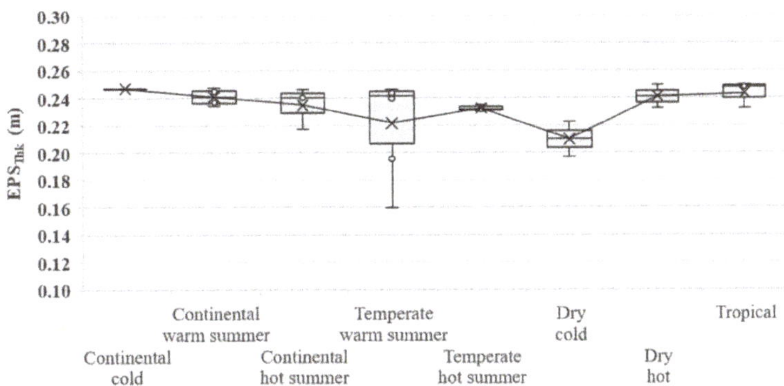


**Figure 17.** Variation in thickness of EPS insulation with climate.

### 3.3.2. Roof Insulation

The optimum solutions show that the insulation thickness of the roof is also higher than the base-case value in most of the climates, as illustrated in Figure 18. Though, it is lower than the EPS thickness of the external walls due to lower thermal conductivity of the rockwool material. The heating load of the location characterizes the rockwool thickness of the roof. Therefore, the roof has a higher rockwool thickness in continental and temperate climates than in dry and tropical climates. The heating load of hot summer climates is lower than warm summer climates. As a result, warm summer locations require larger insulation than the hot summer in continental and temperate climates. Similar to the EPS insulation for the external wall, the dry—cold climate has the minimum mean rockwool thickness of 0.159 m.

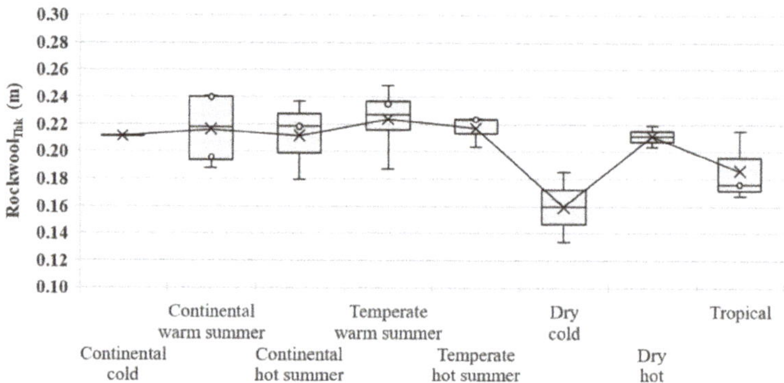

**Figure 18.** Variation in thickness of rockwool insulation with climate.

### 3.3.3. Window Aperture Angle

Figure 19 shows the variation in the optimal window aperture angle with changing climates. Since infiltration through window opening is active only for passive cooling, the aperture angle is more significant in hot climates. The aperture angle is higher in hot regions of continental, temperate, and dry climates. The dry—hot zone has a maximum mean aperture angle of 18.14 degrees. Although the dominant thermal load is cooling in a tropical climate, it also has a higher humidity level throughout the year. As a result, the aperture angle is relatively lower than the dry climate, and it even reduces to 5 degrees in Jakarta, a tropical-monsoon region. The dry—cold zone has the minimum aperture angle compared to other climate zones.

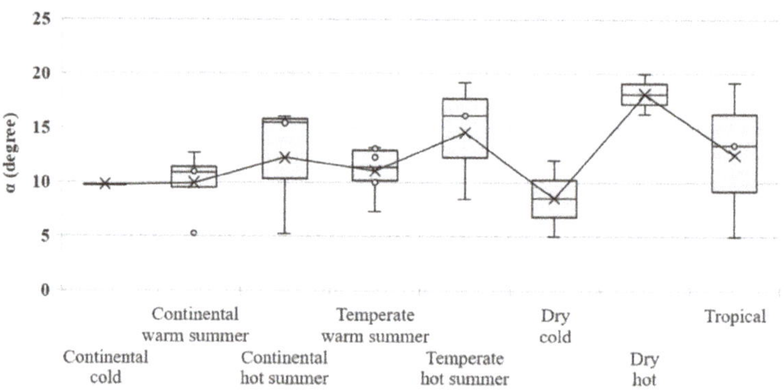

**Figure 19.** Variation in window aperture angle with climate.

### 3.3.4. Window-to-Wall Ratio

WWR is the most imperative element regarding the solar gains in both climates, heating-dominant or cooling-dominant. The box plots of WWR in different climate zones are presented in Figure 20. In general, the optimal solutions ascertain that the heating-dominant regions require higher WWR to maximize the solar gains. Therefore, the continental—cold climate has the maximum WWR, 0.37, continuously decreasing to the mean WWR of 0.27 in the temperate—hot summer climate. On the other hand, in the dry and tropical zone, the goal is to reduce the heat gains of solar radiation. Consequently, the WWR equals 0.2, the lower bound, in optimum solutions.

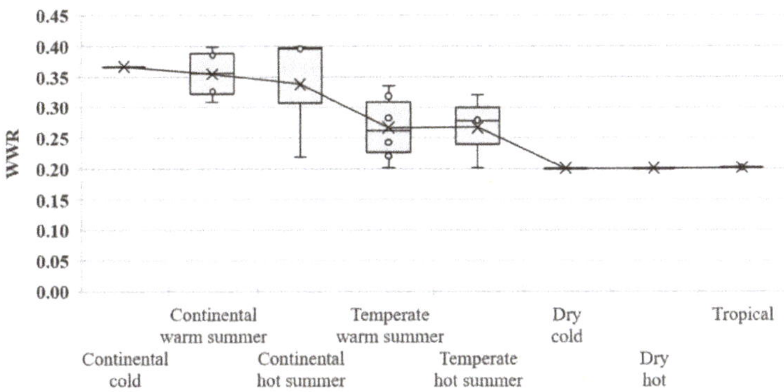

**Figure 20.** Variation in WWR with climate.

### 3.3.5. Solar Radiation for Shading Control

Minimum solar radiation to activate the window shading is another factor to control the solar gains into the household. Figure 21 shows the box plot of IT_H in different climate zones. This value is relatively higher in continental and temperate climates because the dominant thermal load is heating. The continental—cold climate has the maximum value of 279 W for IT_H. On the other hand, window shading activates at low solar radiation, around 250 W, in dry and tropical climates due to high cooling loads.

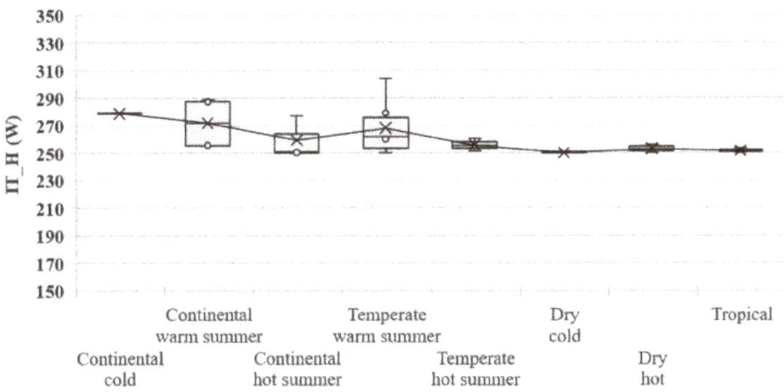

**Figure 21.** Variation in solar radiation for shading control with climate.

### 3.3.6. Window Shading Fraction

This study considers the shading fraction as a function of shading in December, the month of minimum shading fraction, as presented in Figure 6. To minimize the solar gains during summer, the optimal shading fraction in December remains close to the upper bound. Since the cooling load in Dwb (Daocheng) is negligible, the window shading fraction in December ($Shd_{Dec}$) drops to 0.253. For all other locations, it is above 0.326. On average, the shading fraction is 0.326 in December and 0.86 in May.

### 3.3.7. Building Orientation

The building orientation strongly influences the solar gains. Optimization results show that South or North is the optimal orientation in all climates. In most of the locations, the front facade is facing North, as shown in Figure 22. However, the locations in dry and temperate climate zones with minimal heating load have a South-facing optimal building orientation. The optimal orientation is an essential aspect of the building architecture since it increases energy efficiency without additional investment costs. It should be noted that all investigated climates are in the Northern hemisphere. Therefore, these results are only applicable for households having an architecture similar to the case building in the Northern hemisphere.

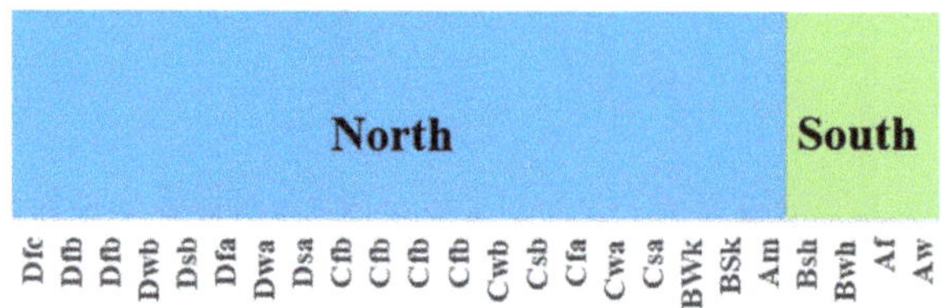


**Figure 22.** Optimal household orientation in the investigated climates.

## 4. Discussion

The bi-objective optimization poses two areas of discussion about the architectural design in different climate zones: energy saving in annual thermal load and the optimal design parameters.

For the case of energy saving through optimization, the continental—warm summer and continental—hot summer zones show 37.8% and 35.8% improvement on average in the annual thermal load, respectively. The temperate climate zone has the average energy-saving potential of 39.17% in warm summer regions and 39.51% in hot summer regions. This energy saving is higher due to the lower heating demand as compared to the continental climate. The design optimization more effectively restricts the transmission losses or gains in the heating-dominant climates. Therefore, energy saving decreases with the increased cooling demand. As a result, energy saving reduces to 34.8% and 31.6% in dry—cold and dry—hot climate zones. The tropical climate zone has the minimum energy saving of 26% in annual thermal load. Interestingly, the design optimization would also have a significant impact on the operational cost of the building. Since the optimization process shows a substantial improvement in the energy performance of the building, it would also decrease the operating cost compared to the base-case building. The operating cost of a building depends upon the type of equipment, energy-supply system, and local energy prices. Thus, the monetary savings from design optimization varies in each location. The current optimization process does not account for the cost of building operation, which is a limitation of this work.

With regard to the design parameters, Table 9 provides a criterion for selecting the energy-optimal ranges according to the climate zone and respective degree days. Although this criterion is based on the investigation of 24 cities in major climate zones, its legitimacy is asserted by achieving the optimal solution after a large number of simulations, i.e., 5000, in each location. However, the adaption of individual parameters is not advised because the design variables are strongly reliant on each other.

The continental—cold climate is represented by maximum HDD. So, it needs a high EPS thickness of 0.247 and rockwool thickness of 0.211, resulting in a low thermal transmittance of 0.15 W/m$^2$ K for the envelope. It is also characterized by large WWR, IT_H, and ShdDec. The continental—warm and continental—hot climates show a similar pattern for design variables, but their ranges drop with a decrease in HDD and increase in CDD. Furthermore, the window aperture angle needs to be increased from the continental—cold to continental—hot climate. Regarding the orientation, a North-facing household is the optimum choice in the continental climate zone.

**Table 9.** Ranges of optimal design parameters and degree days in different climate zones.

| | Continental | | | Temperate | | | Dry | | Tropical | |
|---|---|---|---|---|---|---|---|---|---|---|
| | Cold | Warm Summer | Hot Summer | Warm Summer | Hot Summer | Cold | Hot | | Rainforest/ Savanna | Monsoon |
| HDD18 | 5468 | 3496–3922 | 2320–3523 | 1137–3364 | 1404–2099 | 1182–1693 | 0–348 | 0 | 0 |
| CDD10 | 429 | 378–1305 | 1430–2523 | 835–2204 | 2115–2589 | 3052–3312 | 5382–6910 | 6594–6782 | 6045 |
| $EPS_{Thk}$ (m) | 0.247 | 0.234–0.247 | 0.216–0.245 | 0.160–0.246 | 0.231–0.238 | 0.197–0.223 | 0.232–0.249 | 0.247–0.25 | 0.232 |
| $Rockwool_{Thk}$ (m) | 0.211 | 0.188–0.241 | 0.187–0.239 | 0.187–0.249 | 0.203–0.227 | 0.133–0.185 | 0.203–0.219 | 0.167–1.176 | 0.215 |
| $U_w$ (W/m$^2$ K) | 0.15 | 0.150–0.157 | 0.150–0.169 | 0.150–0.222 | 0.157–0.159 | 0.165–0.184 | 0.149–0.159 | 0.148–0.150 | 0.159 |
| $U_r$ (W/m$^2$ K) | 0.154 | 0.136–0.170 | 0.138–0.178 | 0.132–0.171 | 0.145–0.159 | 0.173–0.230 | 0.148–0.159 | 0.180–0.189 | 0.151 |
| $\alpha$ (degree) | 9.8 | 5.2–11 | 5.2–16.1 | 7.3–13.2 | 8.4–17 | 5–12 | 16.3–20 | 13.4–19.2 | 5 |
| WWR | 0.37 | 0.33–0.4 | 0.2–0.4 | 0.2–0.34 | 0.2–0.32 | 0.2 | 0.2 | 0.2 | 0.2 |
| IT_H (W) | 279 | 255–289 | 251–277 | 250–304 | 251–255 | 250–251 | 250–256 | 251–253 | 250 |
| $Shd_{Dec}$ | 0.329 | 0.253–0.33 | 0.33 | 0.327–0.33 | 0.328–0.33 | 0.33 | 0.33 | 0.33 | 0.33 |
| Orientation | North | North | North | North | North | North | South | South | North |

The dominant thermal load in the temperate zone is space heating. Therefore, it also requires a high level of insulation and consequently lower thermal transmittance of the envelope. Interestingly, the lower limits of insulation materials and U-values are higher in the temperate—hot summer zone than temperate—warm summer zone, and the upper limits are low. The reason is that the lower limit of HDD is large, but the upper limit of HDD is small in this climate zone. Furthermore, the temperate—hot summer zone has a higher CDD. Moreover, the design variables responsible for solar gains are adjusted to minimize the solar heat gain; i.e., the ranges of WWR and IT_H decrease, and the ShdDec range increases in the temperate zone. The optimal orientation is North, the same as in the continental climate zone.

In the dry—cold zone, neither cooling nor heating is the dominant thermal load. As a result, the HDD and CDD are in the same range and have relatively low values. The insulation materials have a smaller thickness range of 0.197–0.223 m for EPS and 0.133–0.185 m for rockwool. Similarly, the U-values of the external walls and roof have relatively higher ranges of 0.165–0.184 W/m$^2$ K and 0.173–0.230 W/m$^2$ K, respectively. The window aperture angle is also smaller compared to the temperate climate. The WWR and IT_H are kept to the minimum, and ShdDec is maximized to restrict the solar gains. The optimal orientation is North in the dry—cold zone. On the other hand, the dry—hot climate is represented by a cooling-dominant thermal load and higher CDD. The optimal solution set is also quite different from that of a dry—cold climate. The thermal transmittance of the envelope is lower than the dry—cold temperate climate zones, and the window aperture angle ranges to its upper limit. The optimal orientation also changes to South in dry-hot climate. Nevertheless, other design variables are the same as in the dry—cold climate.

The tropical zone consists of cooling-dominant locations, and the CDD are above 3000 in all locations. The values of WWR, IT_H, and ShdDec follow the same trend as in other cooling-dominant climates. In Af and Aw climate zones, the EPS insulation is the highest of all climates and thus has the minimum thermal transmittance range, i.e., 0.148–0.150 W/m$^2$ K. However, the thermal transmittance has a comparatively higher range of 0.180–0.189 W/m$^2$ K. The Am climate has lesser CDD than other tropical climates and requires a relatively lower level of envelope insulation. The U-value of the external wall is 0.159 W/m$^2$ K, and it is 0.151 W/m$^2$ K for the roof. The aperture angle is 5 degrees in Am climate due to higher humidity levels throughout the year. The households are South-facing for optimum energy performance in Af and Aw climates, whereas in the Am climate the optimal orientation is North.

The simulation-based performance investigation of buildings is a well-established methodology to make appropriate decisions at the design stage. The experimental validation of building performance is time-consuming and financially infeasible. Nevertheless, the optimal values of the design parameters were compared with the previous studies for validation. Since the building design parameters are dependent on each other, the complete set of design parameters is taken for the explicit comparison. Previous studies used different combinations of envelope parameters, and those studies were conducted for limited climate zones. However, the individual parameters were compared for different climates with the available data from previous studies.

Table 10 shows the comparison of the envelope thermal transmittance of WWR values between the current study and previous studies. Though the optimal values do not exactly match due to differences in the constraints and optimization models, the results are consistent with the previous data. The thermal transmittance in continental climates is low, and WWR has a higher value. In temperate climates, the thermal transmittance of the external walls and roof is higher than the continental region in the current study and the previous studies as well, and the difference between the optimal solutions is very small. The WWR varies between 0.2 and 0.34 in temperate climates. The recommended WWR for temperate climate zones is 0.25. The WWR has approximately the same value in the current and previous studies for temperate—hot summer climates (Cfa and Csa). The thermal transmittances of the envelope and WWR are also coherent with the previous study for the dry—hot climate (Bsk). A study conducted a parametric analysis of the household envelope for thermal transmittance between 0.2 and 0.4 $W/m^2$ K in different climate zones. The optimal values were found to be 0.2 ($W/m^2$ K); i.e., the minimum thermal transmittance in cold and hot climates. In this study, the cold and hot climates are also characterized by lower thermal transmittance of the wall and roof, as shown in Table 9. Although the comparative analysis is given for the limited climate zones, it can be asserted on the basis of the consistency between the current and previous studies that the optimization has produced conclusive results for other climates.

**Table 10.** Comparison of the results with previous studies.

| Climate Zone | Parameters | Optimal Values | |
|---|---|---|---|
| | | **Current Study** | **Previous Studies** |
| Dfa | $U_w$ ($W/m^2$ K) | 0.15 | 0.14 [85] |
| Dwa | $U_w$ ($W/m^2$ K) | 0.154 | 0.12 [86] |
| | WWR | 0.4 | 0.31 |
| Csb | $U_w$ ($W/m^2$ K) | 0.185 | 0.16 [25] |
| | $U_r$ ($W/m^2$ K) | 0.171 | 0.16 |
| | WWR | 0.2 | 0.29 |
| Cfa | $U_w$ ($W/m^2$ K) | 0.159 | 0.19 [87] |
| | $U_r$ ($W/m^2$ K) | 0.146 | 0.18 |
| | WWR | 0.28 | 0.275 |
| Csa | $U_w$ ($W/m^2$ K) | 0.157 | 0.11 [25] |
| | $U_r$ ($W/m^2$ K) | 0.145 | 0.16 |
| | WWR | 0.2 | 0.19 |
| BSk | $U_w$ ($W/m^2$ K) | 0.165 | 0.18 [25] |
| | $U_r$ ($W/m^2$ K) | 0.173 | 0.16 |
| | WWR | 0.2 | 0.23 |
| Temperate | WWR | 0.2–0.34 | 0.25 [8] |
| Cold climate zones | $U_w$ ($W/m^2$ K) | 0.15–0.22 | 0.2 [35] |
| | $U_r$ ($W/m^2$ K) | 0.13–0.17 | 0.2 |
| Hot climate zones | $U_w$ ($W/m^2$ K) | 0.15–0.18 | 0.2 [35] |
| | $U_r$ ($W/m^2$ K) | 0.15–0.23 | 0.2 |

## 5. Conclusions

The present work analyzes a household's passive design parameters and thermal energy demand for its dependence on the climate. TRNSYS and Python-based NSGA-III were used for bi-objective optimization of a single-family household for twenty-four cities in twenty climate zones. The design variables of the optimization problem were insulation thickness of the envelope, window aperture angle, WWR, window shading fraction, radiation-based shading control, and orientation. Annual thermal energy demand and the investment cost of insulation were considered as the objective functions. A MCDM process was implemented through CRITIC and TOPSIS methods to find out the best solution from the PFs. Even though the optimization reduces the thermal load in all investigated climates, it is more effective in the heating-dominant regions. It is observed that the weather conditions strongly influence the passive design parameters. Moreover, the optimal solutions strictly rely on one another and do not indicate remarkable improvement in energy demand if implemented individually.

The optimization results show that thermal insulations of the envelope and WWR are the most perceptive design parameters as they determine the solar gains and transmission gains or losses of the household. In fact, the insulation material is thicker for high thermal loads. Therefore, the EPS thickness on external walls has higher ranges in continental—cold, continental—warm summer, dry—hot, and tropical climate zones. The rockwool thickness on the roof is larger in the heating-dominant locations of continental and temperate climate zones. The dry—cold climate zone is associated with mix climate conditions and requires a lower level of insulation. The windows in the south direction, which are exposed to sunlight for an extended period, try to increase the solar gains in continental and temperate climates. Similarly, the optimal orientation is North in those climates, enabling the façade with the maximum WWR to face South. On the contrary, dry—hot and tropical climate zones are characterized by cooling-dominant loads, minimum WWR, maximum $Shd_{Dec}$, and South (expect tropical-monsoon) as the optimal orientation.

The outcomes of this work provide comprehensive guidelines for the designers to make appropriate decisions about a household's passive design according to the climate. Previous building energy standards in the investigated locations provide only the limiting thermal transmittance values for the building envelope. These results set a benchmark for selecting energy-efficient envelope parameters and respective thermal transmittance ranges in the investigated climates, which can be applied worldwide, eliminating the traditional energy analysis process. This research study considers limited locations in each climate zone and does not perform statistical analysis for each climate. Therefore, further research should be conducted to statistically analyze these design parameters by considering more locations in each climate.

**Author Contributions:** Conceptualization, M.U. and G.F.; methodology, M.U.; software, M.U.; formal analysis, M.U.; writing—original draft, M.U.; supervision, G.F.; writing—review and editing, G.F. All authors have read and agreed to the published version of the manuscript.

**Funding:** This research received no external funding.

**Institutional Review Board Statement:** Not applicable.

**Informed Consent Statement:** Not applicable.

**Data Availability Statement:** The data presented in this study are available within the paper.

**Conflicts of Interest:** The authors declare no conflict of interest.

## References

1. IEA. *Energy Technology Perspectives 2020—Special Report on Clean Energy Innovation*; IEA: Paris, France, 2020. [CrossRef]
2. Global Alliance for Building and Construction. Executive Summary of the 2020 Global Status Report for Buildings and Construction. 2020. Available online: https://globalabc.org/sites/default/files/inline-files/Buildings%20GSR_Executive_Summary%20FINAL_0.pdf (accessed on 1 November 2021).
3. Energy Information Administration (EIA). International Energy Outlook 2019. 2019. Available online: https://www.eia.gov/outlooks/ieo/pdf/ieo2019.pdf (accessed on 1 November 2021).
4. IEA. *Energy Technology Perspectives 2017*; IEA: Paris, France, 2017. Available online: https://www.iea.org/reports/energy-technology-perspectives-2017 (accessed on 1 November 2021).
5. Far, C.; Far, H. Improving energy efficiency of existing residential buildings using effective thermal retrofit of building envelope. *Indoor Built Environ.* **2019**, *28*, 744–760. [CrossRef]

6. Echenagucia, T.M.; Capozzoli, A.; Cascone, Y.; Sassone, M. The early design stage of a building envelope: Multi-objective search through heating, cooling and lighting energy performance analysis. *Appl. Energy* **2015**, *154*, 577–591. [CrossRef]

7. Yu, W.; Li, B.; Jia, H.; Zhang, M.; Wang, D. Application of multi-objective genetic algorithm to optimize energy efficiency and thermal comfort in building design. *Energy Build.* **2015**, *88*, 135–143. [CrossRef]

8. Yong, S.G.; Kim, J.H.; Gim, Y.; Kim, J.; Cho, J.; Hong, H.; Baik, Y.J.; Koo, J. Impacts of building envelope design factors upon energy loads and their optimization in US standard climate zones using experimental design. *Energy Build.* **2017**, *141*, 1–15. [CrossRef]

9. Guo, Y.; Bart, D. Optimization of design parameters for office buildings with climatic adaptability based on energy demand and thermal comfort. *Sustainability* **2020**, *12*, 3540. [CrossRef]

10. Woolley, J.; Schiavon, S.; Bauman, F.; Raftery, P. Side-by-side laboratory comparison of radiant and all-air cooling: How natural ventilation cooling and heat gain characteristics impact space heat extraction rates and daily thermal energy use. *Energy Build.* **2019**, *200*, 68–85. [CrossRef]

11. Lapinskienė, V.; Motuzienė, V.; Džiugaitė-Tumėnienė, R.; Mikučionienė, R. Impact of internal heat gains on building's energy performance. In Proceedings of the 10th International Conference on Environmental Engineering, Vilnius, Lithuania, 27–28 April 2017. [CrossRef]

12. Acar, U.; Kaska, O.; Tokgoz, N. Multi-objective optimization of building envelope components at the preliminary design stage for residential buildings in Turkey. *J. Build. Eng.* **2021**, *42*, 102499. [CrossRef]

13. Dong, Y.; Sun, C.; Han, Y.; Liu, Q. Intelligent optimization: A novel framework to automatize multi-objective optimization of building daylighting and energy performances. *J. Build. Eng.* **2021**, *43*, 102804. [CrossRef]

14. Starke, A.R.; Cardemil, J.M.; Escobar, R.; Colle, S. Multi-objective optimization of hybrid CSP+PV system using genetic algorithm. *Energy* **2018**, *147*, 490–503. [CrossRef]

15. Jalali, Z.; Noorzai, E.; Heidari, S. Design and optimization of form and facade of an office building using the genetic algorithm. *Sci. Technol. Built Environ.* **2020**, *26*, 128–140. [CrossRef]

16. Zhang, A.; Bokel, R.; van den Dobbelsteen, A.; Sun, Y.; Huang, Q.; Zhang, Q. Optimization of thermal and daylight performance of school buildings based on a multi-objective genetic algorithm in the cold climate of China. *Energy Build.* **2017**, *139*, 371–384. [CrossRef]

17. Nasruddin, N.; Sholahudin, S.; Satrio, P.; Mahlia, T.M.I.; Giannetti, N.; Saito, K. Optimization of HVAC system energy consumption in a building using artificial neural network and multi-objective genetic algorithm. *Sustain. Energy Technol. Assess.* **2019**, *35*, 48–57. [CrossRef]

18. Yang, M.-D.; Lin, M.-D.; Lin, Y.-H.; Tsai, K.-T. Multiobjective optimization design of green building envelope material using a non-dominated sorting genetic algorithm. *Appl. Therm. Eng.* **2017**, *111*, 1255–1264. [CrossRef]

19. Li, K.; Pan, L.; Xue, W.; Jiang, H.; Mao, H. Multi-Objective Optimization for Energy Performance Improvement of Residential Buildings: A Comparative Study. *Energies* **2017**, *10*, 245. [CrossRef]

20. Ferdyn-Grygierek, J.; Grygierek, K. Multi-Variable Optimization of Building Thermal Design Using Genetic Algorithms. *Energies* **2017**, *10*, 1570. [CrossRef]

21. Mayer, M.J.; Szilágyi, A.; Gróf, G. Environmental and economic multi-objective optimization of a household level hybrid renewable energy system by genetic algorithm. *Appl. Energy* **2020**, *269*, 115058. [CrossRef]

22. Ghaderian, M.; Veysi, F. Multi-objective optimization of energy efficiency and thermal comfort in an existing office building using NSGA-II with fitness approximation: A case study. *J. Build. Eng.* **2021**, *41*, 102440. [CrossRef]

23. Penna, P.; Prada, A.; Cappelletti, F.; Gasparella, A. Multi-objectives optimization of Energy Efficiency Measures in existing buildings. *Energy Build.* **2015**, *95*, 57–69. [CrossRef]

24. Rabani, M.; Madessa, H.B.; Nord, N. Achieving zero-energy building performance with thermal and visual comfort enhancement through optimization of fenestration, envelope, shading device, and energy supply system. *Sustain. Energy Technol. Assess.* **2021**, *44*, 101020. [CrossRef]

25. Ascione, F.; de Masi, R.F.; de Rossi, F.; Ruggiero, S.; Vanoli, G.P. Optimization of building envelope design for nZEBs in Mediterranean climate: Performance analysis of residential case study. *Appl. Energy* **2016**, *183*, 938–957. [CrossRef]

26. Ferrara, M.; Vallée, J.C.; Shtrepi, L.; Astolfi, A.; Fabrizio, E. A thermal and acoustic co-simulation method for the multi-domain optimization of nearly zero energy buildings. *J. Build. Eng.* **2021**, *40*, 102699. [CrossRef]

27. Chang, S.; Castro-Lacouture, D.; Yamagata, Y. Decision support for retrofitting building envelopes using multi-objective optimization under uncertainties. *J. Build. Eng.* **2020**, *32*, 101413. [CrossRef]

28. Delgarm, N.; Sajadi, B.; Delgarm, S. Multi-objective optimization of building energy performance and indoor thermal comfort: A new method using artificial bee colony (ABC). *Energy Build.* **2016**, *131*, 42–53. [CrossRef]

29. Moghtadernejad, S.; Chouinard, L.E.; Mirza, M.S. Multi-criteria decision-making methods for preliminary design of sustainable facades. *J. Build. Eng.* **2018**, *19*, 181–190. [CrossRef]

30. Iwaro, J.; Mwasha, A.; Williams, R.G.; Zico, R. An Integrated Criteria Weighting Framework for the sustainable performance assessment and design of building envelope. *Renew. Sustain. Energy Rev.* **2014**, *29*, 417–434. [CrossRef]

31. Xu, C.; Ke, Y.; Li, Y.; Chu, H.; Wu, Y. Data-driven configuration optimization of an off-grid wind/PV/hydrogen system based on modified NSGA-II and CRITIC-TOPSIS. *Energy Convers. Manag.* **2020**, *215*, 112892. [CrossRef]

32. Babatunde, M.; Ighravwe, D. A CRITIC-TOPSIS framework for hybrid renewable energy systems evaluation under techno-economic requirements. *J. Proj. Manag.* **2019**, *4*, 109–126. [CrossRef]

33. Salameh, T.; Sayed, E.T.; Abdelkareem, M.A.; Olabi, A.G.; Rezk, H. Optimal selection and management of hybrid renewable energy System: Neom city as a case study. *Energy Convers. Manag.* **2021**, *244*, 114434. [CrossRef]

34. Zhao, J.; Du, Y. Multi-objective optimization design for windows and shading configuration considering energy consumption and thermal comfort: A case study for office building in different climatic regions of China. *Sol. Energy* **2020**, *206*, 997–1017. [CrossRef]

35. Harkouss, F.; Fardoun, F.; Biwole, P.H. Passive design optimization of low energy buildings in different climates. *Energy* **2018**, *165*, 591–613. [CrossRef]

36. Naji, S.; Aye, L.; Noguchi, M. Multi-objective optimisations of envelope components for a prefabricated house in six climate zones. *Appl. Energy* **2021**, *282*, 116012. [CrossRef]

37. Harkouss, F.; Fardoun, F.; Biwole, P.H. Multi-objective optimization methodology for net zero energy buildings. *J. Build. Eng.* **2018**, *16*, 57–71. [CrossRef]

38. Delgarm, N.; Sajadi, B.; Delgarm, S.; Kowsary, F. A novel approach for the simulation-based optimization of the buildings energy consumption using NSGA-II: Case study in Iran. *Energy Build.* **2016**, *127*, 552–560. [CrossRef]

39. Boverket. *The National Board of Housing, Building and Planning's Building Regulations*; BBR 18, BFS 2011:26; The Swedish National Board of Housing, Building and Planning: Stockholm, Sweden, 2011.

40. Mikulits, R.; Wolfgang, T. *OIB-Directives 6: Energy Saving and Thermal Protection*; Austrian Institute of Construction Engineering: Wien, Austria, 2015.

41. Longo, S.S.; Cellura, M.; Cusenza, M.A.; Guarino, F.; Marotta, I. Selecting insulating materials for building envelope: A life cycle approach. *Tec. Ital.-Ital. J. Eng. Sci.* **2021**, *65*, 312–316. [CrossRef]

42. Bienvenido-Huertas, D.; Oliveira, M.; Rubio-Bellido, C.; Marín, D. A comparative analysis of the international regulation of thermal properties in building envelope. *Sustainability* **2019**, *11*, 5574. [CrossRef]

43. Mahboob, M.; Rashid, T.U.; Amjad, M. Assessment of Energy Saving Potential in Residential Sector of Pakistan through Implementation of NEECA and PEC Building Standards. In Proceedings of the 2019 15th International Conference on Emerging Technologies (ICET), Peshawar, Pakistan, 2–3 December 2019; pp. 1–6.

44. Government of Dubai. *A Practice Guide for Building a Sustainable Dubai*; Dubai Green Building System: Dubai, United Arab Emirates, 2020.

45. Building and Construction Authority. *Code for Environmental Sustainability of Buildings*; Building and Construction Authority: Singapore, 2012.

46. Bureau of Energy Efficiency. *Energy Conservation Building Code*; Bureau of Energy Efficiency: Sewa Bhawan, India, 2017.

47. ASHRAE. Standard 90.2-2018, Energy Efficient Design of Low-Rise Residential Buildings. 2018. Available online: https://www.ashrae.org/news/esociety/newly-revised-standard-90-2-includes-new-performance-specifications-more (accessed on 1 November 2021).

48. Pacific Northwest National Laboratory. *Country Report on Building Energy Codes in China*; US Department of Energy: Washington, DC, USA, 2009.

49. Ramin, H.; Karimi, H. Optimum envelope design toward zero energy buildings in Iran. In *E3S Web of Conferences, 6–9 September 2020*; EDP Sciences: Tallinn, Estonia, 2020; p. 16004. [CrossRef]

50. Council, A. Energy efficiency building standards in Japan. *Acesso Em Março* **2007**, 1–9. Available online: http://www.asiabusinesscouncil.org/docs/BEE/papers/BEE_Policy_Japan.pdf (accessed on 1 November 2021).

51. Hansen, C.F.; Hansen, M.L. *Executive Order on the Publication of the Danish Building Regulations 2015 (BR15)*; The Danish Transport and Construction Agency: Copenhagen, Denmark, 2015.

52. Vogdt, F.U.; Walsdorf-Maul, M.; Schwabe, K.; Schaudienst, F.; Baumbach, A. Comparison between calculating methods of energy saving regulations and their economic efficiency. *Czas. Techniczne. Bud.* **2012**, *109*, 423–430.

53. Britain, G. *The Building Regulations 2010: Conservation of Fuel and Power: Approved Document L1A*; NBS: Chicago, IL, USA, 2014.

54. Bordier, R.; Rezaï, N. *Implementing the Energy Performance of Buildings Directive-EPBD Implementation in France*; EASME: Saint-Josse-ten-Noode, Belgium, 2018.

55. Bruke, R.V.; Gil, M.S.; Gonzalez, D.J.; Rodriguez, J.S. *Guía de Aplicación del DB-HE 2019—CTE*; Ministry of Transport, Mobility and Urban Agenda and Urban Agenda: Madrid, Spain, 2020.

56. Dott, R.; Ruschenburg, J.; Ochs, F.; Bony, J.; Haller, M. The Reference Framework for System Simulation of the IEA SHC Task 44/HPP Annex 38—Part B: Buildings and Space Heat Load. 2013. Available online: http://task44.iea-shc.org/data/sites/1/publications/T44A38_Rep_C1_B_ReferenceBuildingDescription_Final_Revised_130906.pdf (accessed on 1 November 2021).

57. TRANSSOLAR Energietechnik. Multizone Building Modeling with Type56 and TRNBuild. In Trnsys 18 Documantation; 2017; Volume 5. Available online: http://www.trnsys.com/ (accessed on 1 November 2021).

58. Parsons, R. (Ed.) *ASHRAE Handbook—Fundamentals*; American Society of Heating Refrigerating and Air-Conditioning Engineers: Atlanta, GA, USA, 1997.

59. ASHRAE Project Committee 90.1, Schedules and Internal Loads for Appendix C. Available online: https://web.ashrae.org/90_1files/ (accessed on 1 November 2021).

60. Parsons, R. (Ed.) *ASHRAE Handbook—Fundamentals*; American Society of Heating Refrigerating and Air-Conditioning Engineers: Atlanta, GA, USA, 2001.

61. Widén, J.; Lundh, M.; Vassileva, I.; Dahlquist, E.; Ellegård, K.; Wäckelgård, E. Constructing load profiles for household electricity and hot water from time-use data-Modelling approach and validation. *Energy Build.* **2009**, *41*, 753–768. [CrossRef]

62. Verein Deutscher Ingenieure. *VDI 4655—Reference Load Profiles of Single-Family and Multi-Family Houses for the Use of CHP Systems*; Verein Deutscher Ingenieure: Düsseldorf, Germany, 2008.

63. The Energy Informatics Group. PRECON Pakistan Residential Electricity Consumption Dataset. 2019. Available online: https://opendata.com.pk/dataset/precon-pakistan-residential-electricity-consumption-dataset (accessed on 1 November 2021).

64. Almohanna, I. Solar Dcathlon Middle East, Team KSU: Project Manual. 2019. Available online: https://www.solardecathlonme.com/2018/storage/reports/KSU-Project-manual.pdf (accessed on 1 November 2021).

65. Gupta, P.; Zan, T.T.T.; Dauwels, J.; Ukil, A. Flow-Based Estimation and Comparative Study of Gas Demand Profile for Residential Units in Singapore. *IEEE Trans. Sustain. Energy* **2019**, *10*, 1120–1128. [CrossRef]

66. Garg, A.; Shukla, P.R.; Maheshwari, J.; Upadhyay, J. An assessment of household electricity load curves and corresponding $CO_2$ marginal abatement cost curves for Gujarat state, India. *Energy Policy* **2014**, *66*, 568–584. [CrossRef]

67. Kubota, T.; Surahman, U.; Higashi, O. A comparative analysis of household energy consumption in Jakarta and Bandung. In Proceedings of the 30th International PLEA Conference: Sustainable Habitat for Developing Societies, Choosing the Way Forward, Gujarat, India, 16–18 December 2014; Volume 2, pp. 260–267.

68. Wang, H.; Fang, H.; Yu, X.; Liang, S. How real time pricing modifies Chinese households' electricity consumption. *J. Clean. Prod.* **2018**, *178*, 776–790. [CrossRef]

69. Hakimi, S.M. Multivariate stochastic modeling of washing machine loads profile in Iran. *Sustain. Cities Soc.* **2016**, *26*, 170–185. [CrossRef]

70. Shiraki, H.; Nakamura, S.; Ashina, S.; Honjo, K. Estimating the hourly electricity profile of Japanese households—Coupling of engineering and statistical methods. *Energy* **2016**, *114*, 478–491. [CrossRef]

71. Trotta, G. An empirical analysis of domestic electricity load profiles: Who consumes how much and when? *Appl. Energy* **2020**, *275*, 115399. [CrossRef]

72. Zimmermann, J.-P.; Evans, M.; Griggs, J.; King, N.; Harding, L.; Roberts, P.; Evans, C. Household Electricity Survey: A Study of Domestic Electrical Product Usage. 2012. Available online: https://www.gov.uk/government/uploads/system/uploads/attachment_data/file/208097/10043_R66141HouseholdElectricitySurveyFinalReportissue4.pdf (accessed on 1 November 2021).

73. Csoknyai, T.; Legardeur, J.; Akle, A.A.; Horváth, M. Analysis of energy consumption profiles in residential buildings and impact assessment of a serious game on occupants' behavior. *Energy Build.* **2019**, *196*, 1–20. [CrossRef]

74. Alberini, A.; Prettico, G.; Shen, C.; Torriti, J. Hot weather and residential hourly electricity demand in Italy. *Energy* **2019**, *177*, 44–56. [CrossRef]

75. Policy, S.E.; Francisco, D.; Freire, M.; Ferreira, C. Residential Sector Energy Consumption at the Spotlight: From Data to Knowledge. 2017. Available online: http://alteracoesclimaticas.ics.ulisboa.pt/wp-content/teses/2017JoaoGouveia.pdf (accessed on 1 November 2021).

76. Kuusela, P.; Norros, I.; Weiss, R.; Sorasalmi, T. Practical lognormal framework for household energy consumption modeling. *Energy Build.* **2015**, *108*, 223–235. [CrossRef]

77. Xiong, J.; Tzempelikos, A. Model-based shading and lighting controls considering visual comfort and energy use. *Sol. Energy* **2016**, *134*, 416–428. [CrossRef]

78. Bre, F.; Silva, A.S.; Ghisi, E.; Fachinotti, V.D. Residential building design optimisation using sensitivity analysis and genetic algorithm. *Energy Build.* **2016**, *133*, 853–866. [CrossRef]

79. ASHRAE. Standard 90.1-2019, Energy Standard for Buildings Except Low-Rise Residential Buildings. 2019. Available online: https://www.ashrae.org/technical-resources/bookstore/standard-90-1 (accessed on 1 November 2021).

80. Kottek, M.; Grieser, J.; Beck, C.; Rudolf, B.; Rubel, F. World Map of the Köppen-Geiger climate classification updated. *Meteorol. Z.* **2006**, *15*, 259–263. [CrossRef]

81. Benítez-Hidalgo, A.; Nebro, A.J.; García-Nieto, J.; Oregi, I.; del Ser, J. jMetalPy: A Python framework for multi-objective optimization with metaheuristics. *Swarm Evol. Comput.* **2019**, *51*, 100598. [CrossRef]

82. Li, H.; Deb, K.; Zhang, Q.; Suganthan, P.N.; Chen, L. Comparison between MOEA/D and NSGA-III on a set of many and multi-objective benchmark problems with challenging difficulties. *Swarm Evol. Comput.* **2019**, *46*, 104–117. [CrossRef]

83. Insulation Materials. Available online: https://www.obi.de/baustoffhalle/daemmstoffe/c/233#/ (accessed on 5 April 2021).

84. ASHRAE. Standard 55, Thermal Environmental Conditions for Human Occupancy. 2020. Available online: https://www.ashrae.org/technical-resources/bookstore/standard-55-thermal-environmental-conditions-for-human-occupancy (accessed on 1 November 2021).

85. Geng, Y.; Han, X.; Zhang, H.; Shi, L. Optimization and cost analysis of thickness of vacuum insulation panel for structural insulating panel buildings in cold climates. *J. Build. Eng.* **2021**, *33*, 101853. [CrossRef]

86. Wang, R.; Lu, S.; Feng, W. Impact of adjustment strategies on building design process in different climates oriented by multiple performance. *Appl. Energy* **2020**, *266*, 114822. [CrossRef]

87. Ascione, F.; Bianco, N.; Mauro, G.M.; Napolitano, D.F. Building envelope design: Multi-objective optimization to minimize energy consumption, global cost and thermal discomfort. Application to different Italian climatic zones. *Energy* **2019**, *174*, 359–374. [CrossRef]

 *sustainability*



*Article*

# Exploring Climate-Change Impacts on Energy Efficiency and Overheating Vulnerability of Bioclimatic Residential Buildings under Central European Climate

**Luka Pajek and Mitja Košir ***

Faculty of Civil and Geodetic Engineering, University of Ljubljana, 1000 Ljubljana, Slovenia; luka.pajek@fgg.uni-lj.si
* Correspondence: mitja.kosir@fgg.uni-lj.si

**Abstract:** Climate change is expected to expose the locked-in overheating risk concerning bioclimatic buildings adapted to a specific past climate state. The study aims to find energy-efficient building designs which are most resilient to overheating and increased cooling energy demands that will result from ongoing climate change. Therefore, a comprehensive parametric study of various passive building design measures was implemented, simulating the energy use of each combination for a temperate climate of Ljubljana, Slovenia. The approach to overheating vulnerability assessment was devised and applied using the increase in cooling energy demand as a performance indicator. The results showed that a B1 heating energy efficiency class according to the Slovenian Energy Performance Certificate classification was the highest attainable using the selected passive design parameters, while the energy demand for heating is projected to decrease over time. In contrast, the energy use for cooling is in general projected to increase. Furthermore, it was found that, in building models with higher heating energy use, low overheating vulnerability is easier to achieve. However, in models with high heating energy efficiency, very high overheating vulnerability is not expected. Accordingly, buildings should be designed for current heating energy efficiency and low vulnerability to future overheating. The paper shows a novel approach to bioclimatic building design with global warming adaptation integrated into the design process. It delivers recommendations for the energy-efficient, robust bioclimatic design of residential buildings in the Central European context, which are intended to guide designers and policymakers towards a resilient and sustainable built environment.

**Keywords:** climate change; bioclimatic design; passive design; energy efficiency; overheating; building resilience; robustness

**Citation:** Pajek, L.; Košir, M. Exploring Climate-Change Impacts on Energy Efficiency and Overheating Vulnerability of Bioclimatic Residential Buildings under Central European Climate. *Sustainability* **2021**, *13*, 6791. https://doi.org/10.3390/su13126791

Academic Editor: Luisa F. Cabeza

Received: 24 May 2021
Accepted: 12 June 2021
Published: 16 June 2021

**Publisher's Note:** MDPI stays neutral with regard to jurisdictional claims in published maps and institutional affiliations.



## 1. Introduction

Since Neolithic times, the building of homes has provided people with a higher degree of flexibility and independence in terms of climate and consequential habitability. Shelters and houses offered their occupants protection from the environment, predators and intruders [1]. Moreover, people were no longer forced to migrate towards flourishing regions with pleasant weather as the seasons passed and the climate changed. Thus, many relatively inhospitable environments were settled. Alongside the habitation of diverse climates, the struggle of builders to either utilise or fight the climatic characteristics of a location had begun. Only the best performing building design ideas were passed on, and thus, the knowledge on climate-adapted buildings was passed on intrinsically from generation to generation. Climate opportunities, together with the occupants' and society's needs and expectations, and the technological know-how about building, form the so-called triquetra of bioclimatic building design [1]. Therefore, the concept of bioclimatic building design is often associated with the harmonisation of climate, comfort, and energy efficiency [2]. The closer the building can follow and respond to the external dynamics, such as temperature, solar radiation and relative humidity, the more efficient it is [3].

Bioclimatic design is an engineering practice usually described through the building's ability to utilise climatic conditions and resources in a particular location to advance its performance. Hence, the goal is that a building and its elements should facilitate occupant's comfort through an energy- and resource-efficient approach by adapting to the location's climatic conditions to the highest reasonable degree [4,5]. In professional circles, the general opinion is that vernacular (i.e., traditional) architecture is perfectly adapted to the climatic characteristics of a specific location, as it is presumed that it has "evolutionarily" adapted to the given climate over the centuries. Therefore, vernacular architecture is often a source of bioclimatic strategies and corresponding passive design measures incorporated into new buildings [1,6,7]. Nowadays, in building design, bioclimatic strategies are regularly accompanied by sophisticated and expensive active systems that can dynamically reduce energy use and increase thermal comfort [8,9].

As indicated above, climate plays a crucial role in bioclimatic building design. While there are large parts of continents with the same climate type, in some parts of the Earth, such as the Alpine-Adriatic region in Europe, many climate types are found in a relatively small area [10]. According to Köppen–Geiger climate classification [11], the prevailing climates in Central Europe are warm temperate (i.e., C) and boreal (i.e., D), fully humid (i.e., f) climates with warm (i.e., b) or cool (i.e., c) summers. Such climate diversity results in specific bioclimatic architecture [12]. In these climates, a residential building designed according to the bioclimatic design paradigm should mainly facilitate passive solar gains, reduce thermal losses during the colder part of the year, and allow heat storage through high thermal mass of the envelope [1]. Furthermore, the thermal response of residential buildings under temperate and boreal climates is typically envelope dominated [13]. Therefore, implementing bioclimatic (i.e., passive) measures on the level of the building envelope might be highly efficient in optimising building heating energy use.

During the last century, evident changes in climate have been noted [14–18], and by the end of the twenty-first century, global temperature is projected to rise by up to 4 °C [19]. In the times of hunter-gatherer societies, people had the option of migrating to other, more pleasant regions in the event of significant climatic changes. Once buildings were added to the equation, migratory behaviour was no longer an attractive option as a climate adaptation strategy because one would leave behind the result of one's hard work—a building. Hence, climate-adapted buildings carry a possible built-in risk concerning climate change. However, according to the Migration and Climate Change Report [20], over 1 billion people are expected to face displacement by 2050 due to climate warming and related ecological threats. In particular, sub-Saharan Africa, South Asia, the Middle East, and North Africa face the most significant number of threats, such as lack of access to food and water and increased natural disasters occurrence [21]. On the other hand, developed regions in Europe and North America are expected to face fewer ecological threats [21]. Nevertheless, not giving them the immunity to broader implications of climate change, such as the impact on urbanised environments and buildings.

A warmer climate will inevitably affect the thermal performance of buildings, even bioclimatic buildings adapted to the current or past climate. Wang et al. [22] warned that there is an increasing need to clarify the challenges posed by climate warming to limit potential thermal discomfort by applying passive building measures. In climates present in Central Europe, the bioclimatic design measures integrated into buildings are based primarily on heating need to achieve comfort during the winter months. Namely, south-oriented windows for passive solar heating, building envelopes with low thermal conductivity and compact building shapes are commonly used in building design [23]. Nevertheless, the projected effects of a warming climate will lead to a risk of overheating for such buildings, especially if the line between a thermally comfortable and a hot environment is thin. Therefore, bioclimatic strategies used in buildings in such locations must be re-evaluated, as emphasised by Pajek and Košir [24]. Numerous studies have been conducted in order to assess the effects of climate change on building energy performance. Berardi and Jafarpur [25] in Toronto, Canada, showed an average decrease of 18–33% for heating and an average increase of 15–126% for cooling energy use by 2070, depending on climate file and building typology. Furthermore, Rodrigues and Fernandes [26] stated that, in residential buildings, a general increase in cooling demand (up to 137%) and a smaller reduction in heating demand (up to 63%) is expected until 2050 in Mediterranean locations, while the current ideal U-values will mainly not cause overheating. Bravo Dias et al. [27] explored climate change implications on passive building design efficiency in 43 most populated cities in the European Union. They concluded that buildings using passive design measures, whose performance is highly climate-dependent, will be particularly affected. For example, in Southern Europe, the shading season will increase by 2.5 months, making shading by overhangs or other fixed elements less effective.

Therefore, the selection of passive design measures should be based on the ability to achieve the highest possible resilience of a building. Martin and Sundley [28] define resilience as a process that involves several criteria, including vulnerability, resistance, robustness, and recoverability. According to Attia et al. [29], overheating vulnerability assessment considering future climate scenarios should be part of the building design process. Such an approach aims to achieve a design solution with less sensitive performance to "noise" in the form of change of the environmental boundary conditions [30]. Even in the animal world, the idea of resilient "building" can be found in ant gardens, which apparently allow the species to be more resilient to climate change than they would be outside of this system [31]. However, to assess the resilience of cities and buildings to climate change, studies of robustness and vulnerability evaluation have been made (see refs. [32–38]). For instance, Fonseca et al. [32] studied the effects of climate change on the energy use of buildings in the United States. They concluded that additional research is needed to provide more robust estimates of the impact of climate change on the building sector. Similarly, Shen and Lior [33] performed a vulnerability analysis on climate change impacts of present renewable energy systems used in net-zero energy buildings. Different authors, namely Moazami et al. [30], Kotireddy et al. [35], and others, presented workflows and methods for building performance robustness assessment to prevent significant variations in energy use. Given these points, Houghton and Castillo-Salgado [39] recommended using green building programs and certifications to help reduce the vulnerability of buildings to climate change.

Finally, the concept of building resilience concerning building energy use should be discussed, particularly in the context of the EU Energy performance of buildings directive (EPBD) [40]. To help enhance the energy performance of buildings, the EPBD also introduced building energy performance certification (EPC). However, in most countries, more than half of all existing residential buildings with registered EPCs have energy class D or lower [41]. On the other hand, the share of newly constructed nearly Zero-Energy Buildings (nZEB), also introduced through EPBD and characterised by high energy efficiency, is increasing. Furthermore, in 2020, the EU Commission presented its strategy to boost the energy renovation for climate neutrality of buildings in the EU [42]. For this reason, the vulnerability of buildings to climate change must be considered.

Bioclimatic principles are often associated with energy-efficient buildings, especially in temperate climates where buildings are primarily heating-dominated but have considerable potential for passive solar heating. Under such climatic conditions, buildings are usually designed to address the heating energy efficiency while overlooking the potential overheating risk during the warmer part of the year. Therefore, passive design measures, such as large equatorially oriented windows, compact building shapes, and highly thermally insulated envelopes, are commonly applied [43]. Nevertheless, it is unclear to what extent such design practices pose a potential lock-in overheating risk under projected climate scenarios. The paper aims at investigating potential solutions to simultaneously achieve high energy efficiency for the heating of bioclimatically designed buildings while at the same time maintaining low vulnerability to a warming climate. The study was conducted for Ljubljana, Slovenia, as a representative of a location with a temperate Central European climate. Energy models of bioclimatic buildings were evaluated against heating and cooling energy use, applying a comprehensive parametric analysis of passive design measures. The study's main objective was to demonstrate a novel approach to the bioclimatic design of buildings, where the adaptation and resistance to a warming climate are integrated into the design process. Hence, the paper presents recommendations for the adoption of resilient bioclimatic building design into practice and legislation.

## 2. Materials and Methods

The study's methodology was developed to enable the reaching of the above-stated objective of the paper. Thus, in principle, the applied methods can be split into four basic steps:

1. Sourcing historical climate data for the location of Ljubljana and preparing future climate data according to climate change projections using the morphing technique (Section 2.1).
2. Building energy model definition with corresponding variable parameters for the conducted parametric analysis (Section 2.2).
3. Definition of the methodology for energy performance evaluation based on the current Slovenian legislation (Section 2.3).
4. Definition of the methodology applied for overheating vulnerability analysis (Section 2.4).

### 2.1. Location and Climate

The study was performed for a Central European climate. As a representative of such climate, the location of Ljubljana (N 46.22, E 14.48, 385 m above sea level) in Slovenia was selected. This location is characterised by a warm temperate, fully humid climate with warm summers (Cfb according to Köppen–Geiger climate classification). The EPW climate file needed for building energy analysis was sourced from the International Weather for Energy Calculation (IWEC) database representing weather data measured between 1982 and 1999. In the paper, this climate data period was labelled as 1981–2010. Furthermore, the EPW of Ljubljana was used to generate projected EPW climate files for the periods 2011–2040, 2041–2070, and 2071–2100. The projected EPW files were generated using the morphing technique (i.e., time series adjustment method) according to the Intergovernmental Panel on Climate Change (IPCC) Special Report on Emissions Scenarios (SRES) A2 climate change scenario [44] and CCWorldWeatherGen tool [45]. The applied morphing technique uses historical climate data based on representative meteorological measurements in conjunction with projected global climate change patterns derived through numerical computer modelling to generate a new set of future projected climate. The use of recorded climate data as a starting point for future projected climate results in temporal continuity and spatial downscaling. The latter might be an issue for building energy simulations if only projections from global climate models are used.

### 2.2. Parametric Analysis

An extensive parametric analysis was carried out in order to study a vast pool of differently designed residential buildings. A single-family house with 162 m$^2$ of net floor area and a volume of 486 m$^3$ was chosen as the groundwork for the analysed energy models. Several building-related input parameters were fixed as constant for all the models considering the EN 16798-1 standard [46], meaningfully limiting the number of total possible combinations. Accordingly, the heating and cooling set-points were set to 21 °C and 26 °C, respectively, while the indoor temperature was controlled via the operative temperature. The summation of infiltration and natural ventilation was set to 0.60 h$^{-1}$ (April till October) and to 0.375 h$^{-1}$ (November till March). Internal heat gains and occupancy schedules were set according to EN 16798-1, Annex C [46]. Our previous analyses [47] have shown that external window shading is a crucial element of high energy performing bioclimatic buildings and was therefore not parametrised. It was set to block direct solar beams from April till October when incident solar radiation on the window was higher than 130 W/m$^2$ and external air temperature higher than 16 °C. The external thermal emissivity of all opaque building elements was set to 0.80.

The following variable input parameters were selected: opaque envelope thermal transmittance ($U_O$), window thermal transmittance ($U_W$) and the paired solar heat gain coefficient (SHGC), window to floor ratio (WFR), window distribution ($W_{dis}$), building shape expressed through shape factor ($f_0$), diurnal heat storage capacity (DHC) of load-bearing construction, external surface solar absorptivity ($\alpha_{sol}$), and summer natural ventilation cooling rate ($NV_C$) (see Table 1).

**Table 1.** Variable input parameters.

| Parameter | Parameter Range |
|---|---|
| $U_O$ [W/m$^2$K] | 0.10–1.00 |
| $U_W$ [W/m$^2$K] (paired SHGC [-]) | 0.60 (0.45)–2.40 (0.75) |
| WFR [%] | 5.0–45.0 |
| $W_{dis}$ [-] | 0.00, 1.00 [a] |
| $f_0$ [m$^{-1}$] | 0.78 (compact), 0.80 (semi-compact), 1.08 (non-compact) |
| DHC [kJ/m$^2$K] [b] | 63 (cross laminated timber), 98 (brick), 146 (concrete/stone) |
| $\alpha_{sol}$ [-] | 0.20–0.80 |
| $NV_C$ [h$^{-1}$] [c] | 0.0–8.0 |
| total number of models | 496,800 |

[a] 0.00 = equal area of windows at all orientations, 1.00 = south-concentrated windows (3.75% of WFR is distributed among all other orientations); [b] DHC is determined according to the principles presented by Bergman et al. [48]; [c] $NV_C$ is applied between April and October when the following conditions are met: internal air temperature is > 24 °C, external air temperature is between 16 and 30 °C, and temperature difference between internal and external air is $\leq$4 K.

Given the above-presented constant and variable building parameters, building energy models were formed in EnergyPlus [49]. Each model was divided into four thermal zones according to each cardinal axis. The jEPlus [50] software was used to conduct the parametric analysis. The annual building energy use for heating ($Q_{NH}$) and cooling ($Q_{NC}$) per square meter of floor area was calculated to evaluate the performance of each building model. Both $Q_{NH}$ and $Q_{NC}$ values represent the necessary thermal energy that needs to be delivered (or extracted in the case of cooling) to the thermodynamic system of a building in order to reach the specified internal thermal conditions. Therefore, these values do not reflect the effects of heating and cooling systems or specific fuels that would be used for running them. For a detailed explanation of the definition of building models, see the paper by Pajek and Košir [51], where the same methodology was used.

### 2.3. Energy Performance Evaluation

The annual energy use for heating ($Q_{NH}$) and cooling ($Q_{NC}$) of each building model was evaluated in relation to the Slovenian Rules on the efficient use of energy in buildings [52], which implements the EPBD requirements at the national level. These rules apply to all new buildings and all buildings being renovated or retrofitted, where at least 25% of the thermal envelope surface is retrofitted. The rules provide the highest allowed $Q_{NH}$ of a residential building per square meter of conditioned floor area, given by Equation (1):

$$Q_{NH} \leq 45 + 60 \times f_0 - 4.4 \times T_L \tag{1}$$

where $Q_{NH}$ is annual building energy use for heating in kWh/m$^2$, $f_0$ is the ratio between the area of the thermal envelope of the building and the net heated volume of the building in m$^{-1}$ (i.e., building shape factor), and $T_L$ is the average annual outdoor air temperature at the location in °C. $T_L$ for Ljubljana (1981–2010) is 10.7 °C [53].

Although the maximum allowed energy for heating depends on the building shape and location, the Rules on the efficient use of energy in buildings [52] limit the $Q_{NC}$ per square meter of the cooled area to 50 kWh/m$^2$, regardless of building shape and location. Table 2 shows the energy use limits, given the three different building shapes used in the study. The compliance of the building energy use with these rules was evaluated for the climate data, representing the period 1981–2010, since these are the climate data used in current energy efficiency analyses in practice.

**Table 2.** Building energy use upper limit according to the Slovenian Rules on the efficient use of energy in buildings by building shape [52] for the location of Ljubljana, Slovenia.

| $f_0$ | $Q_{NH}$ **Limit** | $Q_{NC}$ **Limit** |
|---|---|---|
| 0.78 (compact) | $\leq$44.7 kWh/m$^2$ | |
| 0.80 (semi-compact) | $\leq$45.9 kWh/m$^2$ | $\leq$50.0 kWh/m$^2$ |
| 1.08 (non-compact) | $\leq$62.7 kWh/m$^2$ | |

Furthermore, building models were classified into energy efficiency classes. They were given labels based on the Slovenian EPC classification (Rules on the methodology of production and issuance of energy performance certificates for buildings [54]). According to Slovenian rules, the EPC labels are based only on $Q_{NH}$ value. However, in the conducted study, each model was also labelled according to the $Q_{NC}$ value using the same methodology and criteria as for the $Q_{NH}$. The EPC labels, colour markings, and corresponding building energy use ranges are presented in Table 3.

**Table 3.** Energy Performance Certificate efficiency classification [54].

| Label | Energy Use [kWh/m$^2$] | Label Colour |
|---|---|---|
| A1 | $Q \leq 10$ | |
| A2 | $10 < Q \leq 15$ | |
| B1 | $15 < Q \leq 25$ | |
| B2 | $25 < Q \leq 35$ | |
| C | $35 < Q \leq 60$ | |
| D | $60 < Q \leq 105$ | |
| E | $105 < Q \leq 150$ | |
| F | $150 < Q \leq 210$ | |
| G | $Q > 210$ | |

*2.4. Overheating Vulnerability Analysis*

The vulnerability of building models to overheating was assessed by conducting a robustness analysis presented by Kotireddy et al. [34] using a minimax regret method. In this method, the performance regret for each climate scenario is the difference in performance between a building design and the best performing design in a given scenario. The maximum performance regret of a design across all scenarios is the measure of its robustness. Thus, the most robust design is the design with the lowest maximum performance regret. The minimax regret method can be explained through Equations (2)–(4).

$$R_{max,i} = \max\left(R_{i1}, R_{i2}, \ldots, R_{ij}\right) \tag{2}$$

$$R_{ij} = PI_{ij} - A_j \tag{3}$$

$$A_j = \min\left(PI_{1j}, PI_{2j}, \ldots, PI_{ij}\right) \tag{4}$$

where $R_{max,i}$ is the maximum performance regret of the i-th building model, $R_{ij}$ is the performance regret of the i-th building model in climate scenario j, $A_j$ is the minimum value of the performance indicator in climate scenario j, and $PI_{ij}$ is the performance indicator of the i-th building model in climate scenario j. Here, i = 1–496,800 and j = 1–4 since the performed parametric analysis resulted in 496,800 individual building models simulated through four different climate scenarios. As a performance indicator (i.e., PI), the increase in energy use for cooling (i.e., $\Delta Q_{NC}$) vis-à-vis the $Q_{NC}$ in the 1981–2010 climate was selected and was calculated for each building model in each future climate scenario, namely 2011–2040, 2041–2070, and 2071–2100 climate (see Section 2.1. Location and climate). Then, the building model with the highest climate change vulnerability, and thus the lowest robustness, was identified through Equation (5):

$$V_{max} = max(R_{max,i}) \tag{5}$$

where $V_{max}$ is the most vulnerable design.

Furthermore, the overheating vulnerability score (OV score) was calculated by normalising the performance regret of each building model with the performance regret of the most vulnerable building model. The building model with the lowest OV score (i.e., equal to 0) is the least vulnerable (i.e., the most robust), and the building model with the highest OV score (i.e., equal to 1) is the most vulnerable to climate change in terms of overheating vulnerability.

## 3. Results

### 3.1. Energy Efficiency

The parametrically simulated building energy models were evaluated concerning the compliance with the Slovenian Rules on the efficient use of energy in buildings. This was done to assess the possibility of meeting the requirements of these rules using exclusively the analysed bioclimatic (i.e., passive) design measures without using any active measures, such as mechanical heat recovery ventilation. The conformity with the rules was evaluated for the 1981–2010 period since these are the climate data used in current energy efficiency compliance assessments in Slovenia. The results showed that 15.7% of simulated building models were compliant with the maximum permissible heating energy use (i.e., $Q_{NH}$) criteria (see Table 2). The median $Q_{NH}$ of the energy-rule-compliant building models was 42.7 kWh/m$^2$, and the absolute best-performing model had a $Q_{NH}$ equal to 24.1 kWh/m$^2$. However, the $Q_{NH}$ threshold is related to $f_0$ of a particular building (see Table 2), which resulted in the fact that compliance with the $Q_{NH}$ criteria was easier achieved in the case of a less compact building design. Namely, the criteria were met in 22.5%, 13.5%, and 11.8% of building models with a non-compact (i.e., $f_0 = 1.08$), a semi-compact (i.e., $f_0 = 0.80$), and a compact (i.e., $f_0 = 0.78$) shape, respectively. At this point, caution should be exercised in generalizing the above-stated results. The described phenomenon is a consequence of the methodology used to determine the threshold $Q_{NH}$ (see Equation (1)) given in the Slovenian Rules on the efficient use of energy in buildings and not of better energy response of such building shape. In general, all the models meeting or surpassing the criteria of $Q_{NH}$ have an equal or lower value of $U_O$ than 0.25 W/m$^2$K. The other parameters are normally distributed. The cooling energy use ($Q_{NC}$) criterion (see Table 2) was achieved in all the analysed models since the highest $Q_{NC}$ of simulated models for the 1981–2010 period was 34.1 kWh/m$^2$. The $Q_{NC}$ of the analysed building models is projected to exceed the limit of 50 kWh/m$^2$ for the first time in the 2041–2070 period.

Furthermore, in order to gain a better insight into energy efficiency, the simulated building models in each of the analysed climate periods were classified according to the Slovenian Rules on the methodology of production and issuance of energy performance certificates for buildings (Figure 1). In general, the results in Figure 1 show that using the selected passive design measures results in building models with relatively satisfactory energy efficiency. Although none of the analysed building models was classified into heating energy efficiency classes A1 (i.e., $Q_{NH} < 10$ kWh/m$^2$) and A2 (i.e., $10 < Q_{NH} > 15$ kWh/m$^2$), either under the current or the future climate file, all the other classes (i.e., B1 through G) are represented (Figure 1). Under the influence of the projected climate change, the heating energy efficiency of the analysed buildings is projected to increase over time. The share of building models with higher heating energy efficiency (i.e., classes B1, B2 and C) is increasing. Accordingly, the share of less energy-efficient models is decreasing (i.e., classes D, E, F and G). This means that during the 1981–2010 period, roughly 28% of building models were in class C or higher ($Q_{NH} < 60$ kWh/m$^2$), while for the 2071–2100 period, this share almost doubled to 54%, an increase of 26 percentage points (p.p.). Furthermore, in the 1981–2010 period, only 37 (i.e., 0.01%) building models can be classified under heating energy efficiency label B1 (i.e., $15 < Q_{NH} > 25$ kWh/m$^2$), while this number increases to 13,740 (i.e., 2.77%) cases in the 2071–2100 period. In general, the most extensive changes in the shares of building models in individual heating energy efficiency classes between the 1981–2010 and 2071–2100 periods can be observed for class B2 and class F, an increase of 13 p.p. in the former and a decrease of 12 p.p. in the latter. Moreover, concerning the analysed building model population, it is projected that there will be no more models with a G heating energy efficiency label in the 2041–2070 period and beyond (Figure 1).

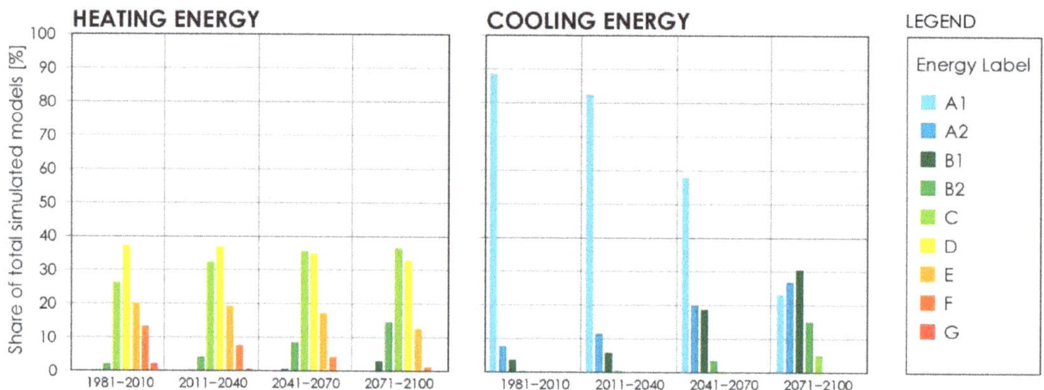


**Figure 1.** Share of total simulated models by heating and cooling energy label for each period.

Taking the 1981–2010 period as a starting point, the $Q_{NH}$ is expected to decrease by 24–39% until the end of the century, with an average decrease of 32%. Table 4 presents the limits (i.e., variance) of building model parameters necessary for achieving a specific heating energy efficiency label. It can be considered that in order to classify one of the analysed building models under the B1 heating energy efficiency label during the 1981–2010 climate, one may choose from a relatively limited pool of choices (i.e., min-max range of a specific parameter). The latter applies to the range of all investigated variable parameters (see Table 4, B1). The other heating energy classes offer more "freedom of choice" concerning the variance of analysed passive design measures.

Furthermore, concerning the cooling energy use of the analysed building models, good cooling energy efficiency can be achieved using passive design measures under the Ljubljana climate. For the 1981–2010 period, the majority (i.e., 89%) of building models can be classified into the A1 cooling energy-efficient label, while the remaining 11% fall at least in class B2 (i.e., $25 < Q_{NC} > 35$ kWh/m$^2$). However, the cooling energy efficiency of the analysed buildings is projected to decrease significantly over time. The share of the most energy-efficient building models (i.e., label A1) is projected to decrease by 66 p.p. between 1981–2010 and 2071–2100 periods with the A2, B1, B2 and C cooling energy efficiency labels increasing proportionally (Figure 1). After the 2041–2070 period, building models classified under labels C (5% in 2071–2100 period) and D (0.01% in 2071–2100 period) appear, which were not present before. Therefore, by the end of the 21st century, the $Q_{NC}$ of each building model is expected to increase by at least 59%, compared to the 1981–2010 period. For some instances, the $Q_{NC}$ increased from zero in 1981–2010 to up to 10 kWh/m$^2$ by the end of the 21st century. Table 5 presents the limits (i.e., variance) of building model parameters necessary for achieving a specific cooling energy efficiency label under the 2071–2100 climate file. In order to maintain the A1 cooling energy efficiency label in the future, the "freedom of choice" (i.e., min-max range) for the values of the varied parameters is not as limited as for heating energy use. Nevertheless, lower than the entire sample average $U_W$, WFR, and $\alpha_{sol}$, and higher than average DHC and $NV_C$ should be used.

**Table 4.** Typical building parameter values by heating energy label using the 1981–2010 climate file (i.e., "current" label).

| Variable Parameter | | Heating Energy Label in the 1981–2010 Period (i.e., "Current" Label) | | | | | | | Entire Sample Average |
|---|---|---|---|---|---|---|---|---|---|
| | | B1 | B2 | C | D | E | F | G | |
| $U_O$ [W/m²K] | mean | 0.10 | 0.10 | 0.16 | 0.34 | 0.63 | 0.90 | 0.99 | 0.43 |
| | min | 0.10 | 0.10 | 0.10 | 0.10 | 0.30 | 0.50 | 0.80 | 0.10 |
| | max | 0.10 | 0.15 | 0.40 | 0.80 | 1.00 | 1.00 | 1.00 | 1.00 |
| $U_W$ [W/m²K] | mean | 0.60 | 0.86 | 1.40 | 1.56 | 1.54 | 1.57 | 1.60 | 1.50 |
| | min | 0.60 | 0.60 | 0.60 | 0.60 | 0.60 | 0.60 | 0.60 | 0.60 |
| | max | 0.60 | 1.80 | 2.40 | 2.40 | 2.40 | 2.40 | 2.40 | 2.40 |
| WFR [%] | mean | 41.2 | 29.4 | 24.5 | 25.2 | 24.6 | 22.8 | 19.7 | 24.6 |
| | min | 35.0 | 10.0 | 5.0 | 5.0 | 5.0 | 5.0 | 5.0 | 5.0 |
| | max | 45.0 | 45.0 | 45.0 | 45.0 | 45.0 | 45.0 | 45.0 | 45.0 |
| $W_{dis}$ [-] | mean | 1.00 | 0.75 | 0.48 | 0.42 | 0.43 | 0.45 | 0.39 | 0.45 |
| | min | 1.00 | 0.00 | 0.00 | 0.00 | 0.00 | 0.00 | 0.00 | 0.00 |
| | max | 1.00 | 1.00 | 1.00 | 1.00 | 1.00 | 1.00 | 1.00 | 1.00 |
| $f_0$ [m⁻¹] | mean | 0.79 | 0.81 | 0.85 | 0.87 | 0.88 | 0.90 | 1.07 | 0.88 |
| | min | 0.78 | 0.78 | 0.78 | 0.78 | 0.78 | 0.78 | 0.80 | 0.78 |
| | max | 0.80 | 1.08 | 1.08 | 1.08 | 1.08 | 1.08 | 1.08 | 1.08 |
| DHC [kJ/m²K] | mean | 146 | 109 | 104 | 102 | 102 | 101 | 100 | 102 |
| | min | 146 | 63 | 63 | 63 | 63 | 63 | 63 | 63 |
| | max | 146 | 146 | 146 | 146 | 146 | 146 | 146 | 146 |
| $\alpha_{sol}$ [-] | mean | 0.75 | 0.55 | 0.52 | 0.51 | 0.50 | 0.46 | 0.34 | 0.50 |
| | min | 0.60 | 0.20 | 0.20 | 0.20 | 0.20 | 0.20 | 0.20 | 0.20 |
| | max | 0.80 | 0.80 | 0.80 | 0.80 | 0.80 | 0.80 | 0.80 | 0.80 |

**Table 5.** Typical building parameter values by cooling energy label using the 2071–2100 climate file.

| Variable Parameter | | Cooling Energy Label in the 2071–2100 Period (i.e., Projected Label) | | | | | | Entire Sample Average |
|---|---|---|---|---|---|---|---|---|
| | | A1 | A2 | B1 | B2 | C | D | |
| $U_O$ [W/m²K] | mean | 0.44 | 0.42 | 0.41 | 0.44 | 0.57 | 0.99 | 0.43 |
| | min | 0.10 | 0.10 | 0.10 | 0.10 | 0.10 | 0.80 | 0.10 |
| | max | 1.00 | 1.00 | 1.00 | 1.00 | 1.00 | 1.00 | 1.00 |
| $U_W$ [W/m²K] | mean | 1.36 | 1.43 | 1.51 | 1.69 | 1.86 | 2.27 | 1.50 |
| | min | 0.60 | 0.60 | 0.60 | 0.60 | 0.60 | 1.80 | 0.60 |
| | max | 2.40 | 2.40 | 2.40 | 2.40 | 2.40 | 2.40 | 2.40 |
| WFR [%] | mean | 13.2 | 20.4 | 29.5 | 35.0 | 38.2 | 44.6 | 24.6 |
| | min | 5.0 | 5.0 | 5.0 | 5.0 | 5.0 | 40.0 | 5.0 |
| | max | 45.0 | 45.0 | 45.0 | 45.0 | 45.0 | 45.0 | 45.0 |
| $W_{dis}$ [-] | mean | 0.49 | 0.46 | 0.37 | 0.46 | 0.52 | 0.92 | 0.45 |
| | min | 0.00 | 0.00 | 0.00 | 0.00 | 0.00 | 0.00 | 0.00 |
| | max | 1.00 | 1.00 | 1.00 | 1.00 | 1.00 | 1.00 | 1.00 |
| $f_0$ [m⁻¹] | mean | 0.90 | 0.89 | 0.88 | 0.84 | 0.83 | 0.79 | 0.88 |
| | min | 0.78 | 0.78 | 0.78 | 0.78 | 0.78 | 0.78 | 0.78 |
| | max | 1.08 | 1.08 | 1.08 | 1.08 | 1.08 | 0.80 | 1.08 |
| DHC [kJ/m²K] | mean | 110 | 106 | 102 | 93 | 79 | 63 | 102 |
| | min | 63 | 63 | 63 | 63 | 63 | 63 | 63 |
| | max | 146 | 146 | 146 | 146 | 146 | 63 | 146 |
| $\alpha_{sol}$ [-] | mean | 0.35 | 0.49 | 0.54 | 0.61 | 0.69 | 0.80 | 0.50 |
| | min | 0.20 | 0.20 | 0.20 | 0.20 | 0.20 | 0.80 | 0.20 |
| | max | 0.80 | 0.80 | 0.80 | 0.80 | 0.80 | 0.80 | 0.80 |
| $NV_C$ [h⁻¹] | mean | 4.6 | 4.0 | 3.9 | 3.6 | 3.1 | 2.8 | 4.0 |
| | min | 0.0 | 0.0 | 0.0 | 0.0 | 0.0 | 0.0 | 0.0 |
| | max | 8.0 | 8.0 | 8.0 | 8.0 | 8.0 | 8.0 | 8.0 |

### 3.2. Climate-Change Vulnerability

The above-presented results indicate that heating energy efficiency is projected to improve over time under the projected climate change scenario. Therefore, the overheating vulnerability analysis for each building model was made according to the heating energy efficiency label attainted under the 1981–2010 climate, as explained in Section 2.3. Figure 2 shows that models with different heating energy efficiency labels also have different overheating vulnerability score (OV score). However, since radiative forcing and global average temperatures are projected to increase over time due to climate change, the overheating risk of buildings is expected to follow that pattern. Consequently, the OV score is highest for buildings evaluated under the 2071–2100 climate (Figure 2).

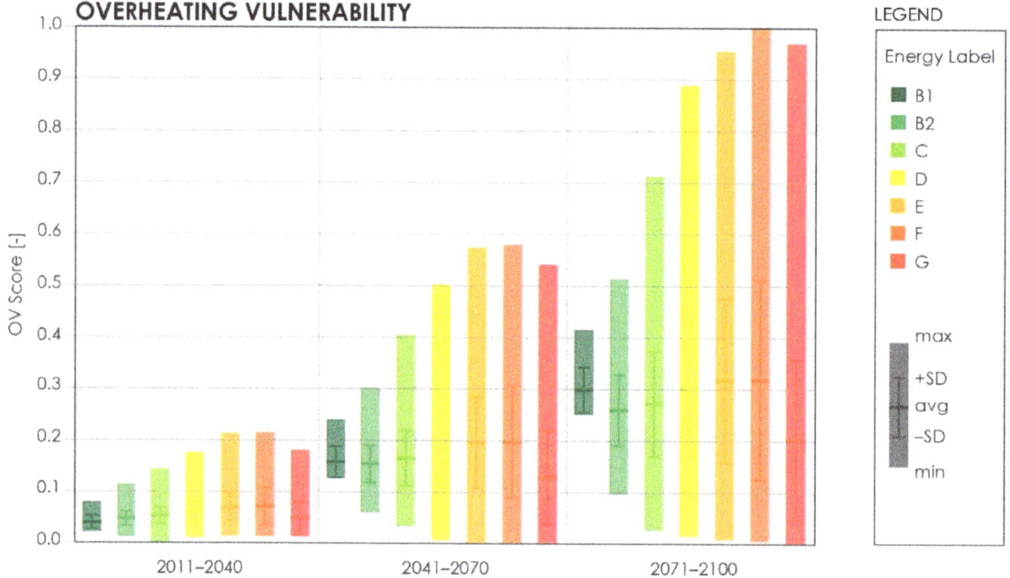


**Figure 2.** Overheating vulnerability score (OV score) of single-family houses in each future climate period. Building models are classified by heating energy label attained according to the 1981–2010 climate file, namely "current" heating energy label.

The average OV score is projected to increase similarly for all the energy labels. Building models classified under the B2 and C heating energy efficiency labels display on average the lowest susceptibility to increasing overheating vulnerability over the studied period. In particular, the average OV score of the B2 label buildings increases by 0.213 from 0.041 in 2011–2040 to 0.256 in 2071–2100. Simultaneously, the min-max range increases substantially from 0.093 in 2011–2040 to 0.413 in 2071–2100. Although the lower average OV score in 2041–2070 and 271–2100 periods are reached for the G labelled buildings, these buildings are also characterised by one of the highest min-max ranges (i.e., 0.971 in 2071–2100). Consequentially, this indicates that they have on average a low overheating risk, although individual building configurations can be very susceptible to it. The OV score min-max range is the narrowest in most heating energy-efficient buildings (i.e., B1 label), meaning that the overheating vulnerability is easier to control for highly heating energy-efficient buildings. Nevertheless, it should be stressed that buildings with the highest heating energy efficiency are generally not characterised by the lowest OV scores. Although in the 2011–2040 period, the B1 label buildings actually have the lowest average OV score (i.e., 0.034), the reached minimum score (i.e., 0.025) is higher than in the case of all other heating energy efficiency labels. The described situation is projected to escalate in the second part of the 21$^{st}$ century when the OV score of the B1 label buildings increases substantially (Figure 2). So much so that in the 2041–2070 period, the B2 and G labelled buildings have a lower average OV score, while in the 2071–2100 period, the B2, C, and G labelled buildings have lower average scores. This indicates that highly heating energy-efficient bioclimatic buildings (i.e., B1 label) are also characterised by substantial locked-in overheating risk. The main reason is that these models have south-concentrated large window areas (i.e., WFR higher than 35%, see Table 4). On the other hand, the maximum OV score of the B1 labelled buildings is the lowest in all periods (Figure 2). Therefore, when using passive design measures for high heating energy efficiency, an overall lower maximum OV score can be expected than in other designs (i.e., B2 to G labelled buildings).

The overall lowest overheating vulnerability score was achieved by a building model having poor thermal insulation ($U_O$ = 1.0 W/m$^2$K, namely 2 cm of thermal insulation), highly thermally insulated windows ($U_W$ = 0.6 W/m$^2$K, SHGC = 0.45), minimal window areas (WFR = 5%), a non-compact shape ($f_0$ = 1.08), high thermal mass (DHC = 146 kJ/m$^2$K), light-coloured external surfaces ($\alpha_{sol}$ = 0.20) and high rates of natural ventilation cooling ($NV_C$ = 8 h$^{-1}$). Its $Q_{NC}$ is projected to increase from 0.0 kWh/m$^2$ in the 1981–2010 period to 3.2 kWh/m$^2$ in 2071–2100. However, the building model is highly energy inefficient from the aspect of heating energy use (i.e., G heating energy efficiency label). On the other hand, the most overheating vulnerable building model is characterised by poor thermal insulation ($U_O$ = 1.0 W/m$^2$K), low thermally insulated windows ($U_W$ = 2.2 W/m$^2$K, SHGC = 0.75), equally distributed extremely large window area (WFR = 45%), a compact shape ($f_0$ = 0.78), high thermal mass (DHC = 146 kJ/m$^2$K), dark-coloured external surfaces ($\alpha_{sol}$ = 0.80) and without natural ventilation cooling ($NV_C$ = 0 h$^{-1}$). Its heating energy efficiency is classified under the F label, while its $Q_{NC}$ is projected to increase by 37.7 kWh/m$^2$, from 12.7 kWh/m$^2$ in the 1981–2010 period to 50.4 kWh/m$^2$ in 2071–2100, an increase of 297%. Table 6 shows typical values of building parameters by OV score percentiles. It can be concluded that, in general, the least prone to overheating (i.e., p05 in Table 6) were building models with above-average $U_O$, $W_{dis}$, $f_0$, DHC, and $NV_C$, and below-average $U_W$, WFR, and $\alpha_{sol}$.

**Table 6.** Typical building parameter values by long-term (2071–2100) overheating vulnerability score (OV score) percentiles.

| Variable Parameter | | Long-Term (2071–2100) OV Score Percentiles | | | | | | Entire Sample Average |
|---|---|---|---|---|---|---|---|---|
| | | p05 | Q1 | Q2 | Q3 | Q4 | p95 | |
| $U_O$ [W/m$^2$K] | mean | 0.49 | 0.42 | 0.41 | 0.38 | 0.51 | 0.74 | 0.43 |
| | min | 0.10 | 0.10 | 0.10 | 0.10 | 0.10 | 0.10 | 0.10 |
| | max | 1.00 | 1.00 | 1.00 | 1.00 | 1.00 | 1.00 | 1.00 |
| $U_W$ [W/m$^2$K] | mean | 1.30 | 1.35 | 1.40 | 1.51 | 1.74 | 1.78 | 1.50 |
| | min | 0.60 | 0.60 | 0.60 | 0.60 | 0.60 | 0.60 | 0.60 |
| | max | 2.40 | 2.40 | 2.40 | 2.40 | 2.40 | 2.40 | 2.40 |
| WFR [%] | mean | 9.6 | 13.8 | 20.8 | 29.6 | 34.0 | 34.2 | 24.6 |
| | min | 5.0 | 5.0 | 5.0 | 5.0 | 5.0 | 5.0 | 5.0 |
| | max | 40.0 | 45.0 | 45.0 | 45.0 | 45.0 | 45.0 | 45.0 |
| $W_{dis}$ [-] | mean | 0.49 | 0.52 | 0.46 | 0.42 | 0.38 | 0.26 | 0.45 |
| | min | 0.00 | 0.00 | 0.00 | 0.00 | 0.00 | 0.00 | 0.00 |
| | max | 1.00 | 1.00 | 1.00 | 1.00 | 1.00 | 1.00 | 1.00 |
| $f_0$ [m$^{-1}$] | mean | 0.94 | 0.89 | 0.89 | 0.87 | 0.85 | 0.85 | 0.88 |
| | min | 0.78 | 0.78 | 0.78 | 0.78 | 0.78 | 0.78 | 0.78 |
| | max | 1.08 | 1.08 | 1.08 | 1.08 | 1.08 | 0.80 | 1.08 |
| DHC [kJ/m$^2$K] | mean | 114 | 108 | 106 | 100 | 95 | 85 | 102 |
| | min | 63 | 63 | 63 | 63 | 63 | 63 | 63 |
| | max | 146 | 146 | 146 | 146 | 146 | 63 | 146 |
| $\alpha_{sol}$ [-] | mean | 0.24 | 0.35 | 0.49 | 0.51 | 0.64 | 0.74 | 0.50 |
| | min | 0.20 | 0.20 | 0.20 | 0.20 | 0.20 | 0.80 | 0.20 |
| | max | 0.80 | 0.80 | 0.80 | 0.80 | 0.80 | 0.80 | 0.80 |
| $NV_C$ [h$^{-1}$] | mean | 4.7 | 4.7 | 4.0 | 3.9 | 3.4 | 3.3 | 4.0 |
| | min | 0.0 | 0.0 | 0.0 | 0.0 | 0.0 | 0.0 | 0.0 |
| | max | 8.0 | 8.0 | 8.0 | 8.0 | 8.0 | 8.0 | 8.0 |

## 4. Discussion

In the bioclimatic design of buildings, the decision-making conditions are diverse, with several design objectives and criteria to be considered, particularly occupant comfort, energy efficiency, and daylighting [55–57]. In practice, trade-offs between these goals are very common, which need to be addressed appropriately. Only the energy efficiency aspect for providing thermal comfort was undertaken as a central part of this study, while occupant thermal comfort, indoor air quality and daylighting were not directly addressed. Therefore, the presented results should be interpreted in the exposed context. Similarly, the results should be understood in the framework of the applied passive design parameters and their value ranges. At the same time, several other design measures, such as evaporative cooling, fixed shading, sunspace, ground heat exchanger cooling, etc., were excluded from the analysis. Their exclusion from the analysis was based on the fact that they are either not common in the design practice (e.g., ground heat exchanger cooling) or ineffective (e.g., evaporative cooling) in the studied climatic context. Under these circumstances, the paper aimed to analyse the energy efficiency and overheating vulnerability of bioclimatic single-family houses in the Central European climate of Slovenia, Ljubljana. The energy efficiency was evaluated according to the annual energy use for heating ($Q_{NH}$) and cooling ($Q_{NC}$) per m$^2$ of building floor area. According to the Slovenian building energy efficiency rules, a B1 heating energy efficiency class was the highest achievable using the selected passive design parameters under the currently applicable climate file (i.e., 1981–2010 period) and the projected future climate scenarios. Nevertheless, a much warmer future climate is projected to improve the heating energy efficiency of such buildings because the energy needed for heating is projected to decrease.

Furthermore, it was highlighted that given the uncertainties of future climate, it is advisable to design buildings for current heating energy efficiency while aiming for low vulnerability to future overheating. Accordingly, Figure 3 displays three conceptual examples of a bioclimatic building designed for the analysed Central European temperate climate of Ljubljana. These three concepts were proposed after the interpretation of the study results. The first building (Figure 3a) corresponds to the B1 label heating energy efficiency with simultaneously the lowest overheating vulnerability score (OV score) of the buildings in the B1 energy label. Next, Figure 3b shows the building design, which meets the B2 label heating energy efficiency with the lowest OV score of the buildings in the B2 energy label. The last building (Figure 3c) is the least overheating vulnerable building design of the buildings that fall into the C label according to the heating energy efficiency. The $Q_{NH}$ value of each exposed building example intensifies from 24.7 kWh/m$^2$ (building B1) to 49.0 kWh/m$^2$ (building C) according to the 1981–2010 climate. At the same time, the $Q_{NC}$ follows the reverse trend. Namely, according to the 2071–2100 climate, the $Q_{NC}$ is highest for building B1 (18.6 kWh/m$^2$) and lowest for building C (4.1 kWh/m$^2$).

Although the best performing concept concerning the heating energy efficiency is the B1 building design (Figure 3a), it has several drawbacks regarding bioclimatic design. According to Potočnik and Košir [58], window size and glazing transmissivity are the dominant parameters to achieve adequate visual and non-visual indoor comfort. Therefore, vast south-concentrated window areas present a significant daylighting related drawback since they would be mainly shaded during summer. In contrast, during the rest of the year, glare might occur while utilising solar gains. On the other hand, building C, shown in Figure 3c, has minimal windows, resulting in potentially inadequate daylighting. It is also less heating energy-efficient than the other two presented design alternatives. Moreover, while using the WFR of 35% (Figure 3a), a natural summer ventilation rate (i.e., $NV_C$) above 4 h$^{-1}$ is recommended to achieve lower overheating vulnerability, which is, in reality, very hard and rarely achievable in residential buildings [59]. Although high-intensity natural ventilation is also preferred in the case of building B2 (Figure 3b), it is not as crucial. The reason is that building B2 has a smaller WFR, and thus solar heat gains and indoor surface temperatures are more governable. In all the best performing three cases, the lowest analysed $U_O$ and $U_W$ were used.

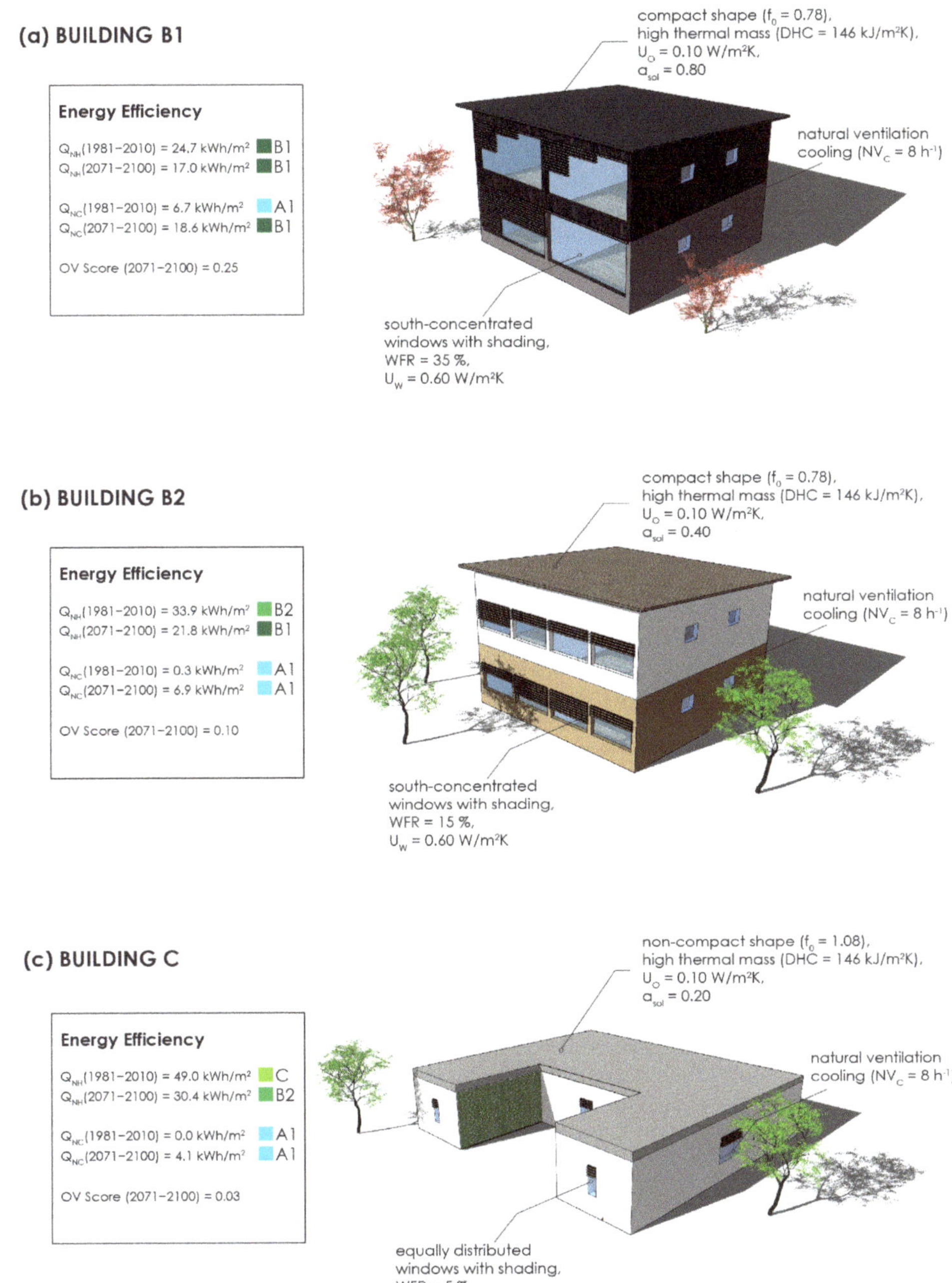



**Figure 3.** Three conceptual examples of bioclimatic building design for the analysed location. Examples represent a building of the most overheating resilient combination of passive measures for a building in: (**a**) B1 heating energy efficiency class; (**b**) B2 heating energy efficiency class; (**c**) C heating energy efficiency class. Each building has a useful floor area equal to 162 m².

Another fact worth noting is that the difference in $Q_{NH}$ between different examples in Figure 3 is projected to halve by the end of the century, while the difference in $Q_{NC}$ is projected to double or triple. Assume both heating and cooling energy use (i.e., $Q_{NH}$ + $Q_{NC}$) of the three buildings are taken together. In this case, it becomes evident that building B1 ($Q_{NH}$ + $Q_{NC}$ = 31.4 kWh/m$^2$) is the best performing in the 1981–2010 period, while building B2 ($Q_{NH}$ + $Q_{NC}$ = 28.7 kWh/m$^2$) is the best performing and building B1 is the worst performing ($Q_{NH}$ + $Q_{NC}$ = 35.6 kWh/m$^2$) in the 2071–2100 period. Furthermore, of the three, building B1 is the only one with higher cumulative heating and cooling energy use in the 2071–2100 period compared to the 1981–2010 period. Therefore, to achieve adequate heating energy efficiency, assure low overheating vulnerability, and at the same time create conditions for adequate daylighting, the combination of passive design measures presented in the case of building B2 (Figure 3b) or similar should be used. Of course, the highlighted findings are limited to the building geometries and envelope configurations considered. Therefore, substantially differently configured buildings may be designed while being aware of their effects on energy use.

Accordingly, it is recommended to use highly thermally insulated building envelopes, especially windows. Furthermore, not too large window areas should be adopted, e.g., WFRs in the range of 10–25%. The windows can be concentrated on the south façade (e.g., window to wall ratio (WWR) between 20 and 60%) for autumn–spring solar harvesting. South concentrated windows also prevent unwanted solar gains in the forenoon and the afternoon during summer. Accordingly, fixed overhangs on the south façade can be used for partial shading. However, in the case of south-concentrated windows, external shading (e.g., blinds) of the entire glazed surface for overheating prevention should be applied. Furthermore, shading operation should be automatically controlled since the overheating risk would be higher if shading devices were manually controlled by occupants [60]. Concerning the building shape, a more compact design is recommended. It is also suggested to use massive construction materials to increase the thermal capacity of the building. Otherwise, the thermal mass should be added in other forms, such as capacitive furniture [61] or phase change materials [62]. Although the B1 heating energy efficiency class can only be achieved using dark coloured external surfaces, it is recommended to use lighter colours (e.g., $\alpha_{sol}$ = 0.40–0.60) that reduce overheating vulnerability. Alternatively, vegetated surfaces (see Figure 3c) [63] or "cool" surface finishes [64] may be used to act as an effective overheating prevention measure. It is advisable to cool spaces using natural ventilation in summer when conditions allow, typically during the night. To this end, cross ventilation or stack ventilation of the building should be made possible by the appropriate arrangement of rooms and openings.

In addition to the presented and proposed passive design measures, additional either active or passive measures could be applied to reduce the energy use of a building. In particular, heating energy efficiency can be further improved by applying the heat recovery mechanical ventilation, improving the airtightness of the envelope, optimising occupant behaviour and similar. Besides, renewable energy sources, such as solar energy through PV or BIPV systems or solar collectors, are advisable [65]. In either case, an emphasis should be placed on long-term overheating vulnerability and not just current heating and cooling energy efficiency. In this way, high resilience and sustainability of the built environment may be achieved, primarily by raising the awareness of designers and policymakers.

## 5. Conclusions

Our civilisation faces the same frustration as the first humans—a struggle to build homes that provide safety and climate independence. As the presented research has demonstrated, the effort continues, while we still have a lot to learn about global warming and its implications for the (energy) performance of the built environment, especially with a limited amount of natural resources. The study successfully demonstrated a novel approach to the bioclimatic design of buildings by attaining current and future energy efficiency while also addressing climate adaptation and overheating resistance. The results of this paper clarify the overall picture concerning the design of bioclimatic residential buildings in the Central European climate. The main conclusions and novelty of the paper can be summarised as:

- The paper demonstrates how to assess overheating vulnerability of bioclimatic buildings. In Central Europe, overheating vulnerability is a significant but often overlooked concern in building design, as designers and policymakers focus primarily on heating energy efficiency. However, overheating vulnerability assessment is required since climate change is projected to negatively affect the cooling energy need of buildings, especially those designed for passive solar energy harvesting during the colder part of the year.
- Recommendations for the energy-efficient resilient bioclimatic building design in Central European temperate climate are given. Such recommendations are needed because residential buildings under this climate are heating-dominated, and with a warming climate comes the risk of overheating. Nevertheless, adapting buildings to current heating energy efficiency requirements while aiming for low vulnerability to future overheating can be achieved with reasonable trade-offs presented in the paper.
- Lastly, the results provide designers and policymakers with information to adopt a resilient bioclimatic building design approach into practice and regulations. A clear path towards the resilience and sustainability of buildings should be defined according to the study findings to preserve resources and mitigate climate change.

**Author Contributions:** Conceptualization, L.P. and M.K.; methodology, L.P. and M.K.; software, L.P.; validation, L.P. and M.K.; formal analysis, L.P.; investigation, L.P.; resources, M.K.; data curation, L.P.; writing—original draft preparation, L.P.; writing—review and editing, M.K.; visualization, L.P.; supervision, M.K. Both authors have read and agreed to the published version of the manuscript.

**Funding:** This research was funded by the Slovenian Research Agency (research core funding No. P2—0158).

**Institutional Review Board Statement:** Not applicable.

**Informed Consent Statement:** Not applicable.

**Data Availability Statement:** The data presented in this study are available on request from the corresponding author. The data are not publicly available as they are not stored on a publicly accessible repository.

**Conflicts of Interest:** The authors declare no conflict of interest.

## References

1. Košir, M. *Climate Adaptability of Buildings: Bioclimatic Design in the Light of Climate Change*; Springer International Publishing: Cham, Switzerland, 2019; ISBN 978-3-030-18455-1.
2. Almusaed, A. *Biophilic and Bioclimatic Architecture: Analytical Therapy for the Next Generation of Passive Sustainable Architecture*; Springer: London, UK, 2011; ISBN 978-1-84996-533-0.
3. Krainer, A. Passivhaus contra bioclimatic design. *Bauphysik* **2008**, 393–404. [CrossRef]
4. Szokolay, S.V. *Introduction to Architectural Science: The Basis of Sustainable Design*, 3rd ed.; Routledge: Oxfordshire, UK, 2014; ISBN 978-0-415-82498-9.
5. Maciel, A.A.; Ford, B.; Lamberts, R. Main influences on the design philosophy and knowledge basis to bioclimatic integration into architectural design—The example of best practices. *Build. Environ.* **2007**, *42*, 3762–3773. [CrossRef]
6. Desogus, G.; Felice Cannas, L.G.; Sanna, A. Bioclimatic lessons from Mediterranean vernacular architecture: The Sardinian case study. *Energy Build.* **2016**, *129*, 574–588. [CrossRef]
7. Oikonomou, A.; Bougiatioti, F. Architectural structure and environmental performance of the traditional buildings in Florina, NW Greece. *Build. Environ.* **2011**, *46*, 669–689. [CrossRef]
8. Yang, L.; Yan, H.; Lam, J.C. Thermal comfort and building energy consumption implications–A review. *Appl. Energy* **2014**, *115*, 164–173. [CrossRef]
9. Halhoul Merabet, G.; Essaaidi, M.; Ben Haddou, M.; Qolomany, B.; Qadir, J.; Anan, M.; Al-Fuqaha, A.; Abid, M.R.; Benhaddou, D. Intelligent building control systems for thermal comfort and energy-efficiency: A systematic review of artificial intelligence-assisted techniques. *Renew. Sustain. Energy Rev.* **2021**, *144*, 110969. [CrossRef]
10. Pajek, L.; Košir, M. Can building energy performance be predicted by a bioclimatic potential analysis? Case study of the alpine-adriatic region. *Energy Build.* **2017**, *139*, 160–173. [CrossRef]

11. Kottek, M.; Grieser, J.; Beck, C.; Rudolf, B.; Rubel, F. World map of the Köppen-Geiger climate classification updated. *Meteorol. Z.* **2006**, *15*, 259–263. [CrossRef]

12. Krainer, A. *Vernacular Buildings in Slovenia*; Architectural Association Graduate School: London, UK, 1993; ISBN 978-0-9525703-6-3.

13. Haggard, K.L.; Bainbridge, D.A.; Aljilani, R.; Goswami, D.Y. *Passive Solar Architecture Pocket Reference Book*; Earthscan: London, UK, 2009; ISBN 978-1-84971-080-0.

14. Houghton, J.T.; Ding, Y.; Griggs, D.J.; Noguer, M.; van der Linden, P.J.; Dai, X.; Maskell, K.; Johnson, C.A. *Climate Change 2001: The Scientific Basis: Contribution of Working Group I to the Third Assessment Report of the Intergovernmental Panel on Climate Change*; Houghton, J.T., Intergovernmental Panel on Climate Change, Eds.; Cambridge University Press: Cambridge, UK, 2001; ISBN 978-0-521-80767-8.

15. Levitus, S.; Antonov, J.I.; Wang, J.; Delworth, T.L.; Dixon, K.W.; Broccoli, A.J. Anthropogenic warming of Earth's climate system. *Science* **2001**, *292*, 267–270. [CrossRef]

16. Zwiers, F.W. Climate Change: The 20-year forecast. *Nature* **2002**, *416*, 690–691. [CrossRef]

17. Houghton, J.T. *Global Warming: The Complete Briefing*, 5th ed.; Cambridge University Press: Cambridge, UK, 2015; ISBN 978-1-107-09167-2.

18. Rubel, F.; Kottek, M. Observed and projected climate shifts 1901–2100 depicted by world maps of the Köppen-Geiger climate classification. *Meteorol. Z.* **2010**, *19*, 135–141. [CrossRef]

19. Collins, M.; Knutti, M.; Arblaster, J.; Dufresne, J.-L.; Fichefet, T.; Friedlingstein, P.; Gao, X.; Gutowski, W.J.; Johns, T.; Krinner, G.; et al. Long-term climate change: Projections, commitments and irreversibility. In *Climate Change 2013: The Physical Science Basis. Contribution of Working Group I to the Fifth Assessment Report of the Intergovernmental Panel on Climate Change*; Climate Change 2013: The Physical Science Basis. Contribution of Working Group I to the Fifth Assessment Report of the Intergovernmental Panel on Climate Change; Cambridge University Press: Cambride, UK; New York, NY, USA, 2013.

20. Brown, O. *IOM Migration Research Series No. 31: Migration and Climate Change*; International Organization for Migration (IOM): Geneva, Switzerland, 2008.

21. IEP. *Institute for Economics & Peace Ecological Threat Register 2020: Understanding Ecological Threats, Resilience and Peace*; Institute for Economics & Peace (IEP): Sydney, Australia, 2020.

22. Wang, S.; Liu, Y.; Cao, Q.; Li, H.; Yu, Y.; Yang, L. Applicability of passive design strategies in China promoted under global warming in past half century. *Build. Environ.* **2021**, *195*, 107777. [CrossRef]

23. La Roche, P. *Carbon-Neutral Architectural Design*, 2nd ed.; CRC Press, Taylor & Francis Group: Boca Raton, FL, USA, 2017; ISBN 978-1-4987-1429-7.

24. Pajek, L.; Košir, M. Implications of present and upcoming changes in bioclimatic potential for energy performance of residential buildings. *Build. Environ.* **2018**, *127*, 157–172. [CrossRef]

25. Berardi, U.; Jafarpur, P. Assessing the impact of climate change on building heating and cooling energy demand in Canada. *Renew. Sustain. Energy Rev.* **2020**, *121*, 109681. [CrossRef]

26. Rodrigues, E.; Fernandes, M.S. Overheating risk in Mediterranean residential buildings: Comparison of current and future climate scenarios. *Appl. Energy* **2020**, *259*, 114110. [CrossRef]

27. Bravo Dias, J.; Soares, P.M.M.; Carrilho da Graça, G. The shape of days to come: Effects of climate change on low energy buildings. *Build. Environ.* **2020**, *181*, 107125. [CrossRef]

28. Martin, R.; Sunley, P. On the notion of regional economic resilience: Conceptualization and explanation. *J. Econ. Geogr.* **2015**, *15*, 1–42. [CrossRef]

29. Attia, S.; Levinson, R.; Ndongo, E.; Holzer, P.; Berk Kazanci, O.; Homaei, S.; Zhang, C.; Olesen, B.W.; Qi, D.; Hamdy, M.; et al. Resilient cooling of buildings to protect against heat waves and power outages: Key concepts and definition. *Energy Build.* **2021**, *239*, 110869. [CrossRef]

30. Moazami, A.; Carlucci, S.; Nik, V.M.; Geving, S. Towards climate robust buildings: An innovative method for designing buildings with robust energy performance under climate change. *Energy Build.* **2019**, *202*, 109378. [CrossRef]

31. Morales-Linares, J.; Corona-López, A.M.; Toledo-Hernández, V.H.; Flores-Palacios, A. Ant-gardens: A specialized ant-epiphyte mutualism capable of facing the effects of climate change. *Biodivers. Conserv.* **2021**, *30*, 1165–1187. [CrossRef]

32. Fonseca, J.A.; Nevat, I.; Peters, G.W. Quantifying the uncertain effects of climate change on building energy consumption across the United States. *Appl. Energy* **2020**, *277*, 115556. [CrossRef]

33. Shen, P.; Lior, N. Vulnerability to climate change impacts of present renewable energy systems designed for achieving net-zero energy buildings. *Energy* **2016**, *114*, 1288–1305. [CrossRef]

34. Kotireddy, R.; Hoes, P.-J.; Hensen, J.L.M. A methodology for performance robustness assessment of low-energy buildings using scenario analysis. *Appl. Energy* **2018**, *212*, 428–442. [CrossRef]

35. Kotireddy, R.; Loonen, R.; Hoes, P.-J.; Hensen, J.L.M. Building performance robustness assessment: Comparative study and demonstration using scenario analysis. *Energy Build.* **2019**, *202*, 109362. [CrossRef]

36. Kotireddy, R.; Hoes, P.-J.; Hensen, J.L.M. Integrating robustness indicators into multi-objective optimization to find robust optimal low-energy building designs. *J. Build. Perform. Simul.* **2019**, *12*, 546–565. [CrossRef]

37. Picard, T.; Hong, T.; Luo, N.; Lee, S.H.; Sun, K. Robustness of energy performance of zero-net-energy (ZNE) homes. *Energy Build.* **2020**, *224*, 110251. [CrossRef]

38. Cantatore, E.; Fatiguso, F. An energy-resilient retrofit methodology to climate change for historic districts. Application in the Mediterranean area. *Sustainability* **2021**, *13*, 1422. [CrossRef]

39. Houghton, A.; Castillo-Salgado, C. Analysis of correlations between neighborhood-level vulnerability to climate change and protective green building design strategies: A spatial and ecological analysis. *Build. Environ.* **2020**, *168*, 106523. [CrossRef]

40. European Union. *EPBD 2018/844/EU Energy Performance of Buildings Directive 2010*; European Union: Brussels, Belgium, 2018.

41. European Commission EU Buildings Factsheets. Available online: https://ec.europa.eu/energy/eu-buildings-factsheets_en (accessed on 19 March 2021).

42. European Commission. *COM 2020 662 Final Communication from the Commission to the European Parliament, the Council, the European Economic and Social Committee and the Committee of the Regions: A Renovation Wave for Europe–Greening Our Buildings, Creating Jobs, Improving Lives*; European Commission: Brussels, Belgium, 2020.

43. Finocchiaro, L.; Georges, L.; Hestnes, A.G. 6–Passive solar space heating. In *Advances in Solar Heating and Cooling*; Woodhead Publishing: Sawston, UK, 2016; pp. 95–116. ISBN 978-0-08-100301-5.

44. *IPCC Intergovernmental Panel on Climate Change AR4 Report*; IPCC: Geneva Switzerland, 2007.

45. University of Southampton, Energy and Climate Change Division CCWorldWeatherGen–Climate Change World Weather File Generator for World-Wide Weather Data. Available online: http://www.energy.soton.ac.uk/ccworldweathergen/ (accessed on 29 June 2020).

46. EN CS. 16798-1. Energy performance of buildings—Ventilation for buildings—Part 1: Indoor environmental input parameters for design and assessment of energy performance of buildings addressing indoor air quality. In *Thermal Environment, Lighting and Acoustics—Module M1-6*; CEN: Brussels, Belgium, 2019.

47. Pajek, L.; Košir, M. Climate change impact on the potential occurrence of overheating in buildings. In Proceedings of the International Conference on Sustainable Built Environment: Smart Building and City for Durability & Sustainability, Seoul, Korea, 12 December 2019; pp. 411–414.

48. Bergman, T.L.; Lavine, A.; DeWitt, D.P.; Incropera, F.P. *Incropera's Principles of Heat and Mass Transfer*, 8th ed.; Wiley: Singapore, 2017; ISBN 978-1-119-38291-1.

49. EnergyPlus. Available online: https://energyplus.net/downloads (accessed on 2 October 2020).

50. JEPlus–An EnergyPlus Simulation Manager for Parametrics. Available online: http://www.jeplus.org/ (accessed on 19 October 2020).

51. Pajek, L.; Košir, M. Strategy for achieving long-term energy efficiency of European single-family buildings through passive climate adaptation. *Appl. Energy* **2021**, *297*, 117116. [CrossRef]

52. Rules on the Efficient Use of Energy in Buildings. *Off. Gaz. Repub. Slov.* **2017**, *52/10*. Available online: https://www.uradni-list.si/glasilo-uradni-list-rs/vsebina/2010-01-2856?sop=2010-01-2856 (accessed on 15 May 2021).

53. Slovenian Environmental Agency Archive–Observed and Measured Meteorological Data in Slovenia. Available online: https://meteo.arso.gov.si/met/sl/archive/ (accessed on 22 March 2021).

54. Rules on the Methodology of Production and Issuance of Energy Performance Certificates for Buildings. *Off. Gaz. Repub. Slov.* **2020**, *92/14*. Available online: https://www.uradni-list.si/glasilo-uradni-list-rs/vsebina/2014-01-3699?sop=2014-01-3699 (accessed on 15 May 2021).

55. Parsaee, M.; Demers, C.M.H.; Hébert, M.; Lalonde, J.-F.; Potvin, A. A photobiological approach to biophilic design in extreme climates. *Build. Environ.* **2019**, *154*, 211–226. [CrossRef]

56. Singh, M.K.; Attia, S.; Mahapatra, S.; Teller, J. Assessment of thermal comfort in existing pre-1945 residential building stock. *Energy* **2016**, *98*, 122–134. [CrossRef]

57. da Guarda, E.L.A.; Gabriel, E.; Domingos, R.M.A.; Durante, L.C.; Callejas, I.J.A.; Sanches, J.C.M.; Rosseti, K. de A.C. Adaptive comfort assessment for different thermal insulations for building envelope against the effects of global warming in the Mid-Western Brazil. *IOP Conf. Ser. Earth Environ. Sci.* **2019**, *329*, 012057. [CrossRef]

58. Potočnik, J.; Košir, M. Influence of geometrical and optical building parameters on the circadian daylighting of an office. *J. Build. Eng.* **2021**, *42*, 102402. [CrossRef]

59. Bekö, G.; Toftum, J.; Clausen, G. Modeling ventilation rates in bedrooms based on building characteristics and occupant behavior. *Build. Environ.* **2011**, *46*, 2230–2237. [CrossRef]

60. Mahdavi, A.; Berger, C.; Amin, H.; Ampatzi, E.; Andersen, R.K.; Azar, E.; Barthelmes, V.M.; Favero, M.; Hahn, J.; Khovalyg, D.; et al. The role of occupants in buildings' energy performance gap: Myth or reality? *Sustainability* **2021**, *13*, 3146. [CrossRef]

61. Chen, Y.; Chen, Z.; Xu, P.; Li, W.; Sha, H.; Yang, Z.; Li, G.; Hu, C. Quantification of electricity flexibility in demand response: Office Building case study. *Energy* **2019**, *188*, 116054. [CrossRef]

62. Zavrl, E.; Zupanc, G.; Stritih, U.; Dovjak, M. Overheating reduction in lightweight framed buildings with application of phase change materials. *Stroj. Vestn. J. Mech. Eng.* **2019**, 3–14. [CrossRef]

63. Lesjak, V.; Pajek, L.; Košir, M. Indirect green façade as an overheating prevention measure. *Građevinar* **2020**, *72*, 569–583. [CrossRef]

64. Fabiani, C.; Pisello, A.L. Cool materials for passive cooling in buildings. In *Urban Microclimate Modelling for Comfort and Energy Studies*; Palme, M., Salvati, A., Eds.; Springer International Publishing: Cham, Switzerland, 2021; pp. 505–537. ISBN 978-3-030-65421-4.

65. Domjan, S.; Arkar, C.; Begelj, Ž.; Medved, S. Evolution of all-glass nearly zero energy buildings with respect to the local climate and free-cooling techniques. *Build. Environ.* **2019**, *160*, 106183. [CrossRef]

 *sustainability*

*Article*

# Retrofitting Building Envelope Using Phase Change Materials and Aerogel Render for Adaptation to Extreme Heatwave: A Multi-Objective Analysis Considering Heat Stress, Energy, Environment, and Cost

Dileep Kumar [1,2], Morshed Alam [1,*] and Jay G. Sanjayan [1]

[1]   Centre for Sustainable Infrastructure and Digital Construction, Department of Civil and Construction Engineering, Swinburne University of Technology, Hawthorn 3122, Australia; dileepkumar@swin.edu.au (D.K.); jsanjayan@swin.edu.au (J.G.S.)
[2]   Department of Mechanical Engineering, Shaheed Zulfiqar Ali Bhutto Campus, Mehran University of Engineering and Technology, Khairpur Mir's 76062, Pakistan
*   Correspondence: mmalam@swin.edu.au

**Citation:** Kumar, D.; Alam, M.; Sanjayan, J.G. Retrofitting Building Envelope Using Phase Change Materials and Aerogel Render for Adaptation to Extreme Heatwave: A Multi-Objective Analysis Considering Heat Stress, Energy, Environment, and Cost. *Sustainability* 2021, 13, 10716. https://doi.org/10.3390/su131910716

Academic Editors: Mitja Košir and Manoj Kumar Singh

Received: 5 August 2021
Accepted: 21 September 2021
Published: 27 September 2021

**Publisher's Note:** MDPI stays neutral with regard to jurisdictional claims in published maps and institutional affiliations.



**Abstract:** Energy retrofitting the existing building stock is crucial to reduce thermal discomfort, energy consumption, and carbon emissions. However, insulating and enhancing the thermal mass of an existing building wall using traditional methods is a very challenging and expensive task. There is a need to develop a material that can be applied easily in an existing occupied building without much interruption to occupants' daily life while also having high thermal resistance and heat storage capacity. This study aimed to investigate a potential building wall retrofit strategy combining aerogel render and Phase change materials (PCM) because aerogel render is highly resistive to heat and PCM has high thermal mass. While a number of studies investigated the thermal and energy-saving performances of aerogel render and PCM separately, no study has been done on the thermal and energy-saving performance of the combination of PCM and aerogel render. In this study, the performance of 12 different retrofit strategies, including aerogel and PCM, were evaluated numerically in terms of heat stress, energy savings, peak cooling, emission, and lifecycle cost using a typical single-story Australian house. The results showed that applying aerogel render and PCM on the outer side of the external walls and PCM and insulation in ceilings is the best option considering all performance indicators and ease of application. Compared to the baseline, this strategy reduced severe discomfort hours by 82% in a free-running building. In an air-conditioned building, it also decreased energy use, peak cooling demand, $CO_2$ emission, and operational energy cost by 40%, 65%, 64%, and 35%, respectively. Although the lifecycle cost savings for this strategy were lower than the "insulated ceiling and rendered wall without PCM" case, the former one was considered the best option for its superior energy, emission, and comfort performance. Parametric analysis showed that 0.025 m is the optimum thickness for both PCM and aerogel render, and the 25 °C melting point PCM was optimum to achieve the best results amongst all performance indicators for a typical Australian house in Melbourne climate.

**Keywords:** building energy retrofitting; phase change materials; aerogel render; heat stress risk; energy savings; emission; lifecycle cost; peak cooling load

## 1. Introduction

The building sector consumes around 30% of total primary energy globally, which is expected to escalate up to 50% by 2050 due to population growth, human lifestyle changes, new technologies, and climate change [1]. Currently, fossil fuels are used to meet around 80% of the world's energy demand, which has an adverse social, economic, and environmental impact [2]. Therefore, the use of renewable energy sources and the adoption of sustainable practices in buildings to minimize fossil fuel consumption are being investigated extensively around the globe [3].

Heat transfer through building envelopes (walls, roofs, windows, and doors) accounts for up to 60% of total heat loss and gain [4], which can be reduced by having insulated building envelopes [3], double and triple glazed windows [5], and thermochromatic windows [6]. A recent review of present authors compared thermal properties and performances of various building insulation materials [7]. It was concluded that a highly insulated building envelope significantly reduces total heating and cooling energy consumption and improves winter thermal comfort in a passive building. However, it resulted in overheating and increased peak cooling demand in a lightweight structure during a heatwave period because of the low heat storage capacity of the insulation and lightweight building materials. Therefore, the building envelope should have higher heat resistance and higher heat storage capacity to reduce heating and cooling energy use in an air-conditioned building and to improve thermal comfort in a passive building [7].

Like many other metropolitan cities, a significant percentage of the residential building stock in Melbourne, Australia were constructed before the introduction of the mandatory five stars (maximum is 10 stars) energy efficiency standard in 2005. In Victoria, approximately 86% of the currently occupied houses were built before 2005 with an average energy efficiency rating of only 1.81, which is very low [8]. Therefore, retrofitting those existing energy inefficient buildings is crucial to reduce energy consumption and harmful greenhouse gas emission from this sector. While insulating a wall in a new dwelling is straightforward, it is more challenging to insulate an existing occupied house unless it is under major renovation where claddings and plasterboards are removed. Sustainability Victoria [9] trialed the pump-in cavity wall insulation method, which resulted in 15.5% energy savings. However, this method is very expensive, and the average payback period was reported to be 29 years. Also, increasing the thermal mass of an existing building using traditional materials (such as bricks and concrete) is impossible. Hence, there is a need to develop a material that can be applied easily in an existing occupied building without much interference and has low thermal conductivity and high thermal storage.

Aerogel-based thermal insulating renders are introduced in the European Union market to insulate existing walls as an alternative to plasterboard and insulation panel [10]. It can be applied easily on the building envelope with limited impacts and interruptions on the occupants' daily life and building functionality [11]. It is a lightweight material with density and thermal conductivity of 150–220 kg/m$^3$ and 0.024–0.027 W/mK, respectively [11–13], depending on the percentage of aerogel granules in the mixture [14]. Aerogel render has higher compressive strength [15,16], low water permeability [17], and is inert to flame. However, aerogel's major drawback is the lower heat storage capacity compared to the conventional insulators and construction materials, as shown in Figure 1 [7]. The lower heat storage capacity causes high indoor temperature fluctuation and summer overheating that have adverse health impacts specifically for older occupants and infants. Therefore, there is a need to improve the heat storage capacity of the aerogel render integrated into the building envelope.

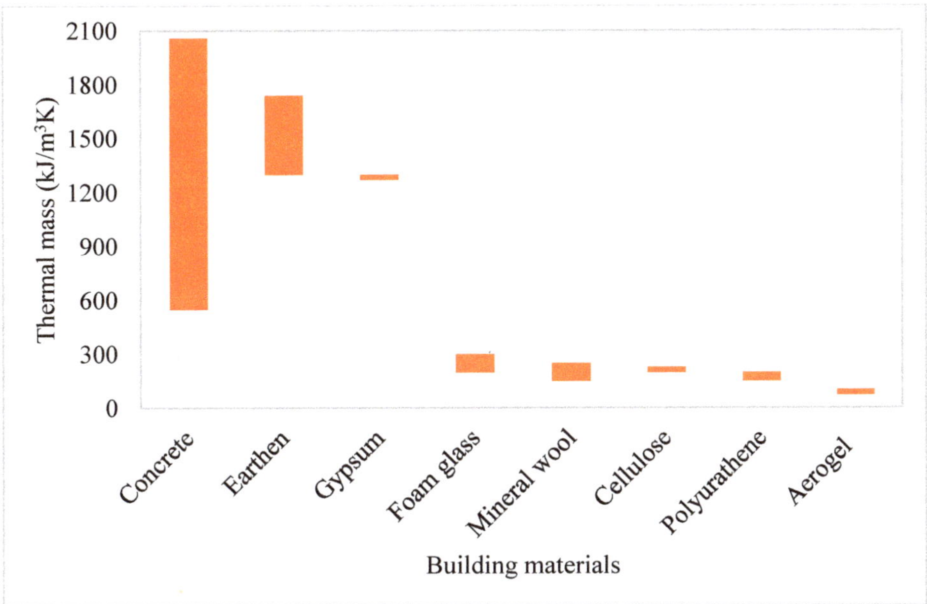


**Figure 1.** Thermal mass of building construction and insulation materials [7].

Integration of Phase change materials (PCM) in building envelope has been shown to increase heat storage capacity significantly [18]. Previous studies successfully integrated micro-encapsulated PCM in building materials, including structural materials [19,20], plaster and mortar [21,22] and insulation [23,24] to improve heat storage capacity and resulting in energy savings in air-conditioned buildings and improving summertime thermal comfort in passive buildings [25]. Hasnat et al. [26] reported a 34% reduction in thermal discomfort hours through the installation of Bio-PCM pouches in the ceilings of a Melbourne house. In other studies, the use of PCM-enhanced geopolymer coating and cement mortar reduced the test hut indoor air temperature up to 2.8 °C [25] and 2.4 °C [27], respectively, in summer, compared to an identical hut containing ordinary cement plaster. Cui et al. [28] developed a thermal energy storage concrete (TESC) using macro-encapsulated lauryl alcohol-lightweight aggregate PCM. The maximum air temperature in a test room (500 mm × 500 mm × 500 mm) containing TESC in the wall and roof was up to 9 °C and 5 °C lower, respectively, compared to no PCM room. Piti et al. [20] incorporated 7.8% Polyethylene glycol type 1450 by weight into a lightweight concrete that increased the heat storage capacity of concrete from 0.92 to 7.7 kJ/kg. Kosny et al. [29] reported that PCM-blended cellulose insulation has similar thermal insulating properties to cellulose insulation up to 30% of PCM addition with a staggering increment in heat storage capacity from 1.04 J/g to 60–80 J/g. The application not only reduced the cooling load (35–40%) but also decreased the heating demand up to 16% in a conventional house located in southern California. Rathore et al. [30] found that the PCM-embedded concrete panel reduces summertime thermal amplitude and time lag by 40.67–59.79% and 7.19–9.18%, respectively, benefitting with cooling energy savings of 0.40 US $/day.

The literature review shows that a PCM combined with aerogel render would be an ideal candidate to retrofit existing buildings because of its ease of application, lower thermal conductivity, and higher thermal mass. While a number of studies investigated the thermal and energy-saving performances of aerogel render and PCM separately, no study has been done on the thermal and energy-saving performance of PCM combined with aerogel render. A multi-objective optimization study could provide important information regarding the optimum PCM melting temperature and optimum PCM and aerogel layer thickness to achieve the desired comfort, energy, environment, and cost performance of a retrofitted building [31–33].

Therefore, this study aims to investigate the performance of building envelope retrofitted with PCM combined with aerogel render in terms of heat stress, energy savings, peak cooling, emission, and lifecycle cost. The specific objectives are:

(1)  To investigate and identify the best retrofit combinations using PCM blanket, aerogel render, and insulation in passive and air-conditioned buildings.

(2)  To determine the optimum PCM temperature, PCM thickness, and aerogel render thickness for the identified best retrofit combination.

In this paper, Section 2 describes the research methodology, including simulation information, metrological parameters (ambient temperature, relative humidity, solar radiation density, and wind speed), material characteristics, and models. It also describes the case study building and proposed retrofit strategies, methods of thermal discomfort assessment, energy use estimation, emission calculations, and lifecycle cost analysis. The comparative analysis of results for different retrofit strategies is shown in Section 3. In Section 4, the study results are discussed further, and the best PCM-aerogel combination is proposed along with optimum phase change temperature and thickness, considering all performance criteria. Finally, Section 5 presents the concluding remarks and future directions.

## 2. Methodology

### 2.1. Case Study Building Description

A typical single-story Australian house was used as a case study to investigate the thermal performance of retrofit strategies. The selected case study building is one of the eight representative Australian houses used to develop the nationwide house energy rating system (NatHERS) in Australia [34]. According to the Australian Building Code Board, the selected single-story house model is one of the two most typical representations of single-story detached houses in Australia. Approximately 72.9% of the Australian dwellings fall in the category of single-story detached houses [35]. The studied house is a four-bedroom, two-bathroom family house with a floor area of 232 m$^2$. Figure 2 [36,37] shows the isometric view and thermal zones of the simulated house, along with the orientation from the north. The thermophysical properties of building materials are given in Table 1. The construction of the building envelopes varies depending on the simulation cases and is presented in the following section.

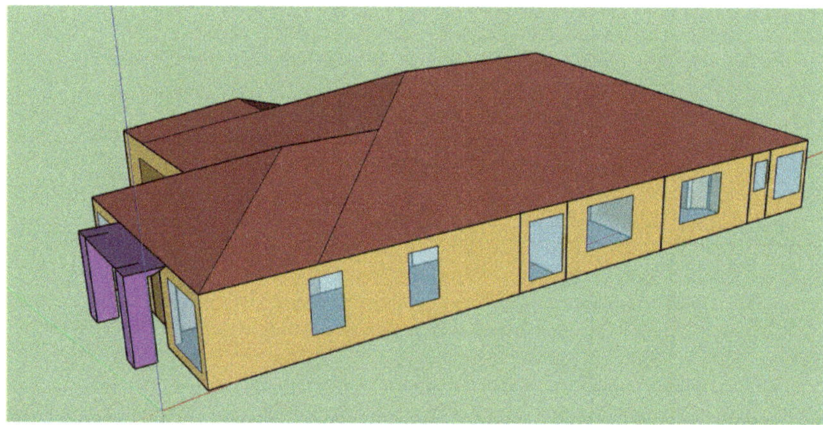

(a)

**Figure 2.** *Cont.*

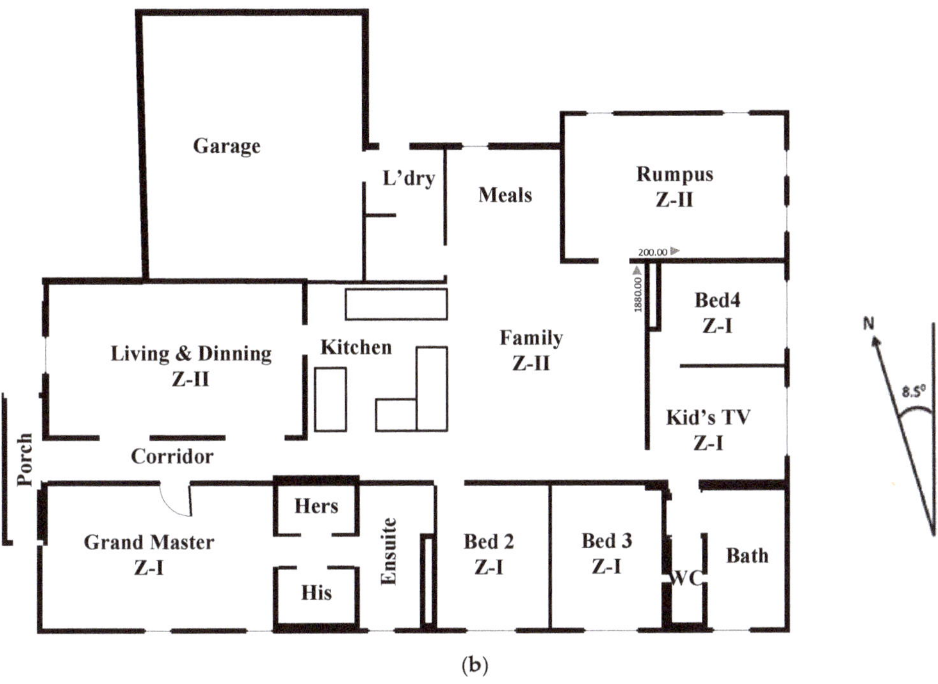


**Figure 2.** The simulated single-story house: (**a**) Isometric view; (**b**) Thermal zones in 2D.

**Table 1.** Thermo-physical properties of building materials.

| Building Materials | Thermo-Physical Properties | | | |
| --- | --- | --- | --- | --- |
| | Thickness (m) | Conductivity (W/m K) | Density (kg/m$^3$) | Specific Heat (J/kg K) |
| Concrete | 0.100 | 1.42 | 2400 | 880 |
| Brick veneer | 0.110 | 0.61 | 1690 | 878 |
| Roof insulation | 0.044 | 0.044 | 12 | 883 |
| Roof tiles | 0.02 | 1.42 | 2400 | 880 |
| plasterboard | 0.013 | 0.17 | 847 | 1090 |
| Carpet | 0.02 | 0.0465 | 104 | 1420 |
| Timber doors | 0.05 | 0.16 | 1122 | 1260 |
| PCM | See Table 4 | 0.2 | 235 | 2400 |
| Aerogel render [36,37] | 0.02 | 0.024 | 100 | 1000 |

## 2.2. Building Energy and Thermal Simulations

The case study building was simulated using building simulation software EnergyPlus v9.2. EnergyPlus v9.2 was developed by the U.S. Department of Energy (DOE). Google Sketchup provided a comprehensive and powerful graphical user interface to EnergyPlus. The simulations were carried out considering the weather file for the year 2009 (the Bureau of Meteorology, Government of Australia), which reported an extreme heatwave in January, as seen in Figure 3 [38]. Melbourne exhibits a temperate oceanic climate, which has a high diurnal temperature swing. Temperate climatic zones are advantageous for PCM application because it allows complete melting/freezing cycle during summer and improves summertime thermal comfort.

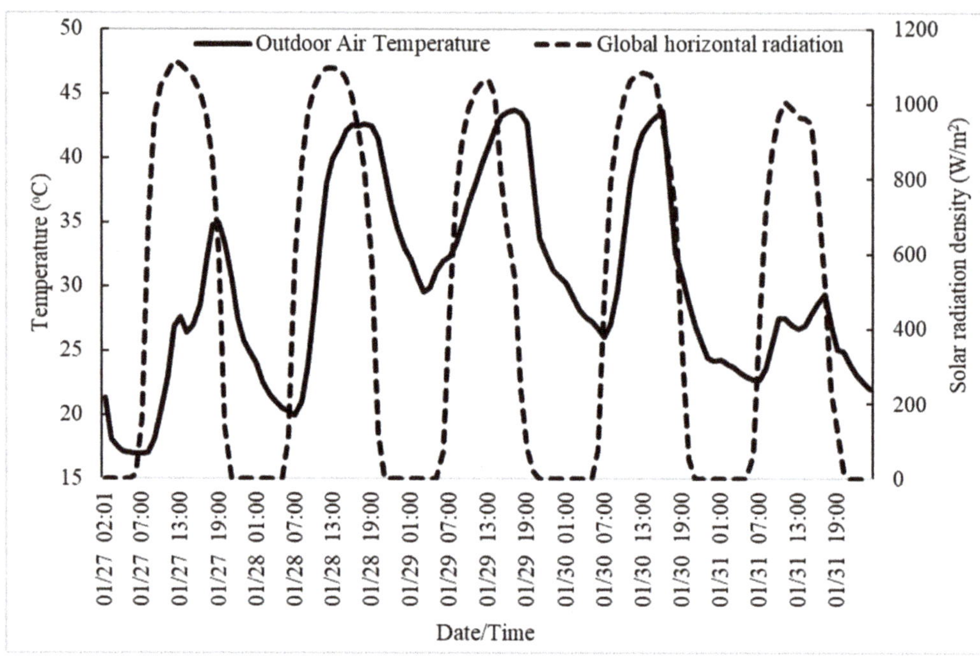

**Figure 3.** Climatic conditions during extreme heatwave period (27–31 January 2009 Melbourne).

The simulations were run using a conduction finite-difference algorithm (ConFD), which allows the simulation of temperature-dependent properties of PCM. In addition, this study used a fully implicit CondFD scheme for dynamic thermal simulation. This scheme accounts for time-dependent phase change phenomenon of PCM using the enthalpy-temperature function.

$$C\rho\Delta x \frac{T_{b+1}^{a+1} - T_b^a}{\Delta x} = \left\{ k_i \frac{\left( T_{b+1}^{a+1} - T_b^{a+1} \right)}{\Delta x} + k_j \frac{\left( T_{a-1}^{a+1} - T_b^{a+1} \right)}{\Delta x} \right\} \tag{1}$$

The specific heat capacity of PCM is temperature-dependent and is updated at every iteration in EnergyPlus according to Equation (2). The effective specific heat capacity of PCM is calculated as:

$$C = \frac{h_i^j - h_i^{j-1}}{T_i^j - T_i^{j-1}} \tag{2}$$

where;

$C$ = specific heat capacity of material (kJ/kg K)
$\rho$ = density of material (kg/m$^3$)
$h$ = specific enthalpy (kJ/kg)
$T$ = Temperature (°C)
$b$ = temperature node, $b - 1$ and $b + 1$ are adjacent inner and outer nodes.
$a + 1$ and $a$ = simulation time and previous time step
$k_i$ and $k_j$ = material's thermal conductivity at a different node.

This study includes BioPCMs having melting point temperature ranges between 20 °C and 32 °C. The thermophysical properties of PCMs are presented in Table 1. Figure 4 shows the enthalpy-temperature graphs of PCMs with different phase change temperatures used in this study [39]. Each PCM has a phase transition range of 4 °C. For example,

PCM24 means it will complete a phase change cycle between 22 °C and 26 °C. While the individual thermal properties of PCM and aerogel layers are known, the thermal properties of PCM-integrated aerogel render are not yet known. The ultimate goal of this project is to develop a PCM-integrated aerogel render for easy retrofitting of an existing building wall. This simulation study was carried out as part of the feasibility study to know how the combination of PCM and aerogel influences building thermal performance and energy consumption. Therefore, for the purpose of this feasibility study and for the sake of simplicity, we assumed PCM and aerogel render as separate layers in this simulation study.

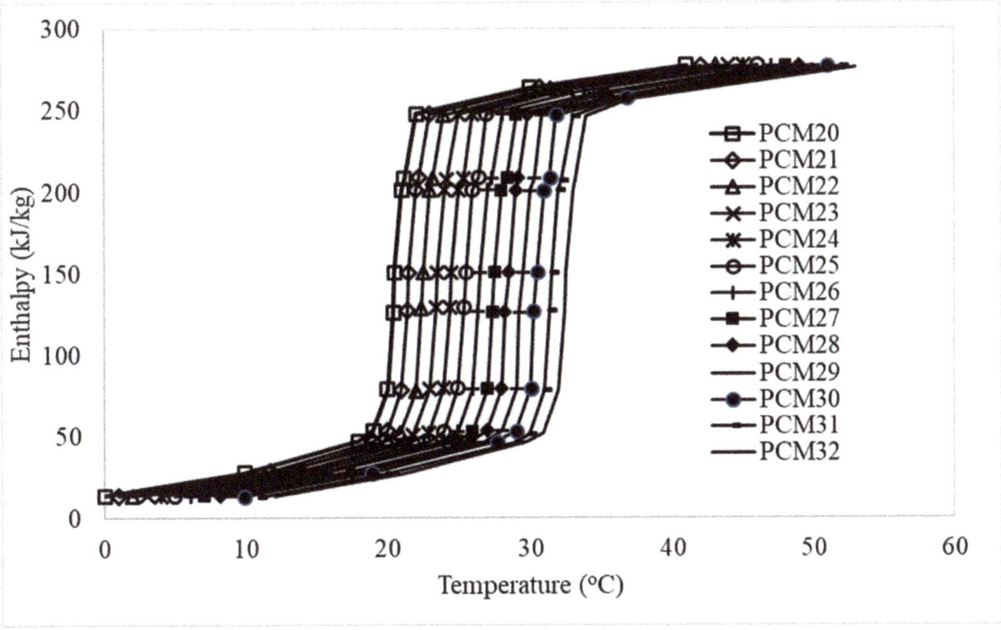

**Figure 4.** Enthalpy-temperature curve of BioPCM [2].

The house was assumed to be occupied by four residents, with different schedules for weekdays and weekends [40]. Figure 5 exhibits the activity level of occupants in different house zones. The GroundHeatTransfer: Slab module of the EnergyPlus software was used to simulate the ground source heat transfer [2]. Moreover, an Effective leakage area model was used to simulate infiltration [41].

Each simulation was conducted twice, considering different retrofit strategies: (1) with and (2) without an HVAC system. The simulations with the HVAC system were used to evaluate the impact of retrofit strategies on total annual energy and peak cooling demand. The simulations without HVAC were used to assess the impact of retrofit strategies on indoor heat stress during a heatwave. The risk of a power outage is very high during the hot summer period, which may leave the HVAC system out of order and pose a significant threat of heat stress to the occupant. Therefore, the retrofit strategies need to be evaluated for both HVAC and no HVAC scenarios. Table 2 gives information about operating, economic parameters, and their respective references.

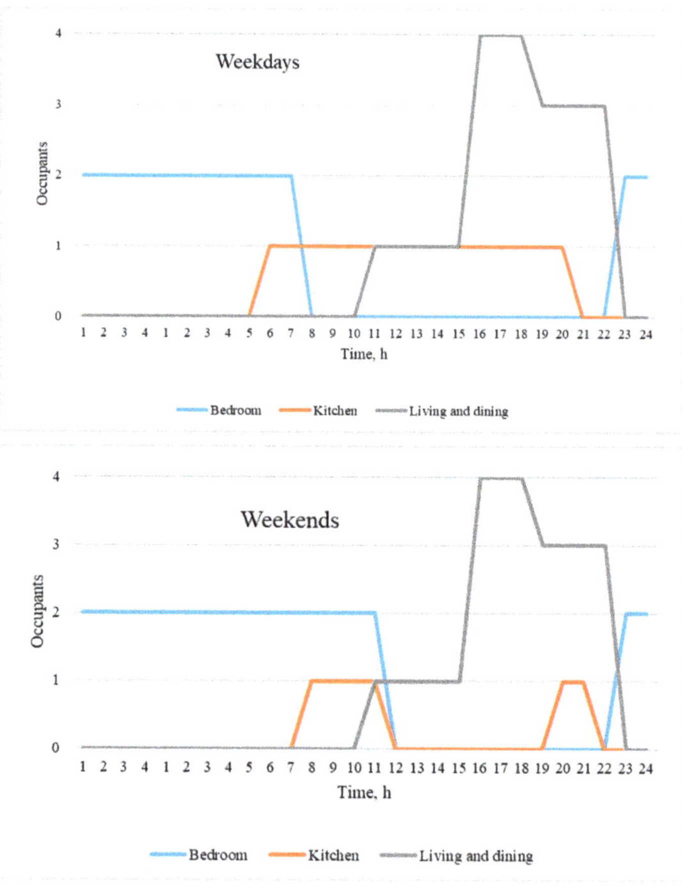


**Figure 5.** Occupants' schedule in a typical Australian house [40].

**Table 2.** Assumption of different operating conditions and economic parameters.

| Operating Conditions | Value | Standard |
| --- | --- | --- |
| Time step | 3 min | Tabares-Velasco et al. [42] |
| Thermostat setpoints | | The house energy rating standards of Australia [43] |
| Heating (°C) | 20 | |
| Cooling (°C) | 24 | (00:00–8:00 and 16:00–24:00 h) |
| People (person) | 4 | |
| Metabolic rate (W/person) | | ASHRAE [44] |
| Writing, seating, standing | 108 | |
| Cooking, cleaning | 171 | See Figure 5 |
| Reading, relaxing | 108 | |
| Lighting (W/m$^2$) | 2.5 | |
| Electric equipment (W/m$^2$) | 1.875 | Australian building code boards [45] |
| Economic Parameter | | |
| Ceiling insulation | 5.93 AUD/m$^2$ | [46] |
| PCM | 4.33 AUD/kg | [47] |
| Aerogel render (AG) | 50–62 AUD/kg | ENERSEN, France [14] |
| Electricity usage rates | 0.31 AUD/kWh | |
| Electricity supply charges | 1.1408 AUD/day | |
| Natural gas usage rate | 0.115 AUD/kWh | Energy Australia [48] |
| Natural gas supply charges | 0.759 AUD/day | |

Table 2. Cont.

| Operating Conditions | Value | Standard |
|---|---|---|
| Electricity Emission Factor | 1.08 kgCO$_{2\text{-eq}}$/kWh | Australian national greenhouse accounts [49] |
| Natural gas emission factor | 3.9 kgCO$_{2\text{-eq}}$/GJ | |
| Ducted cooling system (COP$_{eq}$) | 1.96 | [50] |
| Ducted gas heating system ($\eta$) | 52.5% | [51] |
| Conversion Factor: Electricity | $3.6 \times 10^6$ J/kWh | [52] |
| Heating Value natural gas | $34.526 \times 10^6$ J/m$^3$ | [52] |
| Inflation rate (*i*) | 1.93% | Office of Best Practice Regulation [53] |
| Interest rate (*d*) | 6% | |
| Lifetime (LT) | 40 years | Australian building code boards [45] |

Nosrati and Berardi [17] showed that aerogel render thermal conductivity changes with ambient air relative humidity. To consider this moisture dependency, moisture dependent thermal conductivity data of 90% Aerogel-enhanced plaster data, as reported in [17], were used to vary the thermal conductivity of aerogel in the simulation using energy management system (EMS) object in EnergyPlus.

Figure 6 exhibits a negligible difference in heating and cooling load for hygrothermal and non-hygrothermal simulation because the annual average relative humidity of Melbourne is only 55%, which meagerly changes aerogel render thermal conductivity. The average monthly relative humidity varies from a minimum of 48% in January (Summer) to 72% in June. On the other hand, the relative humidity of around 95% has been shown to impact thermal conductivity significantly [17]. Therefore, in this study, aerogel render was simulated without considering their hygrothermal properties.

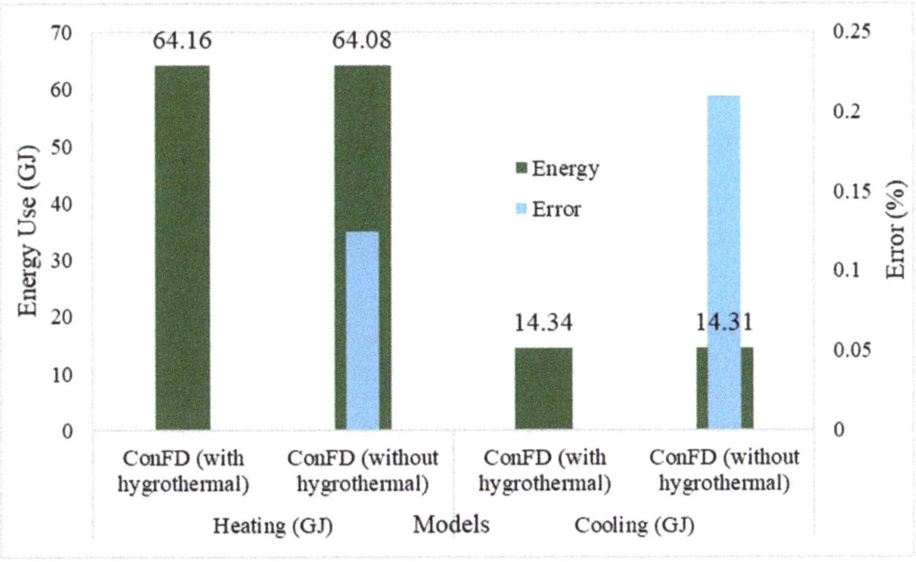


**Figure 6.** Comparison of heating and cooling energy use in building in case of hygrothermal and non-hygrothermal simulation.

Validation of the numerical model is a pre-condition for any simulation-based study. Several studies were conducted by many researchers and the EnergyPlus developer team to validate the EnergyPlus PCM simulation algorithm using analytical (Stefan Problems) [42], comparative testing [54], and field studies [26,55] approaches. For instance, Tabares-Velasco et al. [42] suggested that EnergyPlus is a reliable tool for simulating PCM by comparing the simulation outcomes with experimental investigations. They recommended that the simulation should be conducted considering time step ($\leq$3 min). Moreover, the present author developed and validated the EnergyPlus model of a real duplex house with PCM in a previous study [26]. The single-story house model that was used in this study was developed using a modeling approach similar to that validated duplex house model. This single-story house model was also successfully used to evaluate heat stress conditions in a previous study of the present authors [56]. Therefore, the use of a validated modeling approach provides it with secondary validation.

*2.3. Benchmark Studies*

Table 3 shows the simulation cases with different retrofit strategies. Construction details of the ceiling, internal wall, and external wall are illustrated in Figure 7. Each simulation case was assessed considering heat stress risks, energy-saving potential, emissions, and lifecycle costs. Case 1 is the baseline house without any insulation in the ceiling and walls because this study aims to investigate the retrofitting potential of existing energy-inefficient building stock combining aerogel render with PCM. As mentioned in Section 1, the energy efficiency rating of a significant percentage of existing occupied houses in Melbourne is very low. Therefore, the selection of a baseline case without any insulation in the ceiling and walls is justified.

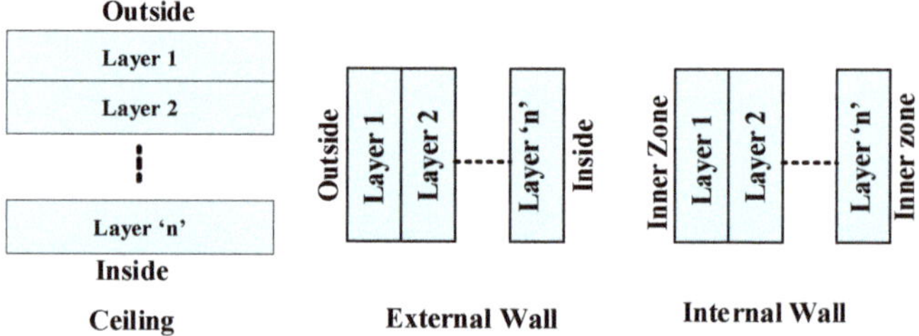


**Figure 7.** Schematics of envelope constructions in different simulation cases.

To allow direct comparison between different cases with PCM, the total amount (kg) of PCM was kept constant in the simulation cases with PCM. This was done by varying the thickness of the PCM layer according to the application surface area.

*2.4. Parametric Studies*

The parametric analysis determined the optimum phase change temperature (OPCT) for each retrofit case with PCMs. PCMs with phase change temperatures ranging from 18 °C to 32 °C were considered, as shown in Figure 4. After the selection of the best retrofit combinations, the second set of parametric studies were carried out to identify optimum aerogel render and PCM thickness. JePlus v2.1 was used together with EnergyPlus to conduct the parametric analysis by varying aerogel render thickness from 0.01 m to 0.05 m and the PCM thickness from 0.005 m to 0.025 m.

**Table 3.** Construction details of the simulation cases.

| Retrofit Cases | Description | Ceiling | | | External Walls | | | | Internal Walls | | | | | | |
|---|---|---|---|---|---|---|---|---|---|---|---|---|---|---|---|
| | | Layer 1 | Layer 2 | Layer 3 | Layer 1 | Layer 2 | Layer 3 | Layer 4 | Layer 1 | Layer 2 | Layer 3 | Layer 4 | Layer 5 | Layer 6 | Layer 7 |
| Case 1 | Baseline (Uninsulated wall and ceiling) | Ceiling Plaster-board | | | Brick Veneer | Wall Plaster-board | | | Wall Plaster-board | Air gap | Wall Plaster-board | | | | |
| Case 2 (Applying render on the outer parts of the external wall) | Rendered wall and uninsulated ceiling | Ceiling Plaster-board | | | Aerogel Render | Brick Veneer | Wall Plaster-board | | Wall Plaster-board | Air gap | Wall Plaster-board | | | | |
| Case 3 | Rendered wall and insulated ceiling | Insulation | Ceiling Plaster-board | | Aerogel Render | Brick Veneer | Wall Plaster-board | | Wall Plaster-board | Air gap | Wall Plaster-board | | | | |
| Case 4 | Rendered wall coupled with PCM and uninsulated ceiling | Ceiling Plaster-board | | | Aerogel Render | PCM | Brick Veneer | Wall Plaster-board | Wall Plaster-board | Air gap | Wall Plaster-board | | | | |
| Case 5 | Rendered wall coupled with PCM and uninsulated ceiling coupled with PCM | PCM | Ceiling Plaster-board | | Aerogel Render | PCM | Brick Veneer | Wall Plaster-board | Wall Plaster-board | Air gap | Wall Plaster-board | | | | |
| Case 6 | Rendered wall coupled with PCM and insulated ceiling | Insulation | Ceiling Plaster-board | | Aerogel Render | PCM | Brick Veneer | Wall Plaster-board | Wall Plaster-board | Air gap | Wall Plaster-board | | | | |
| Case 7 | Rendered wall coupled with PCM and insulated ceiling coupled with PCM | Insulation | PCM | Ceiling Plaster-board | Aerogel Render | PCM | Brick Veneer | Wall Plaster-board | Wall Plaster-board | Air gap | Wall Plaster-board | | | | |
| Case 8 (Applying render on the inner parts of the exterior and interior wall) | Rendered wall and uninsulated ceiling | Ceiling Plaster-board | | | Brick Veneer | Wall Plaster-board | Aerogel Render | | Aerogel Render | Wall Plaster-board | Air gap | Wall Plaster-board | Aerogel Render | | |
| Case 9 | Rendered wall and insulated ceiling | Insulation | Ceiling Plaster-board | | Brick Veneer | Wall Plaster-board | Aerogel Render | | Aerogel Render | Wall Plaster-board | Air gap | Wall Plaster-board | Aerogel Render | | |
| Case 10 | Rendered wall coupled with PCM and uninsulated ceiling | Ceiling Plaster-board | | | Brick Veneer | Wall Plaster-board | PCM | Aerogel Render | Aerogel Render | PCM | Wall Plaster-board | Air gap | Wall Plaster-board | PCM | Aerogel Render |
| Case 11 | Rendered wall coupled with PCM and uninsulated ceiling coupled with PCM | PCM | Ceiling Plaster-board | | Brick Veneer | Wall Plaster-board | PCM | Aerogel Render | Aerogel Render | PCM | Wall Plaster-board | Air gap | Wall Plaster-board | PCM | Aerogel Render |
| Case 12 | Rendered wall coupled with PCM and insulated ceiling | Insulation | Ceiling Plaster-board | | Brick Veneer | Wall Plaster-board | PCM | Aerogel Render | Aerogel Render | PCM | Wall Plaster-board | Air gap | Wall Plaster-board | PCM | Aerogel Render |
| Case 13 | Rendered wall coupled with PCM and insulated ceiling coupled with PCM | Insulation | PCM | Ceiling Plaster-board | Brick Veneer | Wall Plaster-board | PCM | Aerogel Render | Aerogel Render | PCM | Wall Plaster-board | Air gap | Wall Plaster-board | PCM | Aerogel Render |

*2.5. Analysis Methods*

2.5.1. Indoor Heat Stress Risk and Thermal Discomfort

In this study, the thermal discomfort index was used for analyzing the heat stress risk in different retrofit cases. The thermal discomfort index (TDI) is the average of indoor air wet-bulb and dry-bulb temperature, which is estimated using Equation (3) [57].

$$TDI = \frac{T_{drybulb} + T_{wetbulb}}{2} \tag{3}$$

where, $Td$ and $Tw$ denote dry bulb and wet bulb temperature of indoor air. The EnergyPlus model by default calculates the dry-bulb temperature, relative humidity, and barometric pressure of each zone at every time step during the simulation. The wet bulb temperature was calculated using an advanced functionality of EnergyPlus known as EMS application, which uses dry-bulb temperature, relative humidity, and barometric pressure as input [58]. The heat stress risk can be classified as mild, moderate, and severe, observing the behavior of a large population group under different climates. Epstein and Moran established environmental heat stress criteria as tabulated in Table 4 [57].

**Table 4.** Threshold values of thermal discomfort hours [57].

| Discomfort Index (DI) | Classification of Heat Stress |
|---|---|
| DI < 22 | No heat stress is encountered. |
| 22 < DI < 24 | A mild sensation of heat stress. |
| 24 < DI < 28 | Moderate heat stress, people feel very hot, and physical work may be performed with some difficulties. |
| DI > 28 | Heat stress is severe; people engaged in physical work are at increased risk for heat exhaustion and heatstroke. |

2.5.2. Energy Savings

Energy savings (*ES*) is the measure of the percentage of energy consumption reduction in the retrofitted building. It was calculated by using Equation (4) [2].

$$ES = \left( \frac{EC_r - EC_{ret}}{EC_r} \right) \times 100\% \tag{4}$$

where $EC_r$ and $EC_{ret}$ denote energy consumption of reference and retrofitted buildings.

2.5.3. Emission Reduction

In Melbourne, almost 69% of households use the natural gas heater for heating, and 36% of households (highest among other cooling methods) use reverse cycle air-conditioning for cooling [59]. Therefore, this study assumed that the building is equipped with a natural gas heater and split air conditioner to meet the heating demand in winter and cooling demand in summer. The operational GHG emission is the product of heating and cooling energy use and their respective emission factors. Scope 1 emission factor (51.53 kg $CO_2$-e/GJ for Melbourne) was applied to heating demand, while scope 2 emission factor (1.07 kg $CO_2$-e/kWh for Melbourne) was considered for cooling demand [49]. Equation (5) was used to calculate the greenhouse gas emissions:

$$EE_{GHG} = EC_{el} \cdot EF_{el} + EC_{ng} \cdot EF_{ng} \tag{5}$$

where $EC_{el}$ and $EC_{ng}$ denote cooling and heating energy use and their respective emission factors are denoted by $EF_{el}$ and $EF_{ng}$, respectively. The percentage of emission reduction is calculated as.

$$ER = \left( \frac{EE_{GHG,r} - EE_{GHG,ret}}{EE_{GHG,r}} \right) \times 100\% \tag{6}$$

where $EE_{GHG, r}$ and $EE_{GHG, ret}$ denote $GHG$ emission associated with energy use in reference and retrofitted buildings.

### 2.5.4. Lifecycle Cost Analysis

The lifecycle analysis considers the initial investments, operating and maintenance costs up to the disposal, and recovery costs. However, the economic optimization of building envelope walls and roofs excludes maintenance, renewal, and disposal costs. It only includes the investment of proposed alternative and operation energy costs. Operational energy cost varies according to interest and the inflation rate over the expected building lifetime, determined through the present worth factor ($PWF$) over a lifetime. The $PWF$ is calculated using Equation (7) [3].

$$PWF = \sum_{j=1}^{LT} \frac{(1+i)^{j-1}}{(1+d)^j} = \left\{ \begin{array}{ll} \frac{1}{d-i}\left[1 - \left(\frac{1+i}{1+d}\right)^{LT}\right] & if \ d \neq i \\ \frac{LT}{1+i} & if \ d = i \end{array} \right\} \tag{7}$$

where, $i$, $d$, and $LT$ denote inflation rate, interest rate, and the lifetime of a building. The total lifecycle cost is the sum of the annual operation energy cost and the investment cost of retrofit strategies. Which is calculated as [3]:

$$LCC = C_e \cdot PWF + C_i \tag{8}$$

where $C_e$ and $C_i$ denote operation energy cost and initial investment, which are estimated as

$$C_e = \frac{ES_c \ C_{EL}}{COP \ LHV_{EL}} + \frac{ES_h \ C_{NG}}{\eta \ LHV_{NG}} \tag{9}$$

$$C_I = \sum_{i=0}^{n} C_{ins} + C_{render} + C_{PCM} \tag{10}$$

$ES_c$ and $ES_h$ are the cooling and heating energy savings; $C_{EL}$ and $C_{NG}$ are the unit cost of electricity and natural gas; $LHV_{EL}$ and $LVG_{NG}$ are the lower heating value of electricity and natural gas, respectively; $COP$ denotes the co-efficient of performance of non-ducted air conditioning unit; and $\eta$ represents the efficiency of heating system. Moreover, $C_{ins}$, $C_{render}$, and $C_{PCM}$ denote insulation cost, aerogel render cost, and $PCM$ cost, respectively. In this study, a ducted cooling system with a $COP$ of 2.79 with 30% duct losses (equivalent to overall $COP$ of 1.96) was used, which is the minimum energy performance standard (AS/NZS 3823.2) for air-conditioners used in the Australian state [50].

The maximum cost-saving ($CS$) is the difference of lifecycle cost of reference ($LCC_{ref}$) and retrofitted ($LCC_{ret}$) envelope, which is estimated as:

$$CS = LCC_{ref} - LCC_{ret} \tag{11}$$

## 3. Results

### 3.1. Performance of the Retrofitting Strategies in Terms of Heat Stress Risk

The severe discomfort hours corresponding to optimum PCT in the living and bedroom 4 are presented in Figure 8. The living, family, rumpus, and kids tv zones have different occupancy schedules than bedrooms as shown in Figure 5. While discomfort hours were calculated for all zones of the house, results of only two zones were presented here for the sake of brevity. One zone from living type (mostly occupied during daytime) and one zone from bedrooms were selected for the presentation of results.

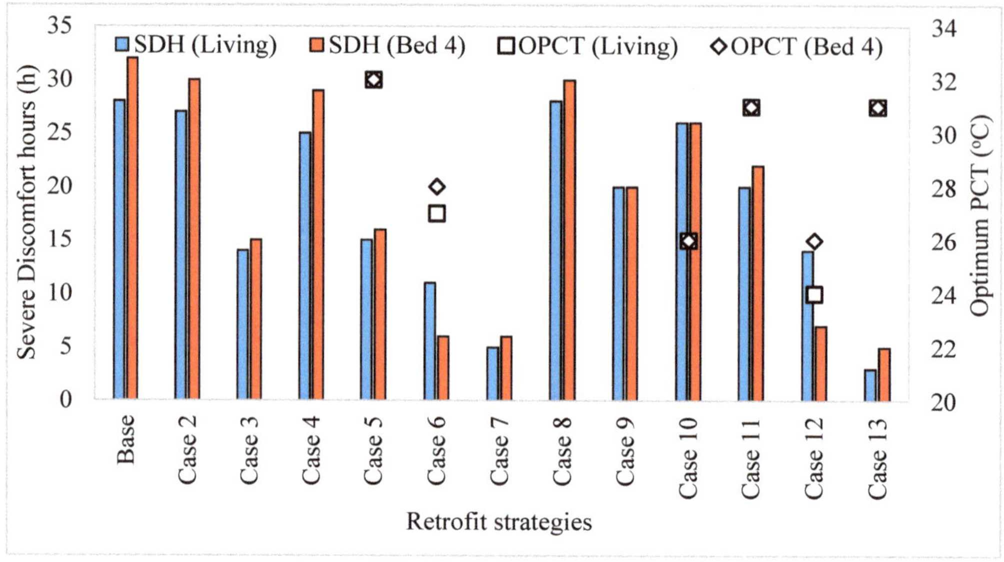



**Figure 8.** Severe discomfort hours corresponding to optimum phase change temperature in living and bedroom.

The discomfort hours in bedroom 4 were slightly higher than in the living room. It could be because of poor cross ventilation in the confined space of Bed 4, which is heated up by the radiation from the eastern sun in the morning. Also, it has a large window to wall ratio compared to the living room. In contrast, the living room has a large internal opening to the corridor, which results in higher cross ventilation. Nevertheless, both zones showed similar severe discomfort hours and optimum phase change temperature corresponding to the retrofit strategies.

Without PCM, the application of ceiling insulation and aerogel rendering on the outer part of the wall (Case 3) was the best combination to reduce discomfort hours. It performed better than the PCM combined wall and ceiling strategies without ceiling insulation (Case 4, 5, 10, and 11). Insulated ceiling mitigated heat stress risk better than the bared ceiling and aerogel rendered wall because heat transfer through the ceiling is higher than the wall [60] and hence, insulation of ceiling significantly reduces heat transfer through the roof than walls. Without ceiling insulation, other retrofitting measures in walls and ceilings were not very effective in minimizing the discomfort hours.

Moreover, in the presence of ceiling insulation, the external wall with aerogel render on its outer part (Case 3) was more effective in minimizing the severe discomfort hours than rendering the internal walls and the inner part of external walls (Case 9). Previous studies reported that increased insulation results in overheating in buildings [61,62], which is somewhat consistent with the current study's findings that incorrect insulation application may lead to higher discomfort hours. In the latter case (Case 9), the heat transfer rate between bedroom 4 and comparatively cooler adjacent family zone decreases due to the application of aerogel render on internal walls. The family zone is comparatively cooler because it has a shaded north window and other zones act as a buffer on three sides.

Furthermore, application of PCM and aerogel render on the outer side of the external walls (Case 6), on internal walls and the interior side of the external walls (Case 12), and the insulated ceiling (Case 7, Case 13) further reduced the severe discomfort hours compared to insulated ceiling case (Case 3 and Case 9). Although the retrofit Case 6 was more effective in minimizing discomfort hours compared to Case 12, the pattern changed with the integration of PCM in ceilings. Figure 8 shows that the integration of PCM and aerogel render on the inner part of the external wall, internal walls, and insulated ceiling (Case 13) was the best strategy to reduce severe discomfort hours. It could be because of the large applied surface area and a thinner layer of PCM that accelerated solidification and melting in Case 13 and absorbed any trapped heat.

The optimum PCM temperature (OPCT) was calculated based on the maximum reduction of severe discomfort hours (SDH). Figure 8 shows that OPCT depends on the application method of PCM. The OPCT was mainly in the range of 22–25 °C when PCMs were applied on the inner and outer parts of the wall. In Case 4, the OPCT was found to be in the range of 24–32 °C, which means there was no change in discomfort hours when PCMs in this temperature range were applied in Case 4. Moreover, in the case of PCM and aerogel render on the outer part of the external wall and in the uninsulated ceiling (Case 4), the OPCT was in the range of 29–32 °C. For Case 7, severe discomfort hours were minimum when phase temperature was within 29–32 °C. No conclusive evidence was found on the impact of PCM position (inner and outer parts of the wall) on OPCT.

### 3.2. Performance of Retrofitting Strategies in Terms of Energy Savings

Figure 9 shows annual heating and cooling energy demand in different simulation cases. The calculated heating loads are generally much higher than the cooling load because Melbourne is in a cool temperate climate zone.

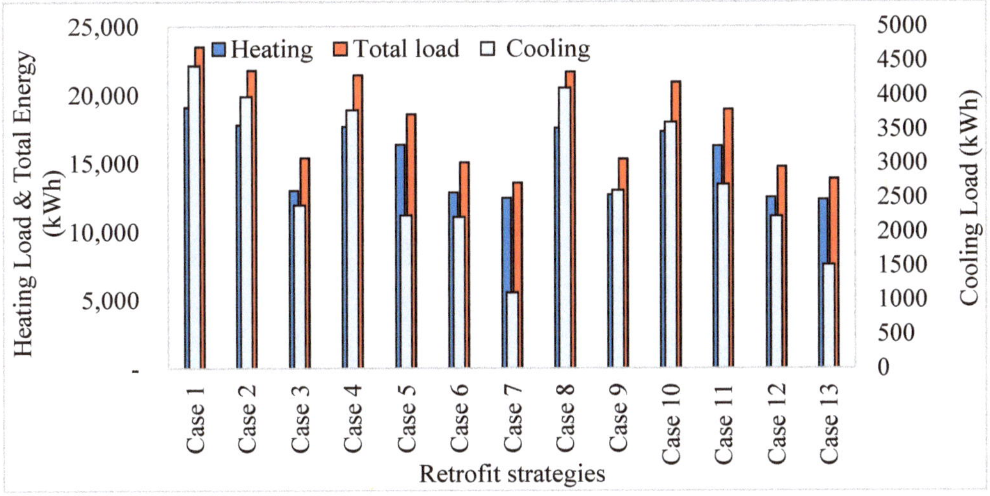


**Figure 9.** Annual cooling, heating, and total loads in reference and retrofitted buildings.

The figure shows that both heating and cooling energy load decreased significantly compared to the reference case (Case 1) with ceiling insulation and aerogel render (Cases 3 and 9). However, aerogel rendering the outer part of the external wall (Case 3) was found to reduce the cooling load slightly higher (47%) than the interior aerogel rendering (41%) (Case 9). On the other hand, the heating load reduction was marginally higher in interior rendering (34%) than exterior rendering (32%). This is in line with the observation in Section 1, where it was reported that rendering the outer side of the external wall minimizes the overheating effect more than the interior rendering. In Cases 5 and 11, the application of PCM in the non-insulated ceiling and aerogel rendered walls significantly reduced the cooling loads (50–53%), which was higher than that of Cases 3 and 9. The reduction in heating loads in Cases 4 and 11 (15–17%) was much lower than that of Cases 3 and 9. However, the total energy consumption in Cases 3 and 9 reduced by 35%, which is much higher than Case 5 and Case 11 (22%), hence they are more preferred options.

Moreover, integration of PCM in insulated ceilings and aerogel rendered walls (Cases 7 and 13) resulted in further significant reduction in cooling load compared to Cases 3 and 9, but the reduction in heating load was marginal. Overall, in Cases 7 and 13, total energy, cooling energy, and heating energy consumption reduced by 41–42%, 69–70%, and 34–36%, respectively, compared to Case 1. Case 13 showed marginally higher heating energy savings but lower cooling energy savings than Case 7. Therefore, in terms of total energy savings, Case 13 is a slightly better option than Case 7. However, as mentioned above, it may be more practical to select Case 7 to avoid interruption in daily activities during retrofitting. The integration of PCM in an insulated ceiling was more efficient than applying PCM only. Therefore, in terms of retrofitting, insulation of the ceiling should come first, and PCM and aerogel render application further reduce the energy use by increasing the heat storage capacity of the building envelope.

Overall, the cooling energy savings potential of different retrofit strategies is much higher than heating energy savings because the PCM layer resists and absorbs heat transfer through the envelope by phase transition (liquefaction and solidification) during summer. In winter, the phase transition activities are very limited due to unfavorable outdoor weather, and the PCM layer mainly resists heat transfer through the envelope being in a solid-state. A different phase transition temperature may be required to enhance the heating energy-saving potential. To further understand this matter, the optimum phase change temperatures in terms of heating, cooling, and total energy savings are presented in Figure 10.

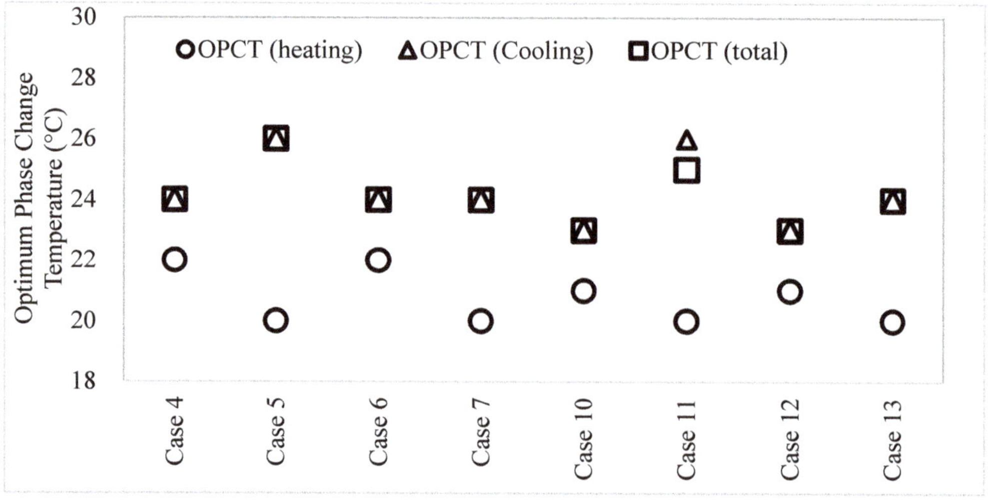


**Figure 10.** Optimum phase change temperatures for heating, cooling, and total energy savings.

Figure 10 shows that the optimum phase change temperature (OPCT) for heating is around 20–22 °C, which is close to the HVAC heating set-point 20 °C of the building. Similarly, the OPCT for cooling is found to be around 23–26 °C, which is close to the HVAC cooling set-point of 24 °C. However, it is not practical to integrate two different phase change materials with optimum temperature for heating and cooling, thereby OPCT is selected based on annual energy use. Figure 10 also shows the OPCT for total energy consumption, which is mostly similar to the OPCT for cooling. Figure 11 shows total annual energy savings for PCM-integrated cases with three different OPCT: OPCT heating, OPCT cooling, and OPCT total. Although OPCT for heating can maximize the heating energy savings, their total energy-saving performance is lower than the OPCT cooling and OPCT total cases. Therefore, OPCT corresponding to cooling or total energy savings is preferred in these cases.

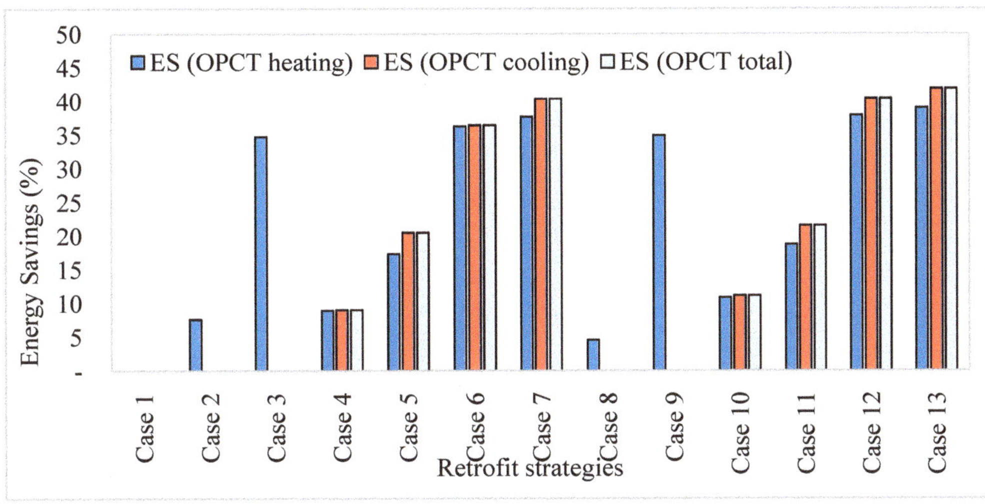



**Figure 11.** Annual energy saving corresponding to optimum phase change temperatures for heating, cooling, and total energy.

Finally, the OPCT corresponding to cooling or total energy savings (23–26 °C) is much lower than the OPCT corresponding to minimum heat stress risk (30–32 °C) for all cases reported in Section 1. Therefore, the selection of OPCT depends on whether a building is free running or mechanically cooled.

### 3.3. Peak Cooling Load Reduction

Figure 12 shows hourly peak cooling loads of different retrofit scenarios. As expected, the peak cooling load in the reference case without any retrofit measures (Case 1) is the highest compared to other cases in the living room of the studied house. However, the figure also shows that the integration of ceiling insulation is crucial to reduce the peak cooling load. In the absence of ceiling insulation, PCM combined with aerogel render (Cases 2, 4, 5, 8, 10, and 11) decreased the peak cooling loads between 5–24%. The integration of ceiling insulation with other retrofit measures significantly reduced the peak cooling load.

Case 3 and Case 9 reduced the peak cooling load by 47% and 45%, respectively, which was further reduced with the application of PCM in the ceiling and walls. The best retrofit scenario for minimizing the peak cooling load was Case 7, with a 65% reduction in peak cooling load.

### 3.4. Performance of Retrofitting Strategies in Terms of Operational Emission

The emission resulting from natural gas consumption and electricity use is illustrated in Figure 13. The emissions of different retrofit strategies with PCM were calculated using OPCT corresponding to the cooling load mentioned in Section 1. Although the building has a comparatively lower cooling load than heating (as shown in Figure 9), the greenhouse gas (GHG) emissions resulting from cooling are higher than emissions associated with heating. Hence, the unit heat produced by natural gas has a lower emission factor than unit electricity used for cooling. Among the retrofitting strategies, aerogel rendered walls (Cases 2 and 8) and aerogel-based render coupled with PCM wall (Cases 4 and 10) have little impact on heating and cooling load emission reduction. However, the application of PCM in the walls and ceiling (Case 5 and Case 11) halved the cooling load emission with a slight decrease in heating load emission, which was further reduced by insulating the ceiling (Cases 6 and 12). The cases with PCM in the insulated ceiling and rendered walls

(Cases 7 and 13) resulted in a maximum 64% total reduction in GHG emission than the base case (Case 1). Out of that 64%, 70% is due to the cooling load emission reduction. Therefore, the PCM application is most beneficial because a significant portion of GHG emissions is associated with the cooling load.

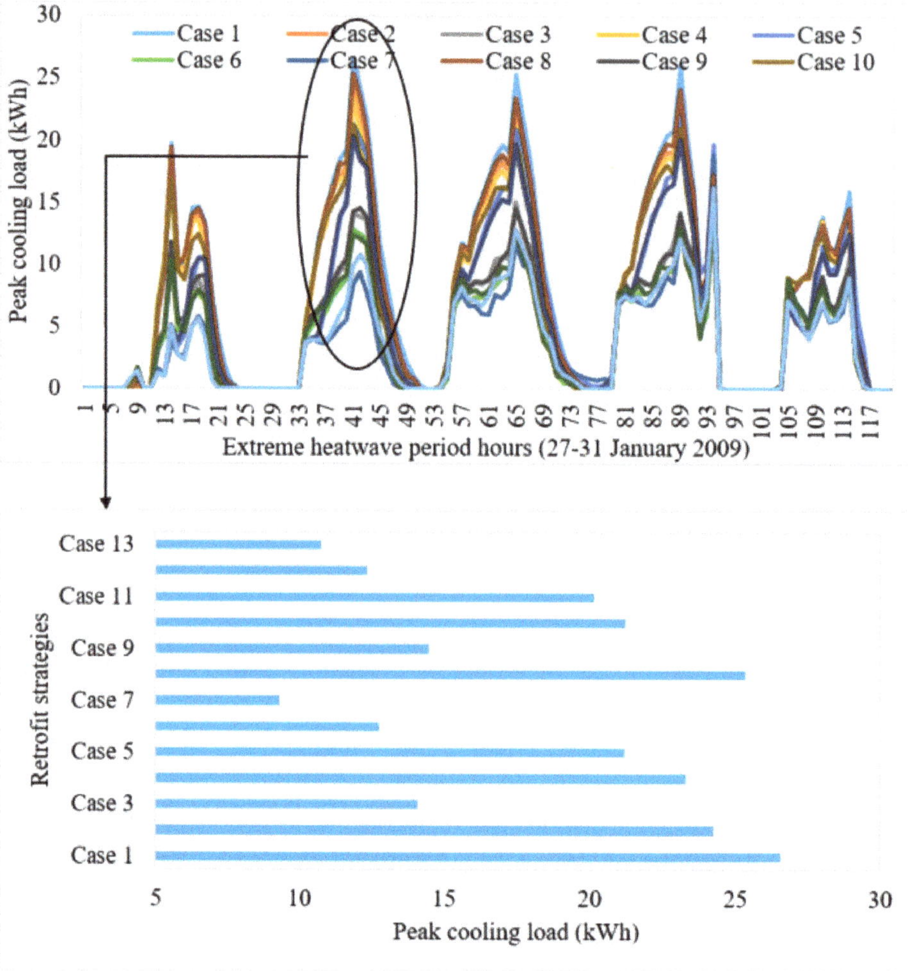



**Figure 12.** Peak cooling load of building under different retrofit strategies.

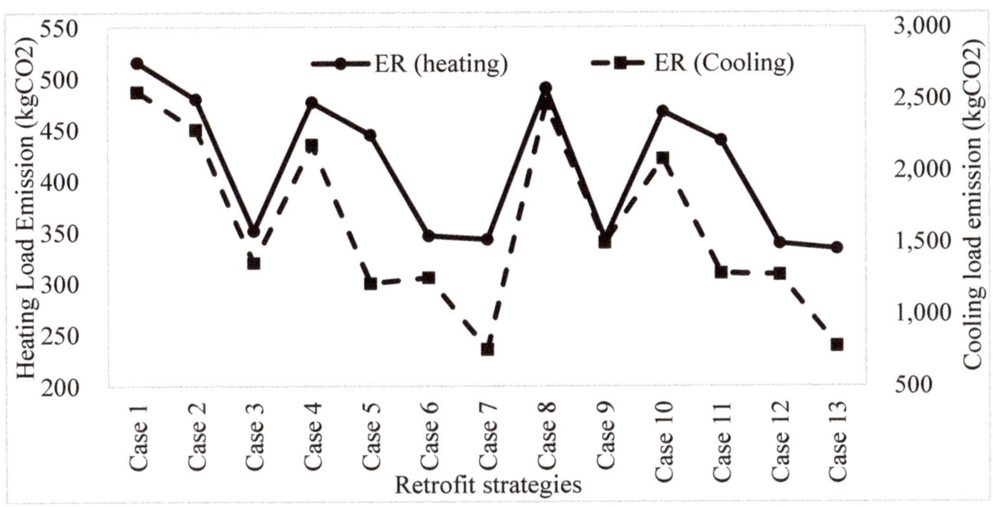


**Figure 13.** $CO_2$ emission associated with natural gas consummation and electricity use in the retrofitted building.

### 3.5. Performance of the Retrofitting Strategies in Terms of Cost

The operational energy cost and initial investment cost of different retrofit strategies are presented in Figure 14, and their respective cost savings are shown in Figure 15. As seen in Figure 14, the operational energy cost (sum of unit electricity cost and natural gas) was reduced by the adding aerogel render, PCM, and insulation. In this study, PCM and insulation quantity were kept constant in all cases because PCM performance varies with its heat storage capacity, which depends on its quantity, while insulation is only applied to the ceiling. The OPCT corresponding to maximum total energy savings was used to evaluate the economic performance of proposed retrofit strategies. The figure shows that applying PCM on the rendered wall and the uninsulated ceiling has an insignificant impact on operation energy cost. However, insulating the ceiling dramatically reduced the operation cost because heat transfer through the roof was higher than the wall [60]. As a consequence, the operation cost dropped to \$3.93–3.96 k/year excluding PCM (Cases 3 and 9), which was further reduced to \$3.83–3.89 k/year and \$3.65–3.73 k/year by applying PCM on the walls (Cases 6–10), and walls and ceiling (Cases 7 and 13), respectively. Although Case 13 was found to have the lowest operational energy cost, the initial investment, in this case, was significantly higher (\$30k) compared to Case 7 (\$18k). Therefore, Case 7 may be preferred from the perspective of lower investment cost and quicker payback period.

Figure 15 shows that the retrofit strategies with insulated ceilings (Case 3, 6, 7, 9, 12, and 13) resulted in positive lifecycle cost savings and can be considered cost-effective. Among all strategies, Case 3 was found to have the highest cost savings over the 40 years lifetime period. The life cycle cost-saving decreases with the integration of PCM and rendering compared to the insulation-only case (Case 3) because of the higher investment cost associated with PCM and render. With aerogel render, PCM, and ceiling insulation, Case 7 resulted in the highest lifecycle cost savings.

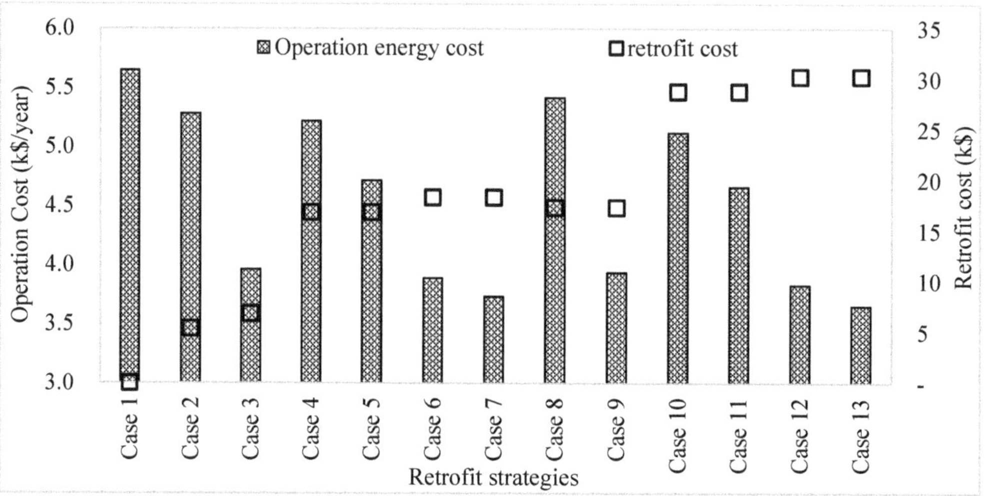



**Figure 14.** Initial investment and annual operation energy cost in different retrofit cases.

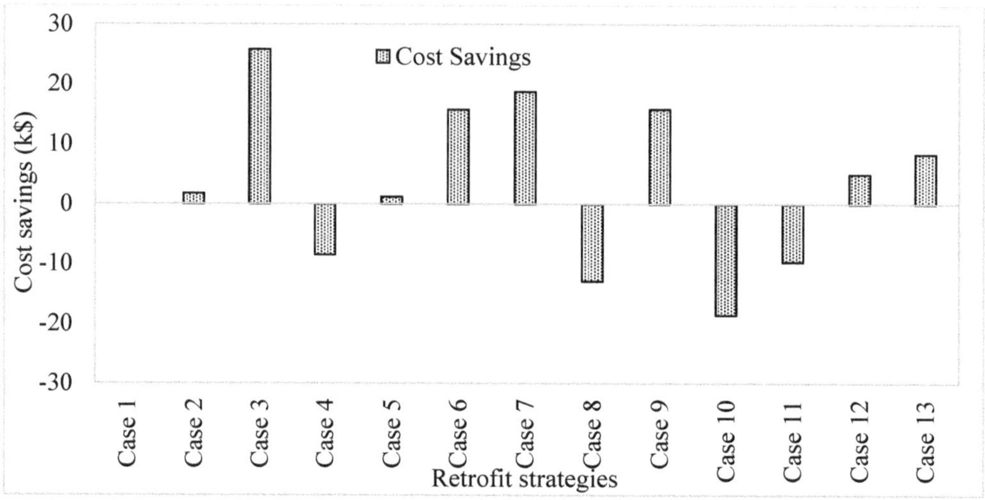



**Figure 15.** Lifecycle cost savings of different retrofit strategies.

## 4. Discussion

The selection of the best retrofit strategy depends on the will of building stakeholders. The private stakeholders are more concerned about thermal comfort and cost savings, while public stakeholders stress on energy-efficient and eco-friendly building design. That is why the application of aerogel renders, PCM, and insulation was evaluated considering the improvement in thermal comfort, increased energy savings, reduction in emission, and maximum lifecycle cost savings. This study assessed 12 different retrofit strategies for mitigating overheating risk in a non-air-conditioned house and minimizing the peak cooling demand, annual energy use, emission, and cost savings in an air-conditioned residential building.

The comparative analysis of results revealed that Case 13 was the best retrofit option to minimize the severe discomfort hours, total heating and cooling energy consumption, and annual operational cost. On the other hand, Case 7 was found to be the best option to minimize peak cooling load during a hot summer period and total $CO_2$ emission associated with heating and cooling load. Also, total lifecycle cost savings were higher for Case 7 than Case 13 because the retrofit investment cost was much higher in the latter case. Therefore, Case 7 may be preferred over Case 13, considering significant lifecycle cost savings, although the performance of Case 7 is marginally lower in terms of discomfort hours, total energy, and operational cost. Furthermore, Case 7 may also be preferred to avoid interrupting occupants' daily life, which is one of the key building energy retrofitting barriers [63]. Finally, the peak cooling load performance of Case 7 was the best among all studied options. As mentioned in 3.3, reducing peak cooling load is very important to eliminate power outages and reduce electricity infrastructure costs that will otherwise be required to meet the peak demand. Hence, Case 7 can be considered as the best retrofit option with PCM.

However, lifecycle cost savings of Case 3 were found to be highest amongst all simulated cases. A comparison between Case 3 and Case 7 showed that lifecycle cost savings of Case 3 is 27% higher than Case 7. However, for the latter case, the severe discomfort hours, total energy consumption, peak cooling load, $CO_2$ emission, and annual operating cost are 64%, 9%, 14%, 36%, and 6% lower than Case 3. Hence, Case 3 can be considered if the cost is the primary selection criteria, as in the case of private stakeholders mentioned above. However, Case 7 could be preferred by public stakeholders with more emphasis on energy-efficient and eco-friendly building design.

*4.1. Impact of Phase Change Temperature on Performance Indicators*

Once Case 7 was selected as the best retrofit strategy, the next key task was selecting the optimum PCM temperature for Case 7, considering the comfort hours, energy savings, peak demand, cost savings, and emissions. Figure 16 shows the performances of Case 7 with different PCM temperatures. The figure shows that the optimum phase change temperature for the minimum severe discomfort hours lies between 29 to 32 °C (also discussed in Section 1). On the other hand, 25 °C PCM results in maximum annual energy savings (40%), emission reduction (63.58%), and lifecycle cost savings ($18.75 k). Furthermore, the peak cooling load was the lowest (9.3 kW) with PCM between 24 and 26 °C. Therefore, if the primary aim is to reduce the thermal discomfort hours in a naturally ventilated (free-running) house during a heatwave period, 29–32 °C should be preferred. However, in an air-conditioned or mixed-mode building, 25 °C PCM is recommended.

However, it should also be noted that although 29–32 °C PCM results in a maximum reduction in discomfort hours during a severe heatwave period, it may not be suitable to increase thermal comfort during the rest of the years. Because approximately 69% of Australians use air conditioner during the hot summer period, 25 °C PCM would be ideal. If there is a power outage during a heatwave, the selection of 25 °C PCM only increases the severe discomfort hours by 3 h during a heatwave period (from 5 to 8 h), but is still 71% lower than the base case scenario mentioned in Figure 8. On the other hand, the use of 29 °C instead of 25 °C in an air-conditioned house increases the peak cooling demand by 6% and decreases the annual energy savings, emission reduction, and cost savings by 2%, 18%, and 28%, respectively. These changes are far more significant than the changes in severe discomfort hours. Therefore, 25 °C PCM can be considered as the optimum PCM for Case 7.

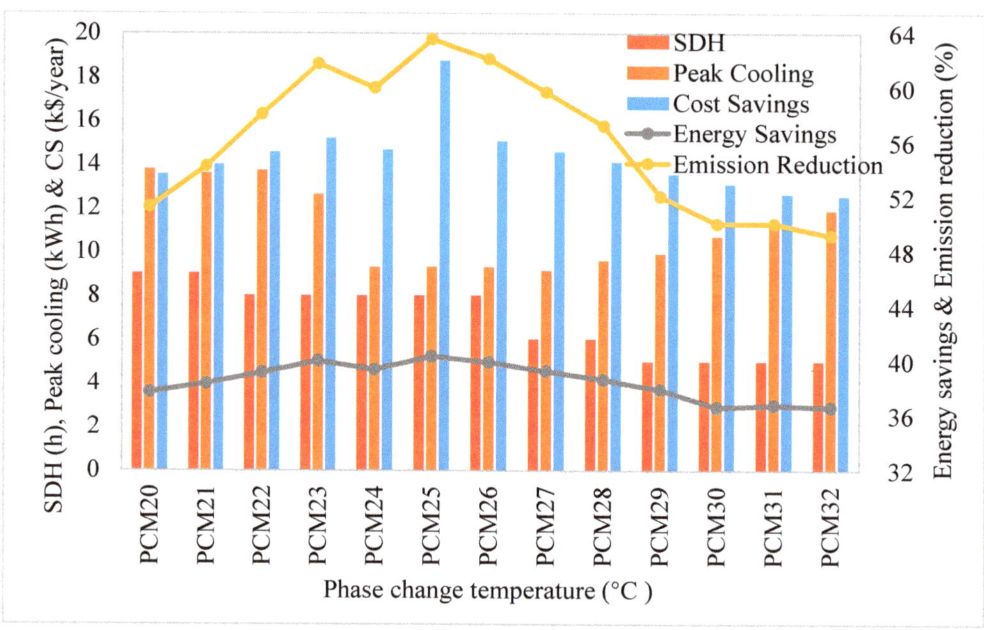



**Figure 16.** Impact of phase change temperature on severe discomfort hours (SDH), peak cooling demand, energy savings, emission reductions, and cost savings.

### 4.2. Impact of PCM and Aerogel Render Thickness on Performance Indicators

Figure 17 shows that severe discomfort hours decrease with increasing render thickness at varying degrees depending on the PCM thickness. Minimum severe discomfort hours can be achieved either by having a combination of thicker render and thinner PCM or with a thinner render and thicker PCM as shown in Figure 17. A thicker render increases resistance to heat transfer from ambient air, and a thicker PCM can absorb more heat from the ambient air. Figure 17 shows that at 0.02 m render thickness, a PCM layer of 0.0225 m is required to achieve minimum discomfort hours. On the other hand, for 0.05 m render thickness, a 0.005 m PCM layer is sufficient to achieve the minimum discomfort hours.

Figure 18 exhibits the influence of aerogel render thickness on annual energy use in an air-conditioned building at different PCM thicknesses. Annual energy use decreases with an increase in both aerogel render and PCM thickness. However, the impact of increasing aerogel render thickness is higher than PCM panel thickness because aerogel render has higher thermal resistance than PCM. Moreover, the building is located in a heating-dominated region where it is desired to retain heat within an occupied space without transferring much into the ambient environment. For instance, the annual energy use was reduced by 550 kWh, with increasing render thickness from 0.01 m to 0.05 m at constant PCM thickness (0.005 m). On the other hand, increasing PCM thickness from 0.005 m to 0.025 m at a constant render thickness of 0.01 m only reduced the energy use by 200 kWh. The thickest layer of PCM (0.025 m) and aerogel render (0.05 m) on the outer part of the wall resulted in the lowest annual energy use because of being heat resistive; aerogel render saves heating energy, whilst, being heat-storage materials, PCM saves cooling energy use. However, this combination also results in the highest investment cost, and therefore, the optimum combination needs to be selected considering the other performance criteria.

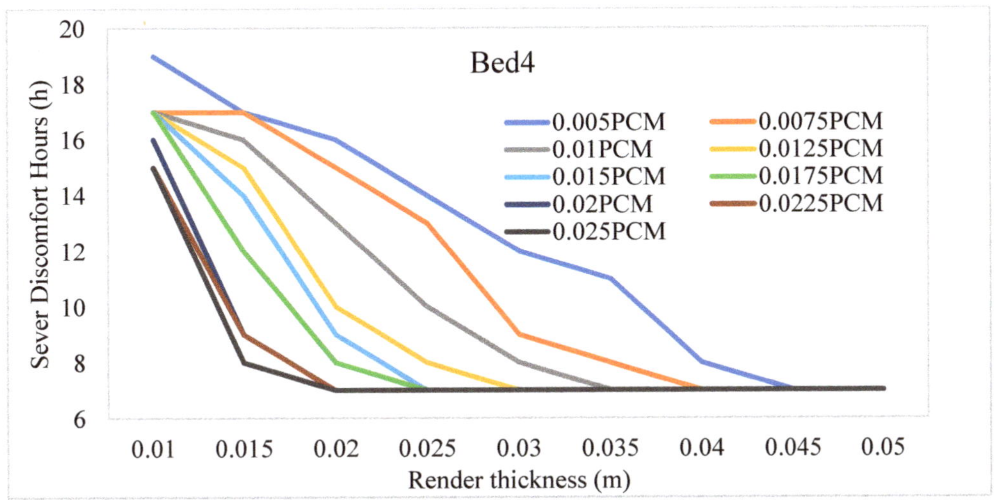


**Figure 17.** Impact of aerogel render thickness on sever discomfort hours considering the different thicknesses of PCM on the exterior side of the wall.

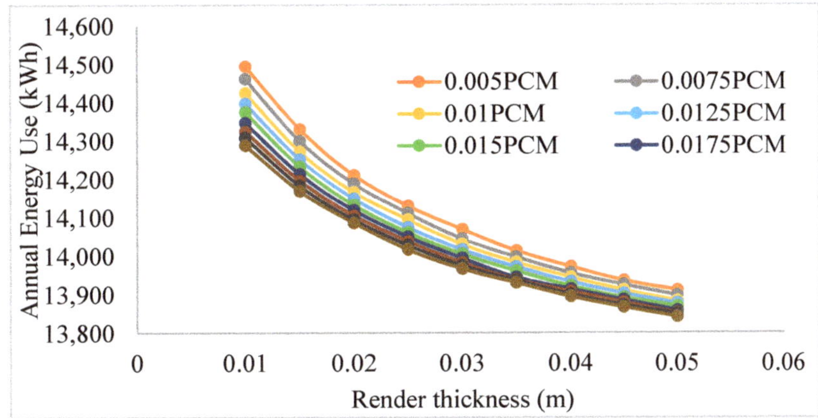


**Figure 18.** Impact of aerogel render thickness on annual energy consumption considering the different thicknesses of PCM on the exterior side of the wall.

Increasing render and PCM thickness also reduced $CO_2$ emission associated with heating and cooling energy use, as shown in Figure 19. Compared to energy, the degree of emission reduction with increasing PCM thickness is comparatively higher at a constant aerogel render thickness. This was because PCM helps to reduce cooling energy demand and the emissions associated with cooling energy use are higher than heating due to their emission factors as discussed in Section 1. The lowest $CO_2$-emission was found for a 0.05 m-thick rendered wall and 0.025 m PCM. An increase in PCM thickness at higher render thickness meagerly reduced $CO_2$ emission compared to lower render thickness.

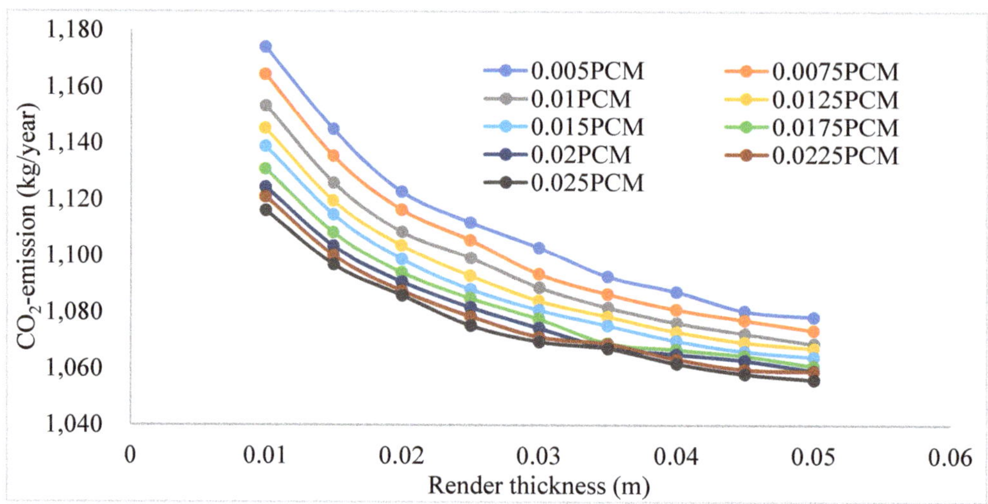


**Figure 19.** Impact of aerogel render thickness on emission considering the different thicknesses of PCM on the exterior side of the wall.

Figure 20 shows variations in life cycle cost savings with different aerogel render and PCM thickness. All combinations of PCM and aerogel render thickness are economically feasible due to positive lifecycle cost savings. However, the savings decrease significantly with the increasing thickness of aerogel render and PCM due to higher initial investment cost. Figure 20 also shows that the cost savings decrease sharply with increasing aerogel render thickness compared to increasing PCM thickness, because the cost of aerogel render is 10 times of PCM. From an economic perspective, a 0.01-m-thick aerogel render and 0.005-m-thick PCM layer should be applied on the outer side of the wall for the highest cost savings among different combinations of PCM and aerogel render thicknesses. However, this combination results in maximum annual energy use and emission, as shown in Figures 18 and 19. Hence, there is a need to find the optimum thickness considering costs, energy, and emission.

Figure 21 shows two Pareto optimization curves created using lifecycle cost, energy consumption, and emission. Pareto font consists of a non-dominated solution where there is no other feasible solution to improve one objective without deteriorating others. The optimum single solution that satisfies the multiple objectives would be selected based on the utopia point criterion. Here, the utopia point represents the point with the lowest lifecycle cost and energy consumption (Figure 21a) and life cycle emission (Figure 21b). The solution is close to the utopia point, which was considered the optimum PCM and aerogel render thickness. The optimum thickness was found to be 0.025 m for both PCM and aerogel render based on both energy consumption and emission. This is different from the best thickness combination identified based on the energy and emission earlier. The optimum solution is highlighted in red in Figure 21 with $84,855 lifecycle cost, 2018 GJ lifecycle energy consumption, and 43 tons of $CO_2$-e emission. This thickness combination results in $24,000 lifecycle cost savings as seen from Figure 21. The identified optimum aerogel-render (0.025 m) and PCM (0.025 m) thickness combination is also suitable to achieve minimum discomfort hours.

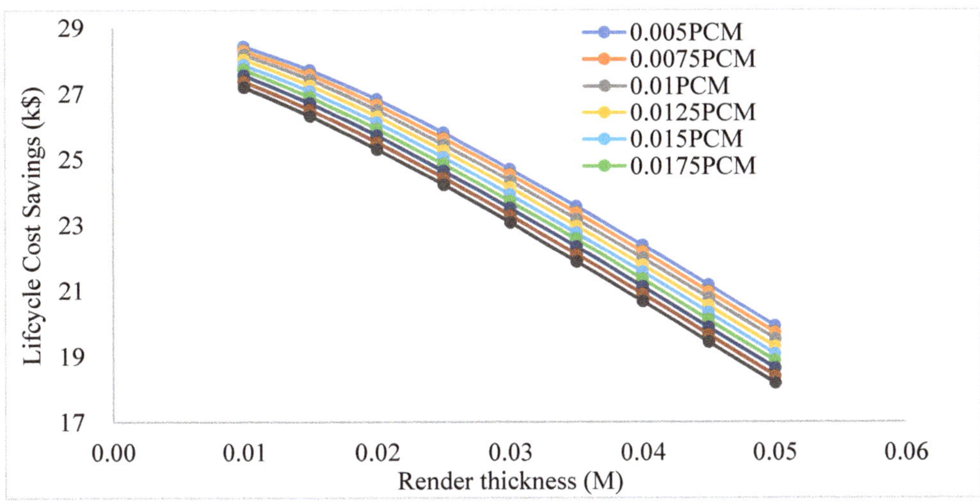



**Figure 20.** Lifecycle cost savings for different aerogel render and PCM thickness on the exterior side of the wall.

This study did not consider supercooling and hysteresis effect of BioPCM due to the lack of available data. Therefore, it may have resulted in some inaccuracies in the energy-saving performance calculation of PCM. PCM with high supercooling may not solidify entirely at night and result in lower cooling energy-saving potential [19]. However, organic PCM generally has a shallow supercooling effect. PCM with high hysteresis improves the thermal performance of PCM walls. Paraffin has low hysteresis with a more negligible difference in melting and solidification curve within 1.2 °C [64].

Moreover, PCM-hysteresis resulted in the mean relative error in the simulated wall's surface temperature and heat flux of 3.5 and 5% compared to PCM without hysteresis [65]. Therefore, the exclusion of PCM-hysteresis may impact heating and cooling energy consumption; however, the impact will be uniform for all simulated cases. Hence, this will not change the critical findings of this study regarding optimum retrofit combinations and optimum PCM temperature, PCM, and aerogel thickness.

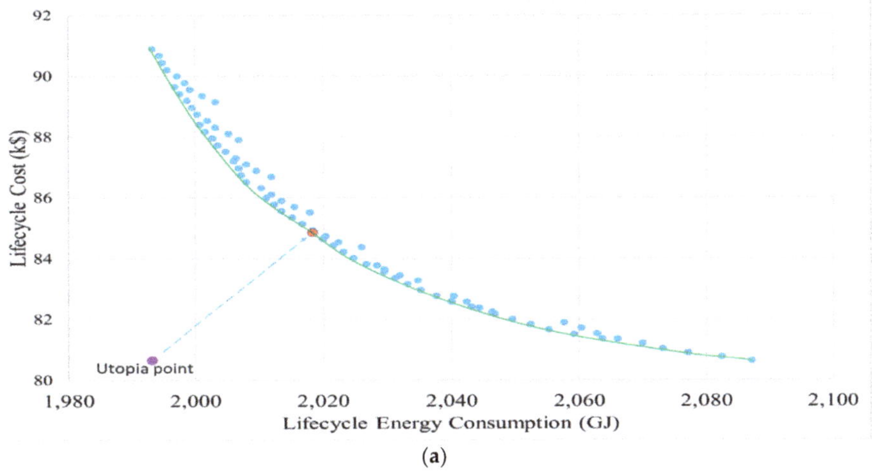



(**a**)

**Figure 21.** *Cont.*

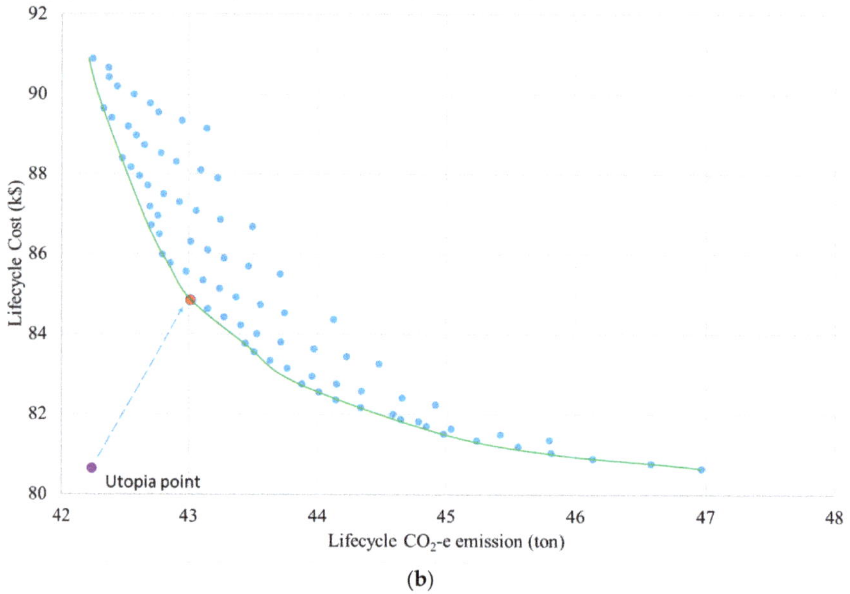



**(b)**

**Figure 21.** Pareto multi-objective optimization curve for (**a**) lifecycle cost vs lifecycle energy consumption, and (**b**) lifecycle cost vs lifecycle CO$_2$-e emission.

## 5. Conclusions

This study numerically investigated 12 different building envelope retrofit strategies, including aerogel render, PCM, and insulation using the building simulation tool EnergyPlus v9.2. The performance of proposed retrofit strategies was evaluated considering overheating risk, energy efficiency, peak cooling load, emission reduction, and cost savings.

The aerogel-based render coupled with PCM outside of external walls and PCM combined with insulated ceilings (Case 7) was found to be the best retrofit strategy considering all performance categories. Compared to the baseline, this strategy reduced severe discomfort hours, total energy consumption, peak cooling load, CO$_2$ emission, and operational energy cost by 82%, 40%, 65%, 64%, and 35%, respectively. Although the lifecycle cost savings of Case 7 are lower than Case 3 (insulated ceiling and externally rendered wall) because of the high investment cost of PCM, the former one can be selected considering its higher environmental performance. Mainly, this would be preferred by public stakeholders where the stress is on energy-efficient and eco-friendly building design. The 25 °C melting point PCM was considered the best option to minimize severe discomfort hours (in non-air-conditioned houses and during the blackout period) during a heatwave period as well as to reduce total energy, emission, cost, and peak cooling load. Parametric analyses showed that the thicker the PCM and aerogel render, the lower is the energy consumption and emission. However, increased PCM and aerogel render thickness decreased lifecycle cost savings due to high investment costs. The optimum thickness for PCM and aerogel render was 0.025 m considering the emission, comfort, energy, and life cycle costs for a typical Australian house in Melbourne climate. This strategy (Case 7) will have a minimum interruption to occupants' daily life while retrofitting because of being applied outside of the external wall.

This study is the first step of a PCM-integrated aerogel render development project. In the future, the findings from this simulation study will be used to develop a PCM-integrated aerogel render. The thermal properties and performance of the developed render will be evaluated experimentally. Then the numerical model developed in this study will be updated to include the properties of PCM-integrated aerogel render, and its performance will be compared against the results presented in this study where PCM and aerogel renders were assumed as separate layers. Furthermore, the thermal performance of the PCM-integrated aerogel render will be evaluated for different climate zones.

**Author Contributions:** Conceptualization, D.K., M.A. and J.G.S.; methodology, D.K.; software, M.A.; validation, M.A.; formal analysis, D.K.; writing—original draft preparation, D.K.; writing—review and editing, M.A.; supervision, M.A, J.G.S.; project administration, J.G.S.; funding acquisition, D.K. All authors have read and agreed to the published version of the manuscript.

**Funding:** This research was funded by Higher Education Commission of Pakistan grant number No. 5-1/HRD/UESTPI(Batch-VI)/6021/2018/HEC.

**Institutional Review Board Statement:** Not Applicable.

**Informed Consent Statement:** Not Applicable.

**Data Availability Statement:** Not applicable.

**Acknowledgments:** Authors would like to thanks the Higher Education Commission (HEC), Pakistan for supporting their research work under the HEC-SUT joint scholarship.

**Conflicts of Interest:** The authors declare no conflict of interest.

## References

1. Berardi, U. A cross-country comparison of the building energy consumptions and their trends. *Resour. Conserv. Recycl.* **2017**, *123*, 230–241. [CrossRef]
2. Sovetova, M.; Memon, S.A.; Kim, J. Thermal performance and energy efficiency of building integrated with PCMs in hot desert climate region. *Sol. Energy* **2019**, *189*, 357–371. [CrossRef]
3. Kumar, D.; Zou, P.X.W.; Memon, R.A.; Alam, M.M.; Sanjayan, J.G.; Kumar, S. Life Cycle Cost Analysis of Building wall and insulation materials. *J. Build.Phys.* **2020**, *43*, 428–455. [CrossRef]
4. Ascione, F.; Bianco, N.; Mauro, G.M.; Napolitano, D.F. Building envelope design: Multi-objective optimization to minimize energy consumption, global cost and thermal discomfort. Application to different Italian climatic zones. *Energy* **2019**, *174*, 359–374. [CrossRef]
5. Li, D.; Zhang, C.; Li, Q.; Liu, C.; Arıcı, M.; Wu, Y. Thermal performance evaluation of glass window combining silica aerogels and phase change materials for cold climate of China. *Appl. Therm.Eng.* **2020**, *165*. [CrossRef]
6. Hoseinzadeh, S. Thermal Performance of Electrochromic Smart Window with Nanocomposite Structure Under Different Climates In Iran. *Micro Nanosyst.* **2019**, *11*, 154–164. [CrossRef]
7. Kumar, D.; Alam, M.; Zou, P.X.W.; Sanjayan, J.G.; Memon, R.A. Comparative analysis of building insulation material properties and performance. *Renew. Sustain. Energy Rev.* **2020**, *131*, 110038. [CrossRef]
8. Sustainablity Victoria Victorian Household Energy Reports, Sustainablity Victoria, Melbourne. 2014. Available online: http://www.sustainability.vic.gov.au/services-and-advice/households/energy-efficiency/toolbox/reports (accessed on 14 October 2020).
9. Victoria, S. Cavity Wall Insulation Retrofit Trial, Melbourne, Australia. 2016. Available online: https://www.sustainability.vic.gov.au/About-us/Research/Household-retrofit-trials (accessed on 14 October 2020).
10. Cuce, E.; Cuce, P.M.; Wood, C.J.; Riffat, S.B. Optimizing insulation thickness and analysing environmental impacts of aerogel-based thermal superinsulation in buildings. *Energy Build.* **2014**, *77*, 28–39. [CrossRef]
11. Berardi, U. Aerogel-enhanced systems for building energy retrofits: Insights from a case study. *Energy Build.* **2018**, *159*, 370–381. [CrossRef]
12. Achard, P.; Rigacci, A.; Echantillac, T.; Bellet, A.; Aulagnier, M.; Daubresse, A. Insulating silica xerogel plaster. WIPO patent WO/083174, 14 July 2011.
13. Ibrahim, M.; Biwole, P.H.; Achard, P.; Wurtz, E. Aerogel-Based Materials for Improving the Building Envelope's Thermal Behavior: A Brief Review with a Focus on a New Aerogel-Based Rendering. In *Energy Sustainability Through (Green Energy)*; Springer: New Delhi, India, 2015; pp. 163–188.
14. Fantucci, S.; Fenoglio, E.; Grosso, G.; Serra, V.; Perino, M.; Marino, V.; Dutto, M. Development of an aerogel-based thermal coating for the energy retrofit and the prevention of condensation risk in existing buildings. *Sci. Technol. Built Environ.* **2019**, *25*, 1178–1186. [CrossRef]
15. Pedroso, M.; Flores-Colen, I.; Silvestre, J.D.; Gomes, M.G.; Silva, L.; Ilharco, L. Physical, mechanical, and microstructural characterisation of an innovative thermal insulating render incorporating silica aerogel. *Energy Build.* **2020**, 109793. [CrossRef]
16. Ibrahimb, M.; Nocentinic, K.; Stipeticd, M.; Dantzd, S.; Caiazzoe, F.G.; Sayegha, H.; Bianco, L. Multi-field and multi-scale characterization of novel super insulating panels/systems based on silica aerogels: Thermal, hydric, mechanical, acoustic, and fire performance. *Build. Environ.* **2019**, *151*, 30–42. [CrossRef]

17. Nosrati, R.H.; Berardi, U. Hygrothermal characteristics of aerogel-enhanced insulating materials under different humidity and temperature conditions. *Energy Build.* **2018**, *158*, 698–711. [CrossRef]
18. Al-Yasiri, Q.; Szabó, M. Incorporation of phase change materials into building envelope for thermal comfort and energy saving: A comprehensive analysis. *J. Build. Eng.* **2021**, *36*, 102122. [CrossRef]
19. Berardi, U.; Gallardo, A.A. Properties of concretes enhanced with phase change materials for building applications. *Energy Build.* **2019**, *199*, 402–414. [CrossRef]
20. Sukontasukkula, P.; Sangpetb, T.; Newlands, M.; Yood, D.-Y.; Tangchirapate, W.; Limkatanyuf, S.; Chindaprasirt, P. Thermal storage properties of lightweight concrete incorporating phase change materials with different fusion points in hybrid form for high temperature applications. *Heliyon* **2020**, *6*, e04863. [CrossRef]
21. Rao, V.V.; Parameshwaran, R.; Ram, V.V. PCM-mortar based construction materials for energy efficient buildings: A review on research trends. *Energy Build.* **2018**, *158*, 95–122. [CrossRef]
22. Cunha, S.; Aguiar, J.B.; Ferreira, V.; Tadeu, A. Influence of Adding Encapsulated Phase Change Materials in Aerial Lime Based Mortars. *Adv. Mater. Res.* **2013**, *687*, 255–261. [CrossRef]
23. Michel, B.; Glouannec, P.; Fuentes, A.; Chauvelon, P. Experimental and numerical study of insulation walls containing a composite layer of PU-PCM and dedicated to refrigerated vehicle. *Appl. Therm. Eng.* **2017**, *116*, 382–391. [CrossRef]
24. Wieprzkowicz, A.; Heim, D. The characteristics of temperature fluctuations in thermal insulation covered with layer of PCM. In *IOP Conference Series: Materials Science and Engineering*; Institute of Physics: Krakow, Poland, 2018; Volume 415, p. 012017. [CrossRef]
25. Bakri, M.M.A.; Jamaludin, A.L.; Abdullah, A.; Razak, R.A.; Hussin, K. *Study on Properties and Morphology of Kaolin Based Geopolymer Coating on Clay Substrates*; Trans Tech Publications Ltd.: Kapellweg, Switzerland, 2014.
26. Jamil, H.; Alam, M.; Sanjayan, J.; Wilson, J. Investigation of PCM as retrofitting option to enhance occupant thermal comfort in a modern residential building. *Energy Build.* **2016**, *133*, 217–229. [CrossRef]
27. Ramakrishnan, S.; Sanjayan, J.; Wang, X. Experimental Research on Using Form-stable PCM-Integrated Cementitious Composite for Reducing Overheating in Buildings. *Buildings* **2019**, *9*, 57. [CrossRef]
28. Cui, H.; Memon, S.A.; Liu, R. Development, mechanical properties and numerical simulation of macro encapsulated thermal energy storage concrete. *Energy Build.* **2015**, *96*, 162–174. [CrossRef]
29. Kośny, J.; Fallahi, A.; Shukla, N.; Kossecka, E.; Ahbari, R. Thermal load mitigation and passive cooling in residential attics containing PCM-enhanced insulations. *Sol. Energy* **2014**, *108*, 164–177. [CrossRef]
30. Rathore, P.K.S.; Shukla, S.K. An experimental evaluation of thermal behavior of the building envelope using macroencapsulated PCM for energy savings. *Renew. Energy* **2019**, *149*, 1300–1313. [CrossRef]
31. Mahmoudan, A.; Samadof, P.; Hosseinzadeh, S.; Garcia, D.A. A multigeneration cascade system using ground-source energy with cold recovery: 3E analyses and multi-objective optimization. *Energy* **2021**, *233*, 121185. [CrossRef]
32. Hoseinzadeh, S.; Ghasemi, M.H.; Heyns, S. Application of hybrid systems in solution of low power generation at hot seasons for micro hydro systems. *Renew. Energy* **2020**, *160*, 323–332. [CrossRef]
33. Gilani, H.A.; Hoseinzadeh, S.; Karimi, H.; Karimi, A.; Hassanzadeh, A.; Garcia, D.A. Performance analysis of integrated solar heat pump VRF system for the low energy building in Mediterranean island. *Renew. Energy* **2021**, *174*, 1006–1019. [CrossRef]
34. Economics, C.f.I. Proposal to Revise Energy Efficiency Requirements of the Building Code of Australia for Residential Buildings, Australian Building Codes Board, Canberra. 2009. Available online: https://www.abcb.gov.au/-/media/Files/Resources/Consultation/RIS-Energy-Efficiency-Residential-Building-Final-Decision-BCA-2010.pdf (accessed on 14 April 2021).
35. ABS Census of Population and Housing. 2016. Available online: https://quickstats.censusdata.abs.gov.au/census_services/getproduct/census/2016/quickstat/036 (accessed on 14 April 2021).
36. Stahl, T.; Brunner, S.; Zimmermann, M. Thermally Insulating Aerogel Based Rendering Materials. Patent WO 2014/090790 Al, 05 November 2014.
37. Ibrahim, M.; Biwole, P.H.; Achard, P.; Wurtz, E.; Ansart, G. Building envelope with a new aerogel-based insulating rendering: Experimental and numerical study, cost analysis, and thickness optimization. *Appl. Energy* **2015**, *159*, 490–501. [CrossRef]
38. Ramakrishnan, S.; Wang, X.; Sanjayan, J.; Wilson, J. Thermal performance of buildings integrated with phase change materials to reduce heat stress risks during extreme heatwave events. *Appl. Energy* **2017**, *194*, 410–421. [CrossRef]
39. Muruganatham, K. Application of Phase Change Materials in Buildings: Field Data vs EnergyPlus Simulation. Master's Thesis, Science Arizona State University, Tempe, AZ, USA, 2010.
40. Harkouss, F.; Fardoun, F.; Biwole, P.H. Passive design optimization of low energy buildings in different climates. *Energy* **2018**, *165*, 591–613. [CrossRef]
41. Alam, M.; Rajeev, P.; Sanjayan, J.; Zou, P.X.W.; Wilson, J. Mitigation of heat stress risks through building energy efficiency upgrade: A case study of Melbourne, Australia. *Aust. J. Civ. Eng.* **2018**, *16*, 64–78. [CrossRef]
42. Tabares-Velasco, P.C.; Christensen, C.; Bianchi, M. Verification and validation of EnergyPlus phase change material model for opaque wall assemblies. *Build. Environ.* **2012**, *54*, 186–196. [CrossRef]
43. ABCB. *Protocol for House Energy Rating Software*; Australian Building Codes Board: Canberra, Australia, 2006.
44. I. ASHRAE. *ASHRAE Handbook: Fundamentals*; American Society of Heating, Refrigeration and Air-Conditioning Engineers: Atlanta, GA, USA, 2009.

45. Alam, M.; Jamil, H.; Sanjayan, J.; Wilson, J. Energy saving potential of phase change materials in major Australian cities. *Energy Build.* **2014**, *78*, 192–201. [CrossRef]

46. Moore, T.; Morrissey, J. Lifecycle costing sensitivities for zero energy housing in Melbourne, Australia. *Energy Build.* **2014**, *79*, 1–11. [CrossRef]

47. Ramakrishnan, S.; Wang, X.; Alam, M.; Sanjayan, J.; Wilson, J. Parametric analysis for performance enhancement of phase change materials in naturally ventilated buildings. *Energy Build.* **2016**, *124*, 35–45. [CrossRef]

48. Current Residential Electricity & Gas Tariffs. 2020. Available online: https://www.originenergy.com.au/content/dam/origin/residential/docs/new-connections/multi-site-pricing-booklet.pdf (accessed on 14 October 2020).

49. *National Greenhouse Accounts Factors: Australian National Greenhouse Accounts*; Department of Climate Change: Canberra, Australia, 2018.

50. Islam, H.; Jollands, M.; Setunge, S.; Ahmed, I.; Haque, N. Life cycle assessment and life cycle cost implications of wall assemblages designs. *Energy Build.* **2014**, *84*, 33–45. [CrossRef]

51. Morrissey, J.; Horne, R.E. Life cycle cost implications of energy efficiency measures in new residential buildings. *Energy Build.* **2011**, *43*, 915–924. [CrossRef]

52. Kumar, D.; Ali, I.; Hakeem, M.; Junejo, A.; Harijan, K. LCC Optimization of Different Insulation Materials and Energy Sources Used in HVAC Duct Applications. *Arab. J. Sci. Eng.* **2019**, *44*, 5679–5696. [CrossRef]

53. OBPR. *Cost-Benefit-Analysis*; Office of the Best Practice Office Regulation, Department of Prime Minister and Cabinet: Canberra, Australia, 2016.

54. Solgi, E.; Hamedani, Z.; Fernando, R.; Kari, B.M. A parametric study of phase change material characteristics when coupled with thermal insulation for different Australian climatic zones. *Build. Environ.* **2019**, *163*, 106317. [CrossRef]

55. Medina, M.A.; King, J.B.; Zhang, M. On the heat transfer rate reduction of structural insulated panels (SIPs) outfitted with phase change materials (PCMs). *Energy* **2008**, *33*, 667–678. [CrossRef]

56. Alam, M.; Sanjayan, J.; Zou, P.X.W.; Stewart, M.G.; Wilson, J. Modelling the correlation between building energy ratings and heat-related mortality and morbidity. *Sustain. Cities Soc.* **2016**, *22*, 29–39. [CrossRef]

57. Epstein, Y.; Moran, D.S. Thermal comfort and the heat stress indices. *Indusrial Health* **2006**, *44*, 388–398. [CrossRef]

58. U.S Department of Energy. *Application Guide for EMS-EnergyPlus™ Version 9.2.0 Documentation*; U.S Department of Energy: Washington, DC, USA, 2019; p. 21.

59. Australian Bureau of Statistics. Environmental Issues: Energy Use and Conservation. 2014. Available online: https://www.abs.gov.au/AUSSTATS/abs@.nsf/Lookup/4602.0.55.001Main+Features1Mar%202014?OpenDocument (accessed on 7 September 2021).

60. Najjar, M.K.; Figueiredo, K.; Hammad, A.W.A.; Tam, V.W.Y.; Evangelista, A.C.J.; Haddad, A. A framework to estimate heat energy loss in building operation. *J. Clean.Prod.* **2019**, *235*, 789–800. [CrossRef]

61. Földváry, V.; Bekö, G.; Langer, S.; Arrhenius, K.; Petráš, D. Effect of energy renovation on indoor air quality in multifamily residential buildings in Slovakia. *Build.Environ.* **2017**, *122*, 363–372. [CrossRef]

62. Tettey, U.Y.A.; Gustavsson, L. Energy savings and overheating risk of deep energy renovation of a multi-storey residential building in a cold climate under climate change. *Energy* **2020**, *202*, 117578. [CrossRef]

63. Alam, M.; Zou, P.X.W.; Stewart, R.A.; Bertone, E.; Sahin, O.; Buntine, C.; Marshall, C. Government championed strategies to overcome the barriers to public building energy efficiency retrofit projects. *Sustain. Cities Soc.* **2019**, *44*, 56–69. [CrossRef]

64. Moreles, E.; Huelsz, G.; Barrios, G. Hysteresis effects on the thermal performance of building envelope PCM-walls. *Build. Simul.* **2018**, *11*, 519–531. [CrossRef]

65. Zastawna-Rumin, A.; Kisilewicz, T.; Berardi, U. Novel Simulation Algorithm for Modeling the Hysteresis of Phase Change Materials. *Energies* **2020**, *13*, 1200. [CrossRef]

 *sustainability*

**MDPI**

*Article*

# Impact of Shape Factor on Energy Demand, CO$_2$ Emissions and Energy Cost of Residential Buildings in Cold Oceanic Climates: Case Study of South Chile

**Manuel Carpio [1,2,*] and David Carrasco [3]**

[1]    Department of Construction Engineering and Management, School of Engineering, Pontificia Universidad Católica de Chile, Santiago 7820436, Chile
[2]    UC Energy Research Center, Pontificia Universidad Católica de Chile, Santiago 7820436, Chile
[3]    Civil Engineer, Valdivia 5090000, Chile; davidcarrascorivas@gmail.com
*    Correspondence: manuel.carpio@ing.puc.cl

**Abstract:** The increase in energy consumption that occurs in the residential sector implies a higher consumption of natural resources and, therefore, an increase in pollution and a degradation of the ecosystem. An optimal use of materials in the thermal envelope, together with efficient measures in the passive architectural design process, translate into lower energy demands in residential buildings. The objective of this study is to analyse and compare, through simulating different models, the impact of the shape factor on energy demand and CO$_2$ emissions depending on the type of construction solution used in the envelope in a cold oceanic climate in South Chile. Five models with different geometries were considered based on their relationship between exposed surface and volume. Additionally, three construction solutions were chosen so that their thermal transmittance gradually complied with the values required by thermal regulations according to the climatic zone considered. Other parameters were equally established for all simulations so that their comparison was objective. Ninety case studies were obtained. Research has shown that an appropriate design, considering a shape factor suitable below 0.767 for the type of cold oceanic climate, implies a decrease in energy demand, which increased when considering architectural designs in the envelope with high values of thermal resistance.

**Keywords:** shape factor; building; thermal envelope; energy demand; CO$_2$ emissions

**Citation:** Carpio, M.; Carrasco, D. Impact of Shape Factor on Energy Demand, CO$_2$ Emissions and Energy Cost of Residential Buildings in Cold Oceanic Climates: Case Study of South Chile. *Sustainability* **2021**, *13*, 9491. https://doi.org/10.3390/su13179491

Academic Editors: Mitja Košir and Manoj Kumar Singh

Received: 2 July 2021
Accepted: 19 August 2021
Published: 24 August 2021

**Publisher's Note:** MDPI stays neutral with regard to jurisdictional claims in published maps and institutional affiliations.



## 1. Introduction

Energy consumption is reflected in the gross domestic product (GDP) of a country. There is a close relationship between GDP and the required electrical energy, which increases every year at the country level and is sustained [1]. The world has created a legal framework to respond to the need to provide energy in the context of sustainable development, given the threats [2]. As a first initiative in the regulatory framework, in 1997, 37 industrialised countries and the European Union established the Kyoto Protocol [3]. A building, especially in the operation stage, can be a great potential consumer of energy, and only using measures and strategies in the design stage which involve insignificant increases in construction costs and significant benefits in energy demand (or energy need [4]) and emission reduction can significantly affect its energy consumption [5].

The energy efficiency of a building depends not only on the thermal properties of the materials in the envelope but also on its shape; the orientation and distribution of spaces, windows and closing ratios; interior temperature; the façade colour and protection against solar radiation [6]. All these parameters influence the passive design of a building, depending on the climatic zone in which it is located. The volumetric impact on a building at the design stage produces better efficiency during the life cycle of the home, reducing energy and natural resource consumption [7,8]. To optimise architectural designs for thermal envelopes, it is essential to study the climate of the building area in detail [9,10].

Compactness is a characteristic of the volume of the building; it is used to adjust the exposed envelope, depending on the useful area, as much as possible [11]. This geometric relationship is represented by the shape factor (*SFv*). Buildings with a high *SFv* value are less compact, resulting in higher heating energy demands in cold climates with poorly sunny winters [12,13]. The impact of *SFv* varies considerably for buildings with different properties in the thermal envelope and weather conditions [14]. A study of the impact of different thermal envelopes in buildings showed that more significant benefits are obtained by using materials with better thermal quality in the envelope when there is more exposed surface per m$^2$ of useful surface [15]. However, for a correct architectural design, a multitude of variables such as orientation, wind and lighting, among others, must be taken into account.

Globally, Chile has an essential representation of cities located in cold oceanic climates [16]. In Chile, one of the most significant increases in energy demand occurred between 2008 and 2014, reaching 18.43% in the commercial, public and residential sectors [17]. Most of this increase (53.7%) was renewable energy from biomass. Furthermore, the imported energy sources are primarily crude oil, natural gas and coal, with 51.8%, 17.0% and 31.2%, respectively. This marked dependence on scarce and non-renewable fuels suggests future problems in energy supply [18]. Additionally, the high carbon emissions resulting from using these fossil fuels generate major environmental and health problems, mainly climate change caused by greenhouse gas emissions [19].

In Chile, the Ordenanza General de Urbanismo y Construcciones (OGUC) (General Ordinance of Urbanism and Constructions) has incorporated the Regulación Térmica (RT) (Thermal Regulation) [20] to be able to classify the energy demand of a building through an energy certification programme, taking into account the different climatic zones (Chilean regulations use the synonym thermal zones). In the specific case of Chile, various investigations were carried out to specify the different climatic zones [21,22], including making projections of future climate change [23] and seeing its application in different fields of study, such as heritage [24]. For cold climates, as in southern Chile, compact buildings with good insulation and infiltration control are recommended. However, care should be taken when using a very low *SFv*, as this can cause ventilation problems and less use of natural light [25].

Considering this background, the objective of this study is to analyse the influence of buildings' thermal envelope *SFv* on energy demand and $CO_2$ emissions in oceanic cold climates by simulating different solutions for optimisation. To carry out this main objective, the following specific objectives were developed: (i) Modelling buildings with different *SFv*; (ii) Applying different architectural designs to the models; (iii) Simulation in different cities in southern Chile; and (iv) Analysing the results. The scope of this study is limited to *SFv* in climatic zones of southern Chile with different constructive solutions. A multitude of additional variables can be studied in future works for a complete analysis of complex architectural designs.

## 2. Materials and Methods

### 2.1. Shape Factor

As the surface of the building in contact with the outside is more significant, there will be more energy exchanges, which may be beneficial or unfavourable in certain types of climates [25], depending on whether the building seeks to conserve the heat inside it or dissipate it to the environment.

*SFv* is a simple equation that relates the enveloping surface to the volume (Equation (1)) [26].

$$SFv = \frac{Se}{V} \tag{1}$$

where the *SFv* is directly related to the heating energy demand in a dwelling, *Se* is the surface area of the exposed envelope and *V* is the habitable volume. The higher the *SFv* (for an identical habitable volume), the higher the heating energy demand of the dwelling.

### 2.2. Climate Data and Climatic Zones

To validate the optimal *SFv* in buildings in cold oceanic climates, as shown in Figure 1, capitals of the southern zone of Chile were chosen—Concepción, Temuco, Valdivia, Puerto Montt, Coyhaique and Punta Arenas. These cities generally represent the climatic characteristics that affect the buildings. According to the study carried out by Sarricolea et al. [16] on the climatic regionalisation of continental Chile, all capitals studied using the Köppen–Geiger climate classification have climate *C* with an oceanic or marine influence. Table 1 shows the climatic zone and climatological station of each of the different cities. In Chile, the regulation that regulates the climatic zones of the country is the OGUC [20], which divides the territory into 7 zones—where Zone 1 is the warmest and Zone 7 is the coldest.

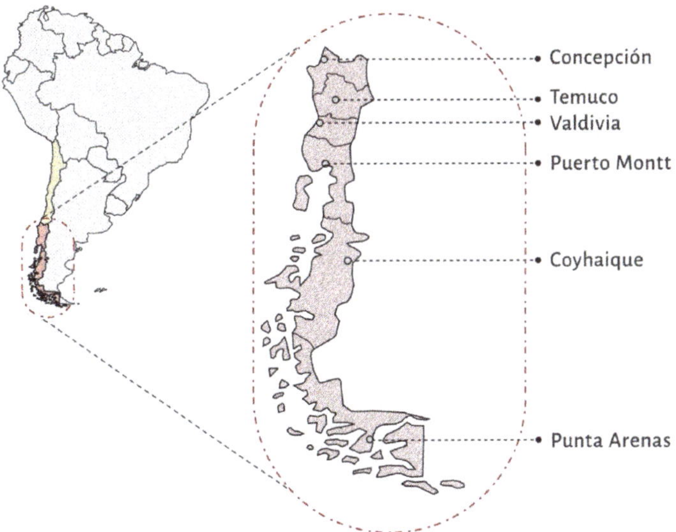


**Figure 1.** Location map of Chile.

**Table 1.** Climatic zones and meteorological station of the regional capitals of southern Chile.

| City | Region | Meteorological Station Data | Climatic Zone OGUC | Temperature—Summer [°C] | Temperature—Winter [°C] | Annual Global Radiation [kWh/m$^2$] |
|---|---|---|---|---|---|---|
| Concepción | Bio-Bio | Global station | 4 | $16.2 \pm 0.7$ | $11.1 \pm 0.3$ | $1729.1 \pm 26.4$ |
| Temuco | Araucanía | Manquehe | 5 | $15.7 \pm 0.7$ | $9.7 \pm 0.6$ | $1552.8 \pm 47.5$ |
| Valdivia | Los Ríos | Pichoy | 5 | $15.4 \pm 0.8$ | $10.0 \pm 0.4$ | $1509.5 \pm 55.9$ |
| Puerto Montt | Los Lagos | El Tepu | 6 | $13.7 \pm 0.5$ | $9.0 \pm 0.2$ | $1335.9 \pm 72.9$ |
| Coyhaique | Aysén del General Carlos Ibáñez del Campo | Teniente V | 7 | $12.4 \pm 1.0$ | $5.4 \pm 0.9$ | $1343.2 \pm 76.5$ |
| Punta Arenas | Magallanes y la Antártida Chilena | Global station | 7 | $10.0 \pm 0.6$ | $4.8 \pm 0.8$ | $1101.3 \pm 29.9$ |

Climatological data for the cities were extracted in *.epw format by the software Meteonorm 7 [27]. Table 1 shows the meteorological stations from which the data were obtained, with radiation periods between 1991 and 2010 and temperature periods between 2000 and 2009.

### 2.3. CO$_2$ Emissions and Energy Costs of Fuels

As shown in Table 2, the theoretical values of CO$_2$ emissions and the lower heating value (LHV) for different fuels were calculated according to data obtained from different official sources [28,29]. Similarly, the cost of using different types of fuels is the price collected from official reports from the Chilean government [30] and other international studies [31–33]. For the present study, only the cost of fuel has been taken into account. The cost of equipment installation and maintenance has not been considered. The equipment used were boilers with a thermal efficiency of 90% and an outlet water temperature of 80 °C for heating [19]. For electricity, instead, an electrical system was used.

**Table 2.** Emissions, LHV and cost of different fuels.

| Fuel | CO$_2$ Emissions [kgCO$_2$/kWh] | LHV | Cost [USD/kWh] |
|---|---|---|---|
| Electricity (Chile) | 0.346 | - | 0.107 |
| Natural gas | 0.204 | 9.771 kWh/m$^3$ | 0.095 |
| Propane gas | 0.254 | 13.131 kWh/m$^3$ | 0.192 |
| Biomass (wood) | Neutral | 2.759 kWh/kg | 0.083 |
| Biomass (pellet) | Neutral | 5.010 kWh/kg | 0.061 |
| Gasoil | 0.287 | 11.939 kWh/kg | 0.082 |

### 2.4. Case Studies

To carry out the present study, five buildings with different *SFv* and three architectural designs in each of the six capitals of the southern regions of Chile were studied, thus obtaining a total of 90 case studies. The buildings and the energy simulations were modelled following the Building Information Modelling (BIM) and Building Performance Analysis (BPA) methodology through Autodesk© software [34] based on the calculation methodology of ISO 52016-1:2017 [4], ISO 52017-1:2017 [35] and ISO 13789:2017 [36]. All models used the same calculation parameters, with 20 m$^2$/person, an 18–22 °C temperature range, 0.5 air renewals/hour, person 1680 Wh daily heat gain and 2.29 Wh/m$^2$ equipment thermal gain. The characteristics of the different case studies are shown below.

#### 2.4.1. Building Geometry

Table 3 and Figure 2 show the five residential building models (M1, M2, M3, M4 and M5) created. Each model varies from the highest to the lowest *SFv*. All models have a square plan with an increase of 10 m of façade between them, a 3 m height between floors and a flat roof.

**Table 3.** *SFv* of the models.

| Model | Floor Dimensions [m × m] | Floor Space [m$^2$] | Number of Floors | Volume [m$^3$] | Exposed Surface [m$^2$] | *SFv* |
|---|---|---|---|---|---|---|
| M1 | 10 × 10 | 100 | 1 | 300 | 320 | 1.067 |
| M2 | 20 × 20 | 400 | 1 | 1200 | 1040 | 0.867 |
| M3 | 40 × 40 | 1600 | 1 | 4800 | 3680 | 0.767 |
| M4 | 30 × 30 | 900 | 2 | 5400 | 2520 | 0.467 |
| M5 | 50 × 50 | 2500 | 3 | 22,500 | 6800 | 0.302 |

However, it is necessary to clarify that the building models used do not correspond to actual buildings. These models are theoretical, and all the buildings have common characteristics, with the *SFv* variable to be compared between them. These theoretical models have the same *SFv* as more common buildings. For example, on the one hand, M3 maintains the same *SFv* as a two-floor building with a 9.3 × 9.3 m floor dimension. On the other hand, M5 maintains the same *SFv* as a six-floor building with a 21 × 21 m floor dimension.

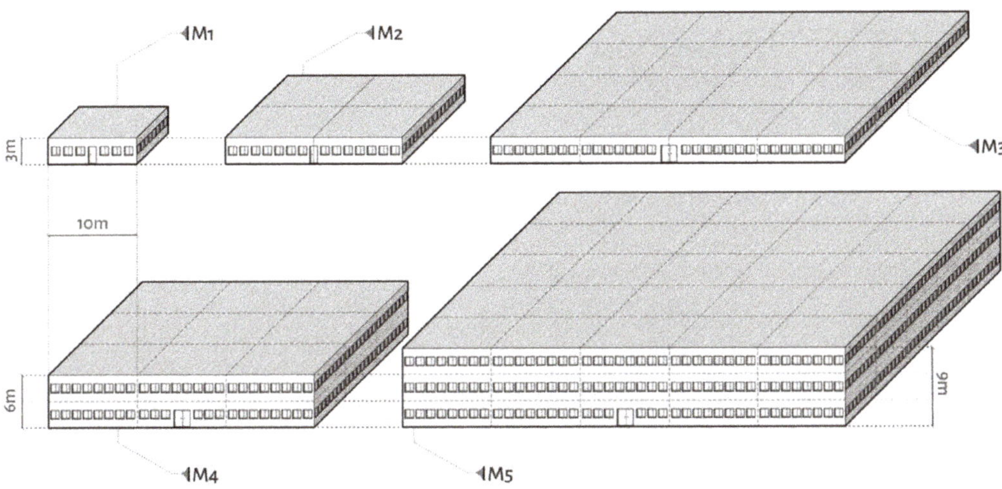



**Figure 2.** Graphic detail of *SFv* models.

### 2.4.2. Architectural Designs for the Thermal Envelopes

Figure 3 and Tables 4 and 5 show the three thermal envelope construction solutions (S1, S2 and S3) for the models described in the previous section. The ratio of window and door area will be 26.67% on all models. According to the material used and the thickness of each layer, each solution has different thermal transmittance (U-value).

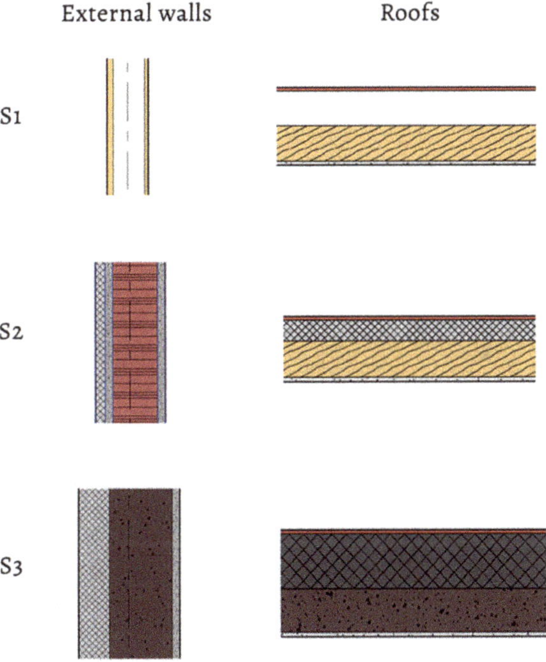



**Figure 3.** Graphic detail of the architectural designs.

**Table 4.** Characteristics of the envelope materials.

| | Material | e [m] | $\lambda$ [W/(m $\times$ K)] | R [(m$^2$ $\times$ K)/W] |
|---|---|---|---|---|
| **S1** | Interior wood lining—Insigne pine 1/2 $\times$ 4″ | 0.013 | 0.104 | 0.122 |
| | Unventilated vertical air chamber | 0.100 | 0.714 | 0.140 |
| | Exterior wood lining—Tinglado dry pine 5 $\times$ 3/4″ | 0.019 | 0.104 | 0.183 |
| | U = 1.625 [W/(m$^2$ $\times$ K)] | 0.132 | | |
| **S2** | Cement mortar | 0.025 | 1.400 | 0.018 |
| | Craft brick—285 $\times$ 143 $\times$ 90 mm—Stonework 20 mm | 0.143 | 0.664 | 0.215 |
| | Cement mortar | 0.025 | 1.400 | 0.018 |
| | Expanded polyethylene with EIFS system—d = 20 kg/m$^3$ | 0.030 | 0.038 | 0.781 |
| | U = 0.832 [W/(m$^2$ $\times$ K)] | 0.223 | | |
| **S3** | Thermal stucco | 0.025 | 0.220 | 0.114 |
| | Reinforced concrete | 0.200 | 1.630 | 0.123 |
| | Expanded polyethylene with EIFS system—d = 20 kg/m$^3$ | 0.100 | 0.038 | 2.604 |
| | U = 0.332 [W/(m$^2$ $\times$ K)] | 0.325 | | |
| **S1** | Plasterboard—d = 700 kg/m$^3$ | 0.013 | 0.260 | 0.048 |
| | Wooden beam—Insigne pine 3 $\times$ 4″ | 0.102 | 0.104 | 0.977 |
| | Non-ventilated vertical air chamber | 0.100 | 0.769 | 0.130 |
| | Fibrocement roof—d = 920 kg/m$^3$ | 0.010 | 0.220 | 0.045 |
| | U = 0.746 [W/(m$^2$ $\times$ K)] | 0.224 | | |
| **S2** | Plasterboard—d = 700 kg/m$^3$ | 0.013 | 0.260 | 0.048 |
| | Wooden beam—Insigne pine 3 $\times$ 4″ | 0.102 | 0.104 | 0.977 |
| | Expanded polyethylene with EIFS system—d = 20 kg/m$^3$ | 0.060 | 0.038 | 1.563 |
| | Fibrocement roof—d = 920 kg/m$^3$ | 0.010 | 0.220 | 0.045 |
| | U = 0.361 [W/(m$^2$ $\times$ K)] | 0.184 | | |
| **S3** | Plasterboard—d = 700 kg/m$^3$ | 0.013 | 0.260 | 0.048 |
| | Reinforced concrete slab | 0.120 | 1.630 | 0.074 |
| | Expanded polyethylene with EIFS system—d = 20 kg/m$^3$ | 0.150 | 0.038 | 3.906 |
| | Fibrocement roof—d = 920 kg/m$^3$ | 0.010 | 0.220 | 0.045 |
| | U = 0.237 [W/(m$^2$ $\times$ K)] | 0.293 | | |

**Table 5.** Doors and windows.

| | | Material | U [W/(m$^2$ $\times$ K)] | Visual Transmittance | Solar Factor |
|---|---|---|---|---|---|
| **S1** | Windows | Single-glazed | 5.736 | 0.90 | 0.86 |
| | Doors | Wood | 3.804 | | |
| **S2** | Windows | Double-glazed | 3.129 | 0.81 | 0.76 |
| | Doors | Hollow wood | 2.326 | | |
| **S3** | Windows | Low emission double-glazed | 2.215 | 0.76 | 0.65 |
| | Doors | Wooden frame—Double-glazed—Glaze against door | 1.936 | | |

## 3. Results

The results obtained in the models for (i) energy demand, (ii) $CO_2$ emissions and (iii) energy cost are shown below.

### 3.1. Energy Demand

Figures 4 and 5 show that the total annual energy demand varied from 37.20 kWh/m$^2$ in M5, located in the city of Concepción with an S3, to 348.98 kWh/m$^2$ in M1, in Punta Arenas, with an S1. Only 2.38% of the total energy is required to cool the building, considering all the architectural designs and climatic zones. Detailed results of heating and cooling demands, in all models with different architectural designs in the six cities, are shown in Tables A1–A6 in the Appendix A.

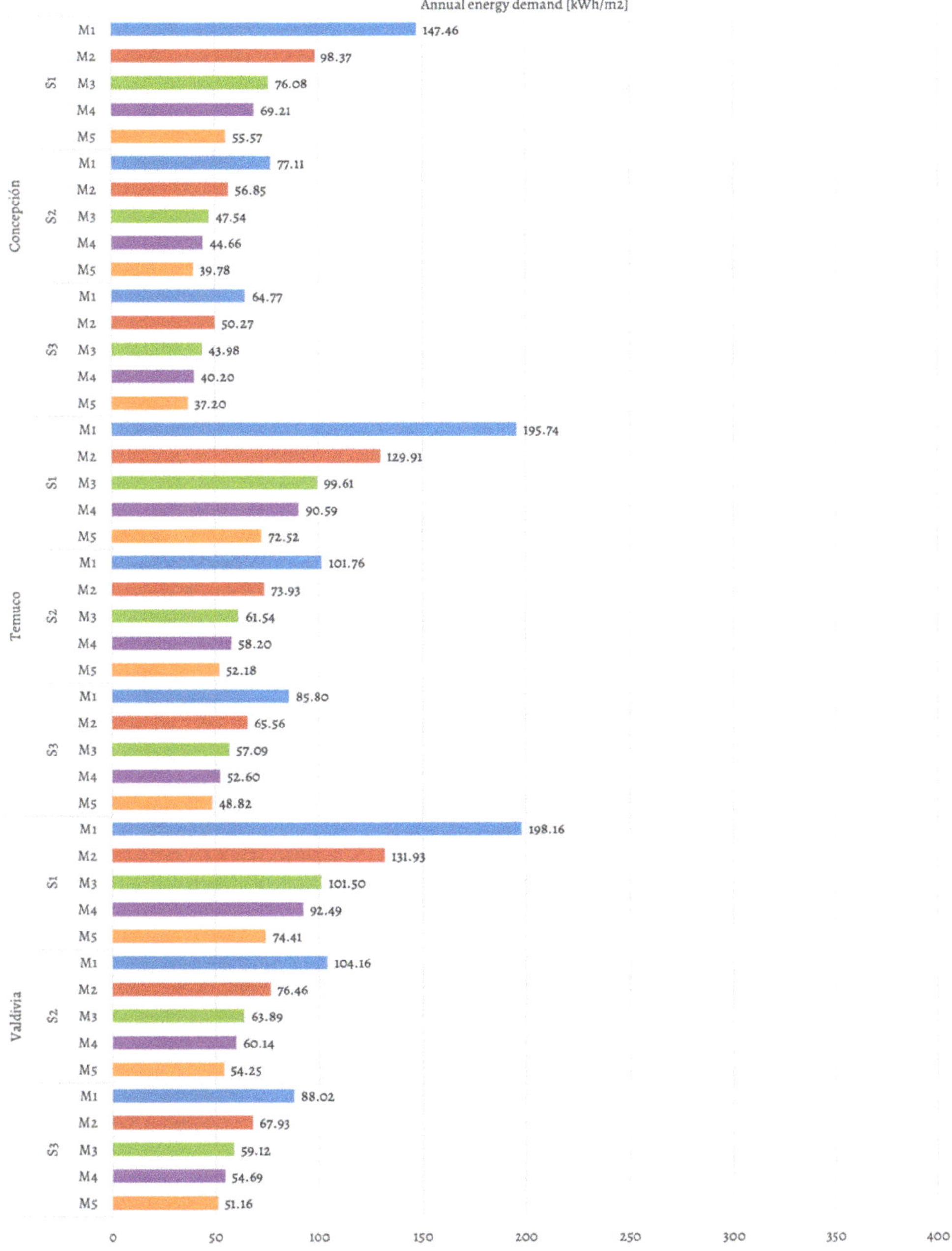



**Figure 4.** Annual energy demand of the models. Concepción, Temuco and Valdivia.

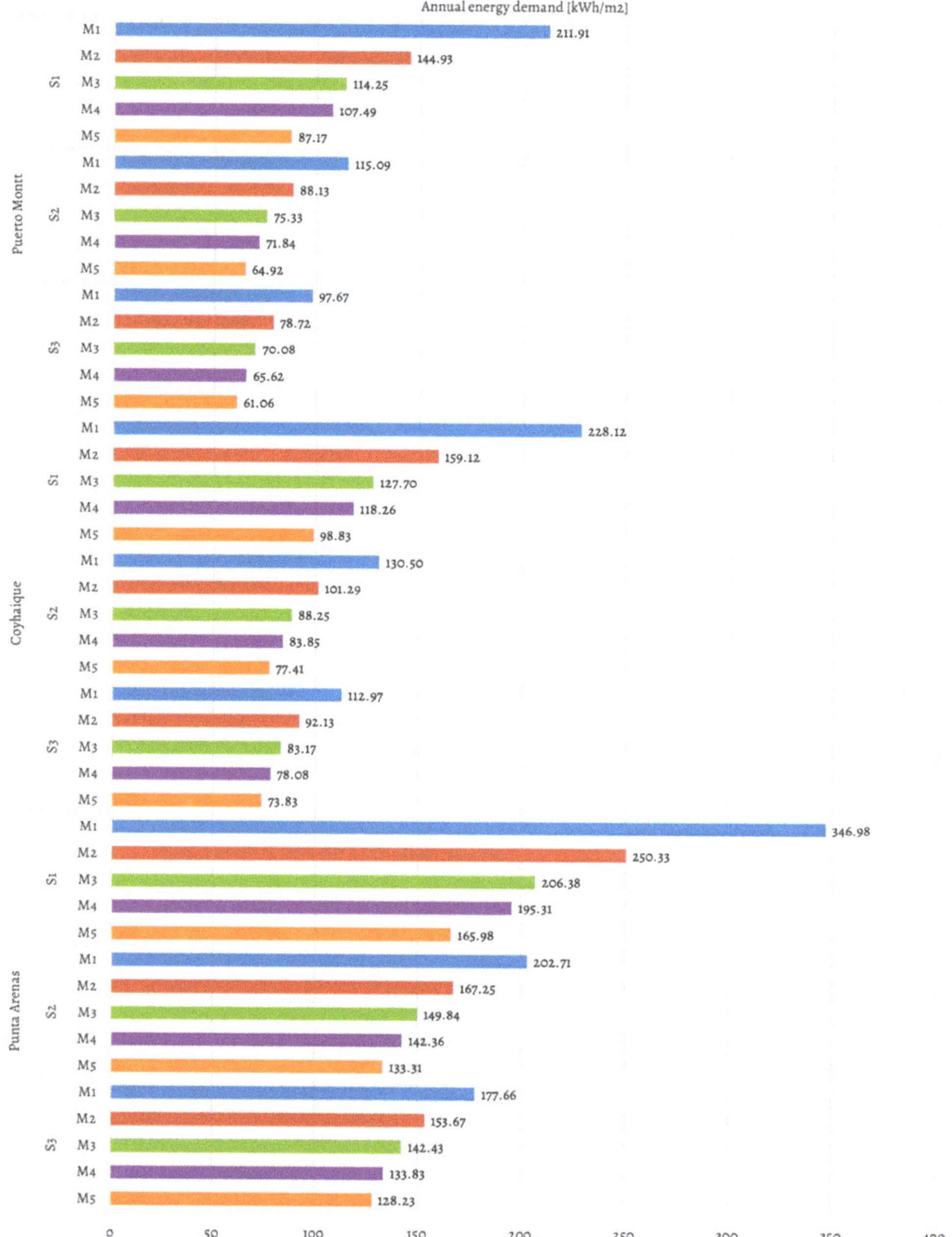



**Figure 5.** Annual energy demand of the models. Puerto Montt, Coyhaique and Punta Arenas.

The results show that the city of Concepción (climatic zone 4) was the one that had the lowest required total energy demand under any of the proposed construction solutions and established models. In contrast, Punta Arenas (climatic zone 7) had the highest total energy demand.

Regarding the design characteristics of the envelope in the different construction solutions considered, S1 is the solution with the highest energy demand in all the models and areas studied, due to the high value of thermal transmittance.

Figure 6 shows the impact of *SFv* on the annual energy demand in each city, depending on the type of construction solution used. The maximum variation in demand (considering M1 and M5 for all cities) was: S1 between 135.30% and 198.70%; S2 between 162.89% and 235.12%; and S3 between 174.29% and 244.71%.

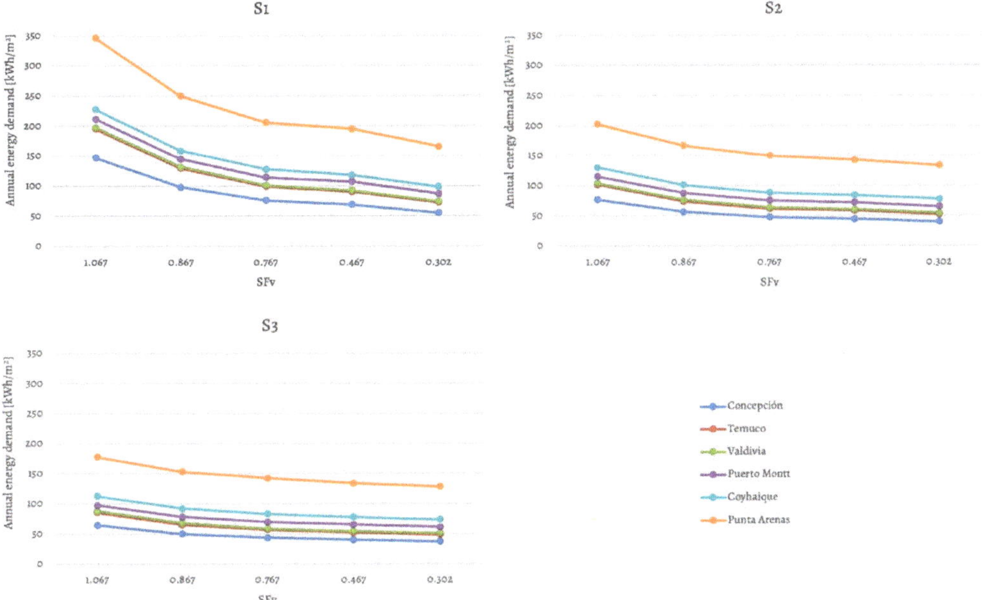

**Figure 6.** Annual energy demand versus *SFv*.

### 3.2. CO$_2$ Emissions

Due to the large amount of data obtained, only the results of S2 will be shown, since it is the most representative of all, considering, in turn, a representative city of each climatic zone—Concepción (4), Valdivia (5), Puerto Montt (6) and Punta Arenas (7).

Figure 7 shows the CO$_2$ emissions generated due to energy demand. The energy used to cool the buildings was assumed as electric for all models. However, the energy source used for heating was variable, based on the values presented in Table 1. For all climatic zones, the least optimal is the exclusive use of electricity, independent of the *SFv* of the dwelling. However, using biomass (wood and pellets) produces low emissions, mainly due to the neutral emission factor [19].

A 160.62 to 235.12% increase in CO$_2$ emissions between climatic zones 4 and 7 was observed when using any heating system. In turn, implementing an S1 to an S3 in the thermal envelope reduced CO$_2$ emissions between 22.74% and 56.67% for all energy options.

### 3.3. Energy Cost

The energy cost of these alternatives is represented in Figure 8. In this figure, the annual cost of heating and cooling the buildings is shown depending on the *SFv* and the alternative used, expressed in USD/m$^2$, based on the values presented in Table 2. The use of propane gas as fuel for heating is the most expensive option of all in any area studied; on the contrary, the use of pellets is the most economical.

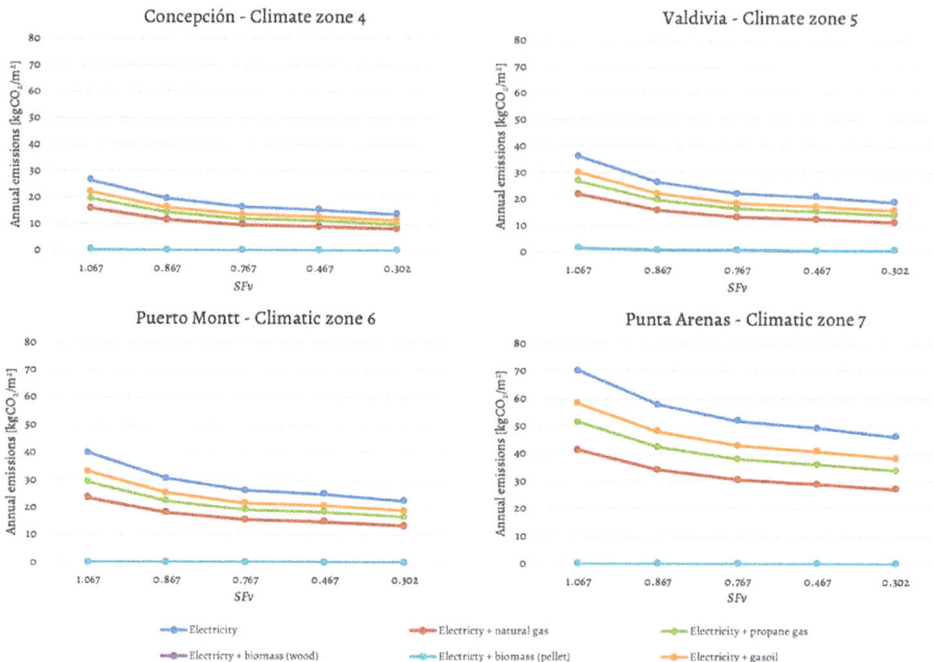


**Figure 7.** Annual emissions depending on the fuel used.

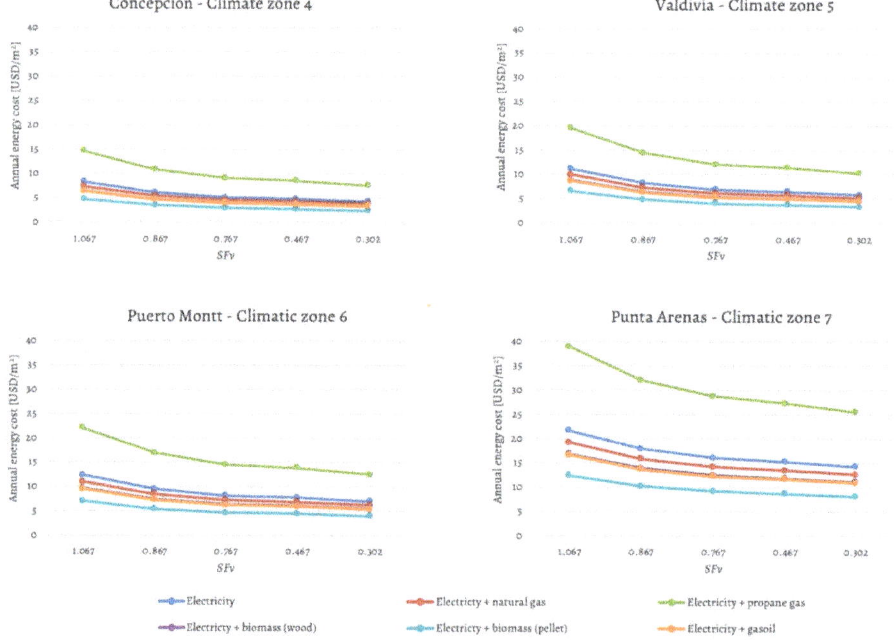


**Figure 8.** Annual energy cost depending on the fuel used.

A 160.43 to 236.25% increase in cost between climate zones 4 and 7, when using any heating system, was similar to what happened in emissions. Implementing an S1 to an S3 in the thermal envelope reduces the cost for all energy options between 22.74% and 56.72%.

Propane gas has had a wide variation in cost between 12.52% and 236.25% for all the climatic zones analysed. The rate of decrease in cost varies depending on the climatic zone. In Concepción its cost drop fluctuates from 6.00 to 26.09% between each *SFv* interval considered; in Punta Arenas, this range was between 4.99% and 17.49%.

The cost of the heating system was reduced between 52.16% and 63.30% using M1 and M5, respectively, when implementing S1, 34.23 to 48.60% with an S2 and 27.82% and 42.65% with an S3.

## 4. Discussion

In the present study, implementing S2 represents a 19.68 to 48.01% decrease in required power demand compared to S1; and a 22.74 to 56.16% decrease in consumption compared to the S3, depending on the city where the building is located and the *SFv*. Comparing our results with the study carried out by Danielski et al. [14], similar results are obtained, where the slope between the total energy demand per square meter and the *SFv* increases when using a construction solution with less thermal resistance in the envelope.

The impact on energy demand from reducing *SFv* was studied in various investigations. In Italy, different energy models, with form factors between 0.54 and 0.78, were analysed in different cities, reaching 34.09 to 43.14% differences in energy demand [37]. In Lithuania, a 33.77% variation in the required energy was obtained by decreasing the *SFv* from 1.35 to 1.17 [8]. In Sweden, heating demand was decreased between 18.00% and 20.00% by reducing the *SFv* from 1.70 to 1.01 in different cities [14].

Additionally, when comparing the energy demand of M1 and M5, 27.82 to 62.95% reductions were reached depending on the city and the construction system considered. The impact of *SFv* was less in the coldest city, Punta Arenas, where consumption only decreased by 27.82 to 52.16%. In contrast, in Concepción, energy savings fluctuated between 42.57% and 62.95%.

Whenever the *SFv* decreases, so do the difference in emissions by improving the architectural design. When the *SFv* is reduced from 1.067 to 0.302: implementing an S1 caused a $CO_2$ decrease between 52.16% and 63.23%; while with an S3, they decrease from 27.82 to 42.64%. With these and similar data from other studies [15], it has been shown that more significant benefits are obtained by improving the thermal resistance of the envelope when there is a higher relation between the exposed surface and the $m^2$ of the surface of the building.

Finally, in Chile there are other studies on the form factor in buildings and its influence on energy demand. For example, Vásquez et al. [38] investigated with the *SFv* of office buildings in the city of Santiago, Chile. They conclude that the *SFv* is essential in architectural design along with other variables such as solar radiation, light, wind or the immediate context.

## 5. Conclusions

This research showed an appropriate design considering a *SFv* suitable for cold oceanic climates, which implied a decrease in energy demand and $CO_2$ emissions. The main conclusions derived from this research are the following:

- The architectural designs with high thermal transmittance values may require from 129.44 to 227.67% of the energy demand of the same building after implementing a solution with a low U-value. Energy demand is widely affected by the weather where housing is located; maximum variations between 135.30 and 244.71% exist for the same *SFv* and architectural design, depending on the city where it is located.
- $CO_2$ emissions depend directly on the climatic zone where the building is located and the fuel used. For all cities, using biomass in heating systems has the lowest emission and cost values, as opposed to what happens when using electricity for

heating. Differences in $CO_2$ emissions from 7.43 to 235.12% can be found between the different climatic zones for the same model. Similarly, the cost of the heating system is reduced by between 31.81 and 32.95% when switching from a fossil fuel, such as propane, to a renewable fuel, such as biomass in the form of pellets.

- Overall, the impact of *SFv*, on both energy demand and $CO_2$ emissions, is greater when architectural designs with a high thermal transmittance value are implemented, reducing energy demand between 22.75% and 56.16%, depending on the area located. Based on the analysis, it is highly recommended to design buildings with a *SFv* below 0.767 for cold oceanic climates, such as in the southern zone of Chile. Among the values shown, energy demand and $CO_2$ emissions tend to stabilise for all the climatic zones and construction solutions studied, with only 9.03% maximum differences in the energy requirement for heating and 10.37% in $CO_2$ emissions.

These results are fully extrapolated to any area with climatic conditions similar to a cold oceanic climate. This study has considered the *SFv* as the main variable, although for a comprehensive architectural design other variables must be taken into account, such solar exposure, wind orientation and passive design characteristics.

**Author Contributions:** Conceptualisation, M.C.; methodology, M.C. and D.C.; software, D.C.; validation, M.C.; formal analysis, M.C. and D.C.; investigation, M.C. and D.C.; resources, M.C.; data curation, D.C.; writing—original draft preparation, M.C. and D.C.; visualisation, M.C. and D.C.; project administration, M.C.; funding acquisition, M.C. All authors have read and agreed to the published version of the manuscript.

**Funding:** This research was funded by Agencia Nacional de Investigación y Desarrollo (ANID) of Chile, through the projects ANID FONDECYT 11160524 and ANID FONDECYT 1201052.

**Institutional Review Board Statement:** Not applicable.

**Informed Consent Statement:** Not applicable.

**Conflicts of Interest:** The authors declare no conflict of interest. The funders had no role in the design of the study; in the collection, analyses, or interpretation of data; in the writing of the manuscript, or in the decision to publish the results.

# Appendix A

Table A1. Energy demand—Concepción.

| S | M | H/C | \<Monthly Energy Demand [kWh/m²]\> | | | | | | | | | | | | \<Annual Energy Demand [kWh/m²]\> | |
|---|---|---|---|---|---|---|---|---|---|---|---|---|---|---|---|---|
| | | | Jan. | Feb. | Mar. | Apr. | May | Jun. | Jul. | Aug. | Sep. | Oct. | Nov. | Dec. | H/C | Total |
| S1 | M1 | Heating | 4.10 | 4.18 | 6.49 | 11.73 | 16.63 | 19.54 | 21.38 | 18.22 | 15.97 | 11.14 | 7.94 | 5.13 | 142.45 | 147.46 |
| | | Cooling | 2.00 | 1.55 | 0.86 | 0.06 | 0.00 | 0.00 | 0.00 | 0.00 | 0.00 | 0.00 | 0.00 | 0.54 | 5.02 | |
| | M2 | Heating | 2.53 | 2.55 | 4.11 | 7.73 | 11.17 | 13.23 | 14.60 | 12.50 | 10.99 | 7.51 | 5.34 | 3.32 | 95.61 | 98.37 |
| | | Cooling | 1.28 | 0.88 | 0.25 | 0.00 | 0.00 | 0.00 | 0.00 | 0.00 | 0.00 | 0.00 | 0.00 | 0.35 | 2.76 | |
| | M3 | Heating | 1.77 | 1.72 | 3.05 | 5.90 | 8.67 | 10.33 | 11.49 | 9.91 | 8.72 | 5.86 | 4.18 | 2.45 | 74.04 | 76.08 |
| | | Cooling | 1.01 | 0.62 | 0.16 | 0.00 | 0.00 | 0.00 | 0.00 | 0.00 | 0.00 | 0.00 | 0.00 | 0.26 | 2.05 | |
| | M4 | Heating | 1.27 | 1.44 | 2.74 | 5.54 | 8.15 | 9.72 | 10.83 | 9.35 | 8.24 | 5.52 | 3.92 | 2.11 | 68.84 | 69.21 |
| | | Cooling | 0.22 | 0.15 | 0.00 | 0.00 | 0.00 | 0.00 | 0.00 | 0.00 | 0.00 | 0.00 | 0.00 | 0.00 | 0.37 | |
| | M5 | Heating | 0.86 | 0.97 | 2.04 | 4.34 | 6.51 | 7.83 | 8.79 | 7.62 | 6.73 | 4.42 | 3.13 | 1.58 | 54.82 | 55.57 |
| | | Cooling | 0.37 | 0.26 | 0.04 | 0.00 | 0.00 | 0.00 | 0.00 | 0.00 | 0.00 | 0.00 | 0.00 | 0.08 | 0.75 | |
| S2 | M1 | Heating | 2.06 | 2.04 | 3.25 | 6.13 | 8.89 | 10.71 | 11.71 | 9.91 | 8.76 | 5.89 | 4.19 | 2.59 | 76.14 | 77.11 |
| | | Cooling | 0.29 | 0.51 | 0.08 | 0.00 | 0.00 | 0.00 | 0.00 | 0.00 | 0.00 | 0.00 | 0.00 | 0.08 | 0.96 | |
| | M2 | Heating | 1.35 | 1.26 | 2.23 | 4.41 | 6.60 | 7.99 | 8.91 | 7.63 | 6.76 | 4.42 | 3.11 | 1.80 | 56.46 | 56.85 |
| | | Cooling | 0.20 | 0.20 | 0.00 | 0.00 | 0.00 | 0.00 | 0.00 | 0.00 | 0.00 | 0.00 | 0.00 | 0.00 | 0.39 | |
| | M3 | Heating | 0.88 | 0.92 | 1.74 | 3.62 | 5.54 | 6.75 | 7.60 | 6.57 | 5.83 | 3.74 | 2.63 | 1.39 | 47.21 | 47.54 |
| | | Cooling | 0.15 | 0.18 | 0.00 | 0.00 | 0.00 | 0.00 | 0.00 | 0.00 | 0.00 | 0.00 | 0.00 | 0.00 | 0.33 | |
| | M4 | Heating | 0.61 | 0.71 | 1.53 | 3.47 | 5.31 | 6.49 | 7.29 | 6.25 | 5.55 | 3.53 | 2.48 | 1.19 | 44.40 | 44.66 |
| | | Cooling | 0.15 | 0.11 | 0.00 | 0.00 | 0.00 | 0.00 | 0.00 | 0.00 | 0.00 | 0.00 | 0.00 | 0.00 | 0.26 | |
| | M5 | Heating | 0.49 | 0.53 | 1.20 | 3.02 | 4.75 | 5.83 | 6.59 | 5.69 | 5.06 | 3.18 | 2.20 | 0.93 | 39.48 | 39.78 |
| | | Cooling | 0.18 | 0.13 | 0.00 | 0.00 | 0.00 | 0.00 | 0.00 | 0.00 | 0.00 | 0.00 | 0.00 | 0.00 | 0.30 | |
| S3 | M1 | Heating | 1.64 | 1.62 | 2.63 | 5.08 | 7.47 | 9.06 | 9.94 | 8.40 | 7.45 | 4.94 | 3.50 | 2.12 | 63.85 | 64.77 |
| | | Cooling | 0.35 | 0.43 | 0.07 | 0.00 | 0.00 | 0.00 | 0.00 | 0.00 | 0.00 | 0.00 | 0.00 | 0.07 | 0.92 | |
| | M2 | Heating | 1.03 | 1.02 | 1.88 | 3.85 | 5.83 | 7.11 | 7.96 | 6.83 | 6.06 | 3.91 | 2.75 | 1.50 | 49.72 | 50.27 |
| | | Cooling | 0.22 | 0.27 | 0.02 | 0.00 | 0.00 | 0.00 | 0.00 | 0.00 | 0.00 | 0.00 | 0.00 | 0.04 | 0.55 | |
| | M3 | Heating | 0.67 | 0.77 | 1.53 | 3.32 | 5.12 | 6.27 | 7.08 | 6.13 | 5.45 | 3.46 | 2.44 | 1.24 | 43.48 | 43.98 |
| | | Cooling | 0.22 | 0.22 | 0.02 | 0.00 | 0.00 | 0.00 | 0.00 | 0.00 | 0.00 | 0.00 | 0.00 | 0.04 | 0.50 | |
| | M4 | Heating | 0.46 | 0.48 | 1.24 | 3.09 | 4.82 | 5.94 | 6.68 | 5.74 | 5.10 | 3.21 | 2.22 | 0.96 | 39.94 | 40.20 |
| | | Cooling | 0.15 | 0.12 | 0.00 | 0.00 | 0.00 | 0.00 | 0.00 | 0.00 | 0.00 | 0.00 | 0.00 | 0.00 | 0.26 | |
| | M5 | Heating | 0.39 | 0.35 | 1.03 | 2.79 | 4.46 | 5.50 | 6.23 | 5.39 | 4.80 | 2.99 | 2.01 | 0.81 | 36.74 | 37.20 |
| | | Cooling | 0.26 | 0.17 | 0.01 | 0.00 | 0.00 | 0.00 | 0.00 | 0.00 | 0.00 | 0.00 | 0.00 | 0.02 | 0.46 | |

S = solution; M = model; H = heating; C = cooling.

**Table A2.** Energy demand—Temuco.

| S | M | H/C | Jan. | Feb. | Mar. | Apr. | May | Jun. | Jul. | Aug. | Sep. | Oct. | Nov. | Dec. | H/C | Total |
|---|---|---|---|---|---|---|---|---|---|---|---|---|---|---|---|---|
| | | | | | | | Monthly Energy Demand [kWh/m²] | | | | | | | | Annual Energy Demand [kWh/m²] | |
| S1 | M1 | Heating | 6.26 | 6.04 | 9.74 | 15.70 | 20.42 | 24.09 | 26.89 | 22.80 | 19.25 | 14.23 | 10.94 | 7.62 | 183.97 | 195.74 |
| | | Cooling | 3.93 | 4.97 | 1.86 | 0.19 | 0.00 | 0.00 | 0.00 | 0.00 | 0.00 | 0.00 | 0.00 | 0.82 | 11.77 | |
| | M2 | Heating | 4.09 | 3.81 | 6.37 | 10.38 | 13.64 | 16.28 | 18.23 | 15.49 | 13.06 | 9.57 | 7.39 | 5.03 | 123.34 | 129.91 |
| | | Cooling | 2.36 | 2.89 | 0.98 | 0.00 | 0.00 | 0.00 | 0.00 | 0.00 | 0.00 | 0.00 | 0.00 | 0.34 | 6.57 | |
| | M3 | Heating | 2.92 | 2.76 | 4.77 | 7.92 | 10.51 | 12.67 | 14.22 | 12.12 | 10.22 | 7.45 | 5.76 | 3.81 | 95.15 | 99.61 |
| | | Cooling | 1.65 | 2.07 | 0.57 | 0.00 | 0.00 | 0.00 | 0.00 | 0.00 | 0.00 | 0.00 | 0.00 | 0.17 | 4.46 | |
| | M4 | Heating | 2.38 | 2.36 | 4.33 | 7.48 | 9.91 | 11.96 | 13.44 | 11.45 | 9.68 | 7.04 | 5.44 | 3.49 | 88.96 | 90.59 |
| | | Cooling | 0.49 | 1.04 | 0.11 | 0.00 | 0.00 | 0.00 | 0.00 | 0.00 | 0.00 | 0.00 | 0.00 | 0.00 | 1.64 | |
| | M5 | Heating | 1.74 | 1.78 | 3.30 | 5.87 | 7.86 | 9.60 | 10.83 | 9.23 | 7.78 | 5.61 | 4.35 | 2.61 | 70.56 | 72.52 |
| | | Cooling | 0.65 | 1.11 | 0.15 | 0.00 | 0.00 | 0.00 | 0.00 | 0.00 | 0.00 | 0.00 | 0.00 | 0.04 | 1.96 | |
| S2 | M1 | Heating | 3.17 | 3.02 | 5.02 | 8.21 | 10.80 | 13.04 | 14.67 | 12.27 | 10.27 | 7.46 | 5.73 | 3.91 | 97.57 | 101.76 |
| | | Cooling | 1.35 | 2.21 | 0.63 | 0.00 | 0.00 | 0.00 | 0.00 | 0.00 | 0.00 | 0.00 | 0.00 | 0.00 | 4.19 | |
| | M2 | Heating | 2.15 | 2.06 | 3.56 | 5.94 | 7.93 | 9.74 | 10.97 | 9.28 | 7.78 | 5.59 | 4.32 | 2.81 | 72.14 | 73.93 |
| | | Cooling | 0.62 | 0.98 | 0.19 | 0.00 | 0.00 | 0.00 | 0.00 | 0.00 | 0.00 | 0.00 | 0.00 | 0.00 | 1.79 | |
| | M3 | Heating | 1.63 | 1.56 | 2.83 | 4.88 | 6.61 | 8.20 | 9.27 | 7.87 | 6.60 | 4.71 | 3.65 | 2.29 | 60.10 | 61.54 |
| | | Cooling | 0.53 | 0.76 | 0.14 | 0.00 | 0.00 | 0.00 | 0.00 | 0.00 | 0.00 | 0.00 | 0.00 | 0.00 | 1.43 | |
| | M4 | Heating | 1.26 | 1.35 | 2.57 | 4.72 | 6.39 | 7.93 | 8.97 | 7.58 | 6.34 | 4.50 | 3.48 | 2.05 | 57.13 | 58.20 |
| | | Cooling | 0.23 | 0.83 | 0.00 | 0.00 | 0.00 | 0.00 | 0.00 | 0.00 | 0.00 | 0.00 | 0.00 | 0.00 | 1.07 | |
| | M5 | Heating | 1.04 | 1.13 | 2.17 | 4.13 | 5.68 | 7.11 | 8.07 | 6.83 | 5.71 | 4.03 | 3.13 | 1.77 | 50.81 | 52.18 |
| | | Cooling | 0.44 | 0.80 | 0.12 | 0.00 | 0.00 | 0.00 | 0.00 | 0.00 | 0.00 | 0.00 | 0.00 | 0.00 | 1.37 | |
| S3 | M1 | Heating | 2.61 | 2.49 | 4.14 | 6.82 | 9.03 | 11.01 | 12.40 | 10.37 | 8.65 | 6.24 | 4.79 | 3.23 | 81.79 | 85.80 |
| | | Cooling | 1.26 | 2.07 | 0.69 | 0.00 | 0.00 | 0.00 | 0.00 | 0.00 | 0.00 | 0.00 | 0.00 | 0.00 | 4.02 | |
| | M2 | Heating | 1.78 | 1.68 | 3.03 | 5.20 | 6.99 | 8.65 | 9.76 | 8.25 | 6.91 | 4.94 | 3.82 | 2.43 | 63.42 | 65.56 |
| | | Cooling | 0.76 | 1.05 | 0.32 | 0.00 | 0.00 | 0.00 | 0.00 | 0.00 | 0.00 | 0.00 | 0.00 | 0.00 | 2.14 | |
| | M3 | Heating | 1.40 | 1.38 | 2.52 | 4.48 | 6.09 | 7.60 | 8.60 | 7.30 | 6.12 | 4.35 | 3.39 | 2.07 | 55.29 | 57.09 |
| | | Cooling | 0.66 | 0.89 | 0.24 | 0.00 | 0.00 | 0.00 | 0.00 | 0.00 | 0.00 | 0.00 | 0.00 | 0.00 | 1.79 | |
| | M4 | Heating | 0.99 | 1.09 | 2.11 | 4.23 | 5.79 | 7.24 | 8.20 | 6.93 | 5.78 | 4.08 | 3.16 | 1.76 | 51.35 | 52.60 |
| | | Cooling | 0.36 | 0.84 | 0.05 | 0.00 | 0.00 | 0.00 | 0.00 | 0.00 | 0.00 | 0.00 | 0.00 | 0.00 | 1.26 | |
| | M5 | Heating | 0.87 | 0.94 | 1.75 | 3.84 | 5.32 | 6.70 | 7.60 | 6.44 | 5.38 | 3.79 | 2.91 | 1.62 | 47.16 | 48.82 |
| | | Cooling | 0.51 | 0.96 | 0.19 | 0.00 | 0.00 | 0.00 | 0.00 | 0.00 | 0.00 | 0.00 | 0.00 | 0.01 | 1.66 | |

S = solution; M = model; H = heating; C = cooling.

**Table A3.** Energy demand—Valdivia.

| S | M | H/C | | Monthly Energy Demand [kWh/m$^2$] | | | | | | | | | | | | Annual Energy Demand [kWh/m$^2$] | |
|---|---|---|---|---|---|---|---|---|---|---|---|---|---|---|---|---|---|---|
| | | | Jan. | Feb. | Mar. | Apr. | May | Jun. | Jul. | Aug. | Sep. | Oct. | Nov. | Dec. | H/C | Total |
| S1 | M1 | Heating | 6.08 | 5.93 | 9.55 | 15.29 | 21.17 | 24.95 | 27.41 | 23.60 | 19.14 | 14.45 | 11.12 | 7.76 | 186.45 | 198.16 |
| | | Cooling | 4.19 | 4.74 | 1.43 | 0.00 | 0.00 | 0.00 | 0.00 | 0.00 | 0.00 | 0.00 | 0.00 | 1.35 | 11.71 | |
| | M2 | Heating | 4.01 | 3.73 | 6.20 | 10.16 | 14.18 | 16.94 | 18.71 | 16.09 | 13.03 | 9.76 | 7.53 | 5.16 | 125.51 | 131.93 |
| | | Cooling | 2.45 | 2.75 | 0.59 | 0.00 | 0.00 | 0.00 | 0.00 | 0.00 | 0.00 | 0.00 | 0.00 | 0.63 | 6.42 | |
| | M3 | Heating | 2.92 | 2.66 | 4.69 | 7.81 | 10.95 | 13.23 | 14.68 | 12.62 | 10.22 | 7.62 | 5.88 | 3.92 | 97.21 | 101.50 |
| | | Cooling | 1.67 | 1.96 | 0.25 | 0.00 | 0.00 | 0.00 | 0.00 | 0.00 | 0.00 | 0.00 | 0.00 | 0.42 | 4.29 | |
| | M4 | Heating | 2.24 | 2.31 | 4.29 | 7.37 | 10.32 | 12.48 | 13.87 | 11.92 | 9.67 | 7.20 | 5.55 | 3.45 | 90.69 | 92.49 |
| | | Cooling | 0.38 | 1.29 | 0.00 | 0.00 | 0.00 | 0.00 | 0.00 | 0.00 | 0.00 | 0.00 | 0.00 | 0.14 | 1.80 | |
| | M5 | Heating | 1.65 | 1.67 | 3.32 | 5.84 | 8.22 | 10.07 | 11.24 | 9.65 | 7.82 | 5.77 | 4.45 | 2.65 | 72.33 | 74.41 |
| | | Cooling | 0.65 | 1.19 | 0.06 | 0.00 | 0.00 | 0.00 | 0.00 | 0.00 | 0.00 | 0.00 | 0.00 | 0.17 | 2.07 | |
| S2 | M1 | Heating | 3.11 | 2.99 | 4.95 | 8.06 | 11.38 | 13.62 | 14.98 | 12.87 | 10.35 | 7.67 | 5.90 | 4.02 | 99.90 | 104.16 |
| | | Cooling | 1.30 | 2.31 | 0.28 | 0.00 | 0.00 | 0.00 | 0.00 | 0.00 | 0.00 | 0.00 | 0.00 | 0.36 | 4.26 | |
| | M2 | Heating | 2.18 | 1.98 | 3.51 | 5.89 | 8.36 | 10.25 | 11.38 | 9.75 | 7.84 | 5.77 | 4.46 | 2.94 | 74.32 | 76.46 |
| | | Cooling | 0.64 | 1.27 | 0.05 | 0.00 | 0.00 | 0.00 | 0.00 | 0.00 | 0.00 | 0.00 | 0.00 | 0.18 | 2.15 | |
| | M3 | Heating | 1.60 | 1.56 | 2.81 | 4.89 | 6.97 | 8.66 | 9.67 | 8.28 | 6.68 | 4.88 | 3.78 | 2.32 | 62.11 | 63.89 |
| | | Cooling | 0.57 | 1.00 | 0.04 | 0.00 | 0.00 | 0.00 | 0.00 | 0.00 | 0.00 | 0.00 | 0.00 | 0.17 | 1.78 | |
| | M4 | Heating | 1.19 | 1.29 | 2.62 | 4.72 | 6.75 | 8.37 | 9.34 | 7.99 | 6.41 | 4.67 | 3.60 | 2.05 | 58.99 | 60.14 |
| | | Cooling | 0.18 | 0.97 | 0.00 | 0.00 | 0.00 | 0.00 | 0.00 | 0.00 | 0.00 | 0.00 | 0.00 | 0.00 | 1.15 | |
| | M5 | Heating | 1.00 | 1.03 | 2.22 | 4.20 | 6.01 | 7.53 | 8.43 | 7.21 | 5.80 | 4.20 | 3.21 | 1.81 | 52.65 | 54.25 |
| | | Cooling | 0.48 | 0.96 | 0.01 | 0.00 | 0.00 | 0.00 | 0.00 | 0.00 | 0.00 | 0.00 | 0.00 | 0.15 | 1.61 | |
| S3 | M1 | Heating | 2.57 | 2.44 | 4.07 | 6.72 | 9.56 | 11.55 | 12.71 | 10.91 | 8.75 | 6.44 | 4.96 | 3.34 | 84.02 | 88.02 |
| | | Cooling | 1.32 | 2.03 | 0.30 | 0.00 | 0.00 | 0.00 | 0.00 | 0.00 | 0.00 | 0.00 | 0.00 | 0.34 | 3.99 | |
| | M2 | Heating | 1.75 | 1.69 | 3.01 | 5.18 | 7.39 | 9.12 | 10.15 | 8.70 | 6.99 | 5.12 | 3.96 | 2.49 | 65.53 | 67.93 |
| | | Cooling | 0.83 | 1.20 | 0.13 | 0.00 | 0.00 | 0.00 | 0.00 | 0.00 | 0.00 | 0.00 | 0.00 | 0.23 | 2.39 | |
| | M3 | Heating | 1.32 | 1.32 | 2.54 | 4.51 | 6.44 | 8.04 | 8.99 | 7.70 | 6.21 | 4.53 | 3.51 | 2.02 | 57.11 | 59.12 |
| | | Cooling | 0.75 | 0.97 | 0.09 | 0.00 | 0.00 | 0.00 | 0.00 | 0.00 | 0.00 | 0.00 | 0.00 | 0.21 | 2.01 | |
| | M4 | Heating | 0.92 | 0.98 | 2.25 | 4.27 | 6.13 | 7.66 | 8.56 | 7.31 | 5.86 | 4.25 | 3.25 | 1.82 | 53.27 | 54.69 |
| | | Cooling | 0.37 | 0.96 | 0.00 | 0.00 | 0.00 | 0.00 | 0.00 | 0.00 | 0.00 | 0.00 | 0.00 | 0.09 | 1.42 | |
| | M5 | Heating | 0.84 | 0.85 | 1.99 | 3.93 | 5.64 | 7.10 | 7.97 | 6.80 | 5.47 | 3.95 | 3.01 | 1.69 | 49.22 | 51.16 |
| | | Cooling | 0.68 | 0.98 | 0.07 | 0.00 | 0.00 | 0.00 | 0.00 | 0.00 | 0.00 | 0.00 | 0.00 | 0.20 | 1.94 | |

S = solution; M = model; H = heating; C = cooling.

**Table A4.** Energy demand—Puerto Montt.

| S | M | H/C | Jan. | Feb. | Mar. | Apr. | May | Jun. | Jul. | Aug. | Sep. | Oct. | Nov. | Dec. | H/C | Total |
|---|---|---|---|---|---|---|---|---|---|---|---|---|---|---|---|---|
| | | | | | | Monthly Energy Demand [kWh/m$^2$] | | | | | | | | Annual Energy Demand [kWh/m$^2$] | |
| S1 | M1 | Heating | 7.11 | 7.06 | 11.37 | 16.92 | 22.88 | 27.66 | 29.23 | 26.72 | 22.28 | 17.13 | 13.66 | 9.45 | 211.46 | 211.91 |
| | | Cooling | 0.06 | 0.39 | 0.00 | 0.00 | 0.00 | 0.00 | 0.00 | 0.00 | 0.00 | 0.00 | 0.00 | 0.00 | 0.45 | |
| | M2 | Heating | 4.71 | 4.66 | 7.52 | 11.40 | 15.62 | 19.05 | 20.19 | 18.57 | 15.44 | 11.83 | 9.38 | 6.35 | 144.71 | 144.93 |
| | | Cooling | 0.00 | 0.22 | 0.00 | 0.00 | 0.00 | 0.00 | 0.00 | 0.00 | 0.00 | 0.00 | 0.00 | 0.00 | 0.22 | |
| | M3 | Heating | 3.63 | 3.56 | 5.77 | 8.88 | 12.28 | 15.06 | 16.01 | 14.82 | 12.30 | 9.43 | 7.43 | 4.93 | 114.09 | 114.25 |
| | | Cooling | 0.00 | 0.17 | 0.00 | 0.00 | 0.00 | 0.00 | 0.00 | 0.00 | 0.00 | 0.00 | 0.00 | 0.00 | 0.17 | |
| | M4 | Heating | 3.36 | 3.18 | 5.42 | 8.38 | 11.58 | 14.19 | 15.11 | 14.01 | 11.65 | 8.93 | 7.03 | 4.64 | 107.49 | 107.49 |
| | | Cooling | 0.00 | 0.00 | 0.00 | 0.00 | 0.00 | 0.00 | 0.00 | 0.00 | 0.00 | 0.00 | 0.00 | 0.00 | 0.00 | |
| | M5 | Heating | 2.58 | 2.42 | 4.24 | 6.72 | 9.38 | 11.60 | 12.37 | 11.55 | 9.56 | 7.31 | 5.72 | 3.69 | 87.13 | 87.17 |
| | | Cooling | 0.00 | 0.05 | 0.00 | 0.00 | 0.00 | 0.00 | 0.00 | 0.00 | 0.00 | 0.00 | 0.00 | 0.00 | 0.05 | |
| S2 | M1 | Heating | 3.65 | 3.63 | 5.88 | 9.06 | 12.45 | 15.49 | 16.30 | 14.95 | 12.26 | 9.23 | 7.30 | 4.90 | 115.09 | 115.09 |
| | | Cooling | 0.00 | 0.00 | 0.00 | 0.00 | 0.00 | 0.00 | 0.00 | 0.00 | 0.00 | 0.00 | 0.00 | 0.00 | 0.00 | |
| | M2 | Heating | 2.65 | 2.64 | 4.29 | 6.77 | 9.47 | 11.85 | 12.56 | 11.66 | 9.61 | 7.27 | 5.69 | 3.68 | 88.13 | 88.13 |
| | | Cooling | 0.00 | 0.00 | 0.00 | 0.00 | 0.00 | 0.00 | 0.00 | 0.00 | 0.00 | 0.00 | 0.00 | 0.00 | 0.00 | |
| | M3 | Heating | 2.19 | 2.14 | 3.55 | 5.71 | 8.07 | 10.13 | 10.79 | 10.11 | 8.31 | 6.31 | 4.90 | 3.10 | 75.33 | 75.33 |
| | | Cooling | 0.00 | 0.00 | 0.00 | 0.00 | 0.00 | 0.00 | 0.00 | 0.00 | 0.00 | 0.00 | 0.00 | 0.00 | 0.00 | |
| | M4 | Heating | 1.98 | 1.90 | 3.38 | 5.47 | 7.74 | 9.78 | 10.40 | 9.69 | 7.95 | 5.97 | 4.64 | 2.93 | 71.84 | 71.84 |
| | | Cooling | 0.00 | 0.00 | 0.00 | 0.00 | 0.00 | 0.00 | 0.00 | 0.00 | 0.00 | 0.00 | 0.00 | 0.00 | 0.00 | |
| | M5 | Heating | 1.70 | 1.61 | 2.94 | 4.92 | 7.00 | 8.87 | 9.46 | 8.87 | 7.26 | 5.48 | 4.24 | 2.58 | 64.92 | 64.92 |
| | | Cooling | 0.00 | 0.00 | 0.00 | 0.00 | 0.00 | 0.00 | 0.00 | 0.00 | 0.00 | 0.00 | 0.00 | 0.00 | 0.00 | |
| S3 | M1 | Heating | 3.01 | 3.01 | 4.88 | 7.62 | 10.55 | 13.26 | 13.94 | 12.82 | 10.47 | 7.84 | 6.18 | 4.08 | 97.67 | 97.67 |
| | | Cooling | 0.00 | 0.00 | 0.00 | 0.00 | 0.00 | 0.00 | 0.00 | 0.00 | 0.00 | 0.00 | 0.00 | 0.00 | 0.00 | |
| | M2 | Heating | 2.31 | 2.29 | 3.75 | 5.99 | 8.45 | 10.64 | 11.29 | 10.51 | 8.64 | 6.52 | 5.08 | 3.24 | 78.72 | 78.72 |
| | | Cooling | 0.00 | 0.00 | 0.00 | 0.00 | 0.00 | 0.00 | 0.00 | 0.00 | 0.00 | 0.00 | 0.00 | 0.00 | 0.00 | |
| | M3 | Heating | 1.98 | 1.90 | 3.24 | 5.29 | 7.51 | 9.46 | 10.08 | 9.48 | 7.78 | 5.90 | 4.58 | 2.87 | 70.08 | 70.08 |
| | | Cooling | 0.00 | 0.00 | 0.00 | 0.00 | 0.00 | 0.00 | 0.00 | 0.00 | 0.00 | 0.00 | 0.00 | 0.00 | 0.00 | |
| | M4 | Heating | 1.71 | 1.61 | 2.97 | 4.98 | 7.10 | 9.01 | 9.59 | 8.96 | 7.33 | 5.50 | 4.26 | 2.59 | 65.62 | 65.62 |
| | | Cooling | 0.00 | 0.00 | 0.00 | 0.00 | 0.00 | 0.00 | 0.00 | 0.00 | 0.00 | 0.00 | 0.00 | 0.00 | 0.00 | |
| | M5 | Heating | 1.47 | 1.47 | 2.66 | 4.63 | 6.62 | 8.40 | 8.97 | 8.43 | 6.90 | 5.20 | 4.01 | 2.30 | 61.06 | 61.06 |
| | | Cooling | 0.00 | 0.00 | 0.00 | 0.00 | 0.00 | 0.00 | 0.00 | 0.00 | 0.00 | 0.00 | 0.00 | 0.00 | 0.00 | |

S = solution; M = model; H = heating; C = cooling.

Table A5. Energy demand—Coyhaique.

| S | M | H/C | Monthly Energy Demand [kWh/m$^2$] | | | | | | | | | | | | Annual Energy Demand [kWh/m$^2$] | |
| --- | --- | --- | --- | --- | --- | --- | --- | --- | --- | --- | --- | --- | --- | --- | --- | --- |
| | | | Jan. | Feb. | Mar. | Apr. | May | Jun. | Jul. | Aug. | Sep. | Oct. | Nov. | Dec. | H/C | Total |
| S1 | M1 | Heating | 3.57 | 4.50 | 7.70 | 15.05 | 27.17 | 33.15 | 37.66 | 29.82 | 22.12 | 14.53 | 10.14 | 5.41 | 210.82 | 228.12 |
| | | Cooling | 6.51 | 6.79 | 0.96 | 0.00 | 0.00 | 0.00 | 0.00 | 0.00 | 0.00 | 0.00 | 0.07 | 2.98 | 17.30 | |
| | M2 | Heating | 2.22 | 2.98 | 5.23 | 10.38 | 18.99 | 23.41 | 26.63 | 21.23 | 15.85 | 10.30 | 7.18 | 3.68 | 148.08 | 159.12 |
| | | Cooling | 4.26 | 4.29 | 0.61 | 0.00 | 0.00 | 0.00 | 0.00 | 0.00 | 0.00 | 0.00 | 0.00 | 1.88 | 11.04 | |
| | M3 | Heating | 1.64 | 2.26 | 4.11 | 8.27 | 15.23 | 18.90 | 21.53 | 17.30 | 13.03 | 8.40 | 5.88 | 2.87 | 119.40 | 127.70 |
| | | Cooling | 3.19 | 3.24 | 0.47 | 0.00 | 0.00 | 0.00 | 0.00 | 0.00 | 0.00 | 0.00 | 0.00 | 1.40 | 8.29 | |
| | M4 | Heating | 1.41 | 1.99 | 3.66 | 7.77 | 14.34 | 17.81 | 20.33 | 16.35 | 12.33 | 7.93 | 5.49 | 2.54 | 111.95 | 118.26 |
| | | Cooling | 2.42 | 2.64 | 0.26 | 0.00 | 0.00 | 0.00 | 0.00 | 0.00 | 0.00 | 0.00 | 0.00 | 0.98 | 6.31 | |
| | M5 | Heating | 1.12 | 1.60 | 2.92 | 6.37 | 11.88 | 14.87 | 17.01 | 13.72 | 10.42 | 6.65 | 4.60 | 2.02 | 93.18 | 98.83 |
| | | Cooling | 2.12 | 2.27 | 0.36 | 0.00 | 0.00 | 0.00 | 0.00 | 0.00 | 0.00 | 0.00 | 0.00 | 0.90 | 5.65 | |
| S2 | M1 | Heating | 1.78 | 2.45 | 4.20 | 8.44 | 15.64 | 19.26 | 21.93 | 17.15 | 12.72 | 8.21 | 5.75 | 2.92 | 120.45 | 130.50 |
| | | Cooling | 3.93 | 4.12 | 0.43 | 0.00 | 0.00 | 0.00 | 0.00 | 0.00 | 0.00 | 0.00 | 0.00 | 1.57 | 10.05 | |
| | M2 | Heating | 1.28 | 1.78 | 3.17 | 6.48 | 12.16 | 15.21 | 17.35 | 13.87 | 10.44 | 6.64 | 4.67 | 2.19 | 95.22 | 101.29 |
| | | Cooling | 2.37 | 2.46 | 0.32 | 0.00 | 0.00 | 0.00 | 0.00 | 0.00 | 0.00 | 0.00 | 0.00 | 0.91 | 6.06 | |
| | M3 | Heating | 1.03 | 1.52 | 2.67 | 5.60 | 10.55 | 13.28 | 15.19 | 12.25 | 9.33 | 5.91 | 4.15 | 1.86 | 83.35 | 88.25 |
| | | Cooling | 1.86 | 2.02 | 0.27 | 0.00 | 0.00 | 0.00 | 0.00 | 0.00 | 0.00 | 0.00 | 0.00 | 0.74 | 4.90 | |
| | M4 | Heating | 0.88 | 1.30 | 2.32 | 5.31 | 10.14 | 12.78 | 14.62 | 11.70 | 8.83 | 5.55 | 3.83 | 1.59 | 78.85 | 83.85 |
| | | Cooling | 1.82 | 2.07 | 0.32 | 0.00 | 0.00 | 0.00 | 0.00 | 0.00 | 0.00 | 0.00 | 0.00 | 0.79 | 5.00 | |
| | M5 | Heating | 0.78 | 1.20 | 2.09 | 4.83 | 9.29 | 11.75 | 13.48 | 10.84 | 8.25 | 5.18 | 3.57 | 1.43 | 72.68 | 77.41 |
| | | Cooling | 1.85 | 1.84 | 0.33 | 0.00 | 0.00 | 0.00 | 0.00 | 0.00 | 0.00 | 0.00 | 0.00 | 0.70 | 4.73 | |
| S3 | M1 | Heating | 1.48 | 2.06 | 3.56 | 7.22 | 13.50 | 16.74 | 19.07 | 14.92 | 11.07 | 7.09 | 4.97 | 2.42 | 104.10 | 112.97 |
| | | Cooling | 3.48 | 3.56 | 0.43 | 0.00 | 0.00 | 0.00 | 0.00 | 0.00 | 0.00 | 0.00 | 0.00 | 1.40 | 8.87 | |
| | M2 | Heating | 1.08 | 1.60 | 2.80 | 5.83 | 11.02 | 13.84 | 15.80 | 12.65 | 9.55 | 6.05 | 4.25 | 1.96 | 86.43 | 92.13 |
| | | Cooling | 2.26 | 2.28 | 0.31 | 0.00 | 0.00 | 0.00 | 0.00 | 0.00 | 0.00 | 0.00 | 0.00 | 0.85 | 5.70 | |
| | M3 | Heating | 0.93 | 1.35 | 2.40 | 5.25 | 9.92 | 12.52 | 14.33 | 11.57 | 8.85 | 5.60 | 3.93 | 1.66 | 78.32 | 83.17 |
| | | Cooling | 1.86 | 1.96 | 0.30 | 0.00 | 0.00 | 0.00 | 0.00 | 0.00 | 0.00 | 0.00 | 0.00 | 0.73 | 4.85 | |
| | M4 | Heating | 0.77 | 1.18 | 2.08 | 4.88 | 9.41 | 11.91 | 13.64 | 10.92 | 8.27 | 5.18 | 3.53 | 1.39 | 73.18 | 78.08 |
| | | Cooling | 1.87 | 1.94 | 0.35 | 0.00 | 0.00 | 0.00 | 0.00 | 0.00 | 0.00 | 0.00 | 0.00 | 0.73 | 4.90 | |
| | M5 | Heating | 0.71 | 1.12 | 1.89 | 4.57 | 8.85 | 11.23 | 12.89 | 10.38 | 7.92 | 4.95 | 3.36 | 1.30 | 69.18 | 73.83 |
| | | Cooling | 1.82 | 1.81 | 0.33 | 0.00 | 0.00 | 0.00 | 0.00 | 0.00 | 0.00 | 0.00 | 0.00 | 0.68 | 4.65 | |

S = solution; M = model; H = heating; C = cooling.

**Table A6.** Energy demand—Punta Arenas.

| S | M | H/C | Jan. | Feb. | Mar. | Apr. | May | Jun. | Jul. | Aug. | Sep. | Oct. | Nov. | Dec. | H/C | Total |
|---|---|---|---|---|---|---|---|---|---|---|---|---|---|---|---|---|
| | | | | | | | Monthly Energy Demand [kWh/m$^2$] | | | | | | | | Annual Energy Demand [kWh/m$^2$] | |
| S1 | M1 | Heating | 14.15 | 13.53 | 21.10 | 29.19 | 39.38 | 46.34 | 48.17 | 42.56 | 32.21 | 26.08 | 20.01 | 14.25 | 346.98 | 346.98 |
| | | Cooling | 0.00 | 0.00 | 0.00 | 0.00 | 0.00 | 0.00 | 0.00 | 0.00 | 0.00 | 0.00 | 0.00 | 0.00 | 0.00 | |
| | M2 | Heating | 10.39 | 9.72 | 15.10 | 20.85 | 27.99 | 32.80 | 34.20 | 30.59 | 23.59 | 19.40 | 15.13 | 10.57 | 250.33 | 250.33 |
| | | Cooling | 0.00 | 0.00 | 0.00 | 0.00 | 0.00 | 0.00 | 0.00 | 0.00 | 0.00 | 0.00 | 0.00 | 0.00 | 0.00 | |
| | M3 | Heating | 8.75 | 8.03 | 12.44 | 17.02 | 22.71 | 26.50 | 27.70 | 25.07 | 19.66 | 16.43 | 13.02 | 9.03 | 206.38 | 206.38 |
| | | Cooling | 0.00 | 0.00 | 0.00 | 0.00 | 0.00 | 0.00 | 0.00 | 0.00 | 0.00 | 0.00 | 0.00 | 0.00 | 0.00 | |
| | M4 | Heating | 8.28 | 7.59 | 11.77 | 16.06 | 21.42 | 25.00 | 26.15 | 23.69 | 18.64 | 15.66 | 12.44 | 8.62 | 195.31 | 195.31 |
| | | Cooling | 0.00 | 0.00 | 0.00 | 0.00 | 0.00 | 0.00 | 0.00 | 0.00 | 0.00 | 0.00 | 0.00 | 0.00 | 0.00 | |
| | M5 | Heating | 7.13 | 6.44 | 9.97 | 13.54 | 17.97 | 20.95 | 21.95 | 20.04 | 15.93 | 13.59 | 10.93 | 7.54 | 165.98 | 165.98 |
| | | Cooling | 0.00 | 0.00 | 0.00 | 0.00 | 0.00 | 0.00 | 0.00 | 0.00 | 0.00 | 0.00 | 0.00 | 0.00 | 0.00 | |
| S2 | M1 | Heating | 8.06 | 7.66 | 12.11 | 17.06 | 23.13 | 27.49 | 28.50 | 25.15 | 18.74 | 15.19 | 11.55 | 8.07 | 202.71 | 202.71 |
| | | Cooling | 0.00 | 0.00 | 0.00 | 0.00 | 0.00 | 0.00 | 0.00 | 0.00 | 0.00 | 0.00 | 0.00 | 0.00 | 0.00 | |
| | M2 | Heating | 6.99 | 6.40 | 9.96 | 13.76 | 18.39 | 21.55 | 22.51 | 20.41 | 15.97 | 13.43 | 10.64 | 7.23 | 167.25 | 167.25 |
| | | Cooling | 0.00 | 0.00 | 0.00 | 0.00 | 0.00 | 0.00 | 0.00 | 0.00 | 0.00 | 0.00 | 0.00 | 0.00 | 0.00 | |
| | M3 | Heating | 6.45 | 5.78 | 8.96 | 12.18 | 16.13 | 18.81 | 19.70 | 18.09 | 14.42 | 12.42 | 10.03 | 6.87 | 149.84 | 149.84 |
| | | Cooling | 0.00 | 0.00 | 0.00 | 0.00 | 0.00 | 0.00 | 0.00 | 0.00 | 0.00 | 0.00 | 0.00 | 0.00 | 0.00 | |
| | M4 | Heating | 5.96 | 5.39 | 8.42 | 11.62 | 15.53 | 18.22 | 19.06 | 17.37 | 13.68 | 11.60 | 9.26 | 6.25 | 142.36 | 142.36 |
| | | Cooling | 0.00 | 0.00 | 0.00 | 0.00 | 0.00 | 0.00 | 0.00 | 0.00 | 0.00 | 0.00 | 0.00 | 0.00 | 0.00 | |
| | M5 | Heating | 5.70 | 5.08 | 7.91 | 10.79 | 14.33 | 16.76 | 17.56 | 16.14 | 12.87 | 11.10 | 8.98 | 6.10 | 133.31 | 133.31 |
| | | Cooling | 0.00 | 0.00 | 0.00 | 0.00 | 0.00 | 0.00 | 0.00 | 0.00 | 0.00 | 0.00 | 0.00 | 0.00 | 0.00 | |
| S3 | M1 | Heating | 7.06 | 6.67 | 10.53 | 14.89 | 20.18 | 23.98 | 24.88 | 22.07 | 16.53 | 13.46 | 10.31 | 7.09 | 177.66 | 177.66 |
| | | Cooling | 0.00 | 0.00 | 0.00 | 0.00 | 0.00 | 0.00 | 0.00 | 0.00 | 0.00 | 0.00 | 0.00 | 0.00 | 0.00 | |
| | M2 | Heating | 6.46 | 5.87 | 9.13 | 12.58 | 16.79 | 19.67 | 20.56 | 18.72 | 14.73 | 12.48 | 9.95 | 6.74 | 153.67 | 153.67 |
| | | Cooling | 0.00 | 0.00 | 0.00 | 0.00 | 0.00 | 0.00 | 0.00 | 0.00 | 0.00 | 0.00 | 0.00 | 0.00 | 0.00 | |
| | M3 | Heating | 6.18 | 5.50 | 8.51 | 11.53 | 15.24 | 17.77 | 18.62 | 17.15 | 13.73 | 11.91 | 9.68 | 6.63 | 142.43 | 142.43 |
| | | Cooling | 0.00 | 0.00 | 0.00 | 0.00 | 0.00 | 0.00 | 0.00 | 0.00 | 0.00 | 0.00 | 0.00 | 0.00 | 0.00 | |
| | M4 | Heating | 5.63 | 5.06 | 7.91 | 10.88 | 14.51 | 17.01 | 17.81 | 16.30 | 12.89 | 11.02 | 8.84 | 5.96 | 133.83 | 133.83 |
| | | Cooling | 0.00 | 0.00 | 0.00 | 0.00 | 0.00 | 0.00 | 0.00 | 0.00 | 0.00 | 0.00 | 0.00 | 0.00 | 0.00 | |
| | M5 | Heating | 5.52 | 4.89 | 7.60 | 10.35 | 13.72 | 16.03 | 16.81 | 15.49 | 12.39 | 10.76 | 8.74 | 5.94 | 128.23 | 128.23 |
| | | Cooling | 0.00 | 0.00 | 0.00 | 0.00 | 0.00 | 0.00 | 0.00 | 0.00 | 0.00 | 0.00 | 0.00 | 0.00 | 0.00 | |

S = solution; M = model; H = heating; C = cooling.

## References


1. Chile Plan de Acción de Eficiencia Energética 2020, Ministerio de Energía; Santiago de Chile. 2013. Available online: https://www.amchamchile.cl/UserFiles/Image/Events/octubre/energia/plan-de-accion-de-eficiencia-energetica2020.pdf (accessed on 10 June 2021).
2. International Energy Agency. *World Energy Investment Outlook*; International Energy Agency: Paris, France, 2021; Volume 23.
3. United Nations. *Kyoto protocol to the United Nations Framework Convention on Climate Change*; United Nations: New York, NY, USA, 1997.
4. ISO 52016-1:2017–Energy Performance of Buildings–Energy Needs for Heating and Cooling, Internal Temperatures and Sensible and Latent Heat Loads–Part 1: Calculation Procedures. Available online: https://www.iso.org/standard/65696.html (accessed on 30 June 2017).
5. Bustamante, W.; Cepeda, R.; Martínez, P.; Santa María, H. Eficiencia energética en vivienda social: Un desafío posible. In *Camino Al Bicentenario: Propuestas Para Chile*; Concurso Políticas Públicas: Santiago, Chile, 2009; pp. 253–283. ISBN 9789561413931.
6. Givoni, B. Conservation and the use of integrated-passive energy systems in architecture. *Energy Build.* **1981**, *3*, 213–227. [CrossRef]
7. Parasonis, J.; Keizikas, A.; Kalibatiene, D. The relationship between the shape of a building and its energy performance. *Arch. Eng. Des. Manag.* **2012**, *8*, 246–256. [CrossRef]
8. Parasonis, J.; Keizikas, A. Possibilities to Reduce the Energy Demand for Multistory Residential Buildings. In Proceedings of the 10th International Conference, Vilnius, Lithuania, 19–21 May 2010; pp. 989–993.
9. Carpio, M.; Jódar, J.; Rodriguez, M.L.; Zamorano, M. A proposed method based on approximation and interpolation for determining climatic zones and its effect on energy demand and $CO_2$ emissions from buildings. *Energy Build.* **2015**, *87*, 253–264. [CrossRef]
10. Verichev, K.; Zamorano, M.; Fuentes-Sepúlveda, A.; Cárdenas, N.; Carpio, M. Adaptation and mitigation to climate change of envelope wall thermal insulation of residential buildings in a temperate oceanic climate. *Energy Build.* **2021**, *235*, 110719. [CrossRef]
11. Parasonis, J.; Keizikas, A.; Endriukaitytė, A.; Kalibatienė, D. Architectural Solutions to Increase the Energy Efficiency of Buildings. *J. Civ. Eng. Manag.* **2012**, *18*, 71–80. [CrossRef]
12. Aksoy, U.T.; Inalli, M. Impacts of some building passive design parameters on heating demand for a cold region. *Build. Environ.* **2006**, *41*, 1742–1754. [CrossRef]
13. Depecker, P.; Menezo, C.; Virgone, J.; Lepers, S. Design of buildings shape and energetic consumption. *Build. Environ.* **2001**, *36*, 627–635. [CrossRef]
14. Danielski, I.; Fröling, M.; Joelsson, A. The Impact of the Shape Factor on Final Energy Demand in Residential Buildings in Nordic Climates. In Proceedings of the World Renewable Energy Forum, Denver, CL, USA, 13–17 May 2012; Volume 7.
15. Carpio, M.; García-Maraver, A.; Ruiz, D.P.; Martín-Morales, M. Impact of the envelope design of residential buildings on their acclimation energy demand, $CO_2$ emissions and energy rating. *WIT Trans. Ecol. Environ.* **2014**, *186*, 387–398.
16. Sarricolea, P.; Herrera-Ossandon, M.; Meseguer-Ruiz, Ó. Climatic regionalisation of continental Chile. *J. Maps* **2017**, *13*, 66–73. [CrossRef]
17. Comisión Nacional de Energía Balance Energético–Chile. Available online: https://www.cne.cl/ (accessed on 30 July 2021).
18. Marcos Martín, F. *Biocombustibles Sólidos de Origen Forestal*; AENOR: Madrid, Spain, 2001; ISBN 9788481432725.
19. Carpio, M.; Zamorano, M.; Costa, M. Impact of using biomass boilers on the energy rating and $CO^2$ emissions of Iberian Peninsula residential buildings. *Energy Build.* **2013**, *66*, 732–744. [CrossRef]
20. *Chile Ordenanza General de Urbanismo y Construcciones (OGUC)*; Ministerio de Vivienda y Urbanismo: Santiago, Chile, 2009.
21. Verichev, K.; Carpio, M. Climatic zoning for building construction in a temperate climate of Chile. *Sustain. Cities Soc.* **2018**, *40*, 352–364. [CrossRef]
22. Verichev, K.; Zamorano, M.; Carpio, M. Assessing the applicability of various climatic zoning methods for building construction: Case study from the extreme southern part of Chile. *Build. Environ.* **2019**, *160*, 106165. [CrossRef]
23. Verichev, K.; Zamorano, M.; Carpio, M. Effects of climate change on variations in climatic zones and heating energy consumption of residential buildings in the southern Chile. *Energy Build.* **2020**, *215*, 109874. [CrossRef]
24. Prieto, A.J.; Verichev, K.; Carpio, M. Heritage, resilience and climate change: A fuzzy logic application in timber-framed masonry buildings in Valparaíso, Chile. *Build. Environ.* **2020**, *174*, 106657. [CrossRef]
25. Agencia Chilena de Eficiencia Energética. Manual de Gestor Energético–Sector Construcción 2014. p. 252. Available online: http://old.acee.cl/eficiencia-energetica/guias (accessed on 10 June 2021).
26. Agencia Chilena De Eficiencia Energética. *Guia De Diseño Para La Eficiencia Energetica En La Vivienda Social 2009*; Agencia Chilena De Eficiencia Energética: Santiago, Chile, 2009; p. 203.
27. Meteotest Meteonorm 7 2019. Available online: http://old.acee.cl/576/articles-61341_doc_pdf.pdf (accessed on 10 June 2021).
28. Chile Factores de Emisión, Ministerio de Energía. Available online: https://www.energia.gob.cl/ (accessed on 30 July 2021).
29. Spain Plan de Energías Renovables 2011–2020. Ministerio de Industria, Turismo y Comercio Gobierno de España, IDEA. 2011; pp. 1–824. Available online: https://www.miteco.gob.es/es/cambio-climatico/legislacion/documentacion/PER_2011-2020_VOL_I_tcm30-178649.pdf (accessed on 10 June 2021).

30. Comisión Nacional de Energía Anuario Estadístico de Energía 2005–2015. 2015. Available online: https://www.cne.cl/wp-content/uploads/2016/07/AnuarioCNE2015_vFinal-Castellano.pdf (accessed on 10 June 2021).
31. Romero, J. Cuantificación, Caracterización Y Análisis De La Comercialización De Leña En Puerto Williams, Isla Navarino, XII Region. 2007. Available online: http://dspace.utalca.cl/handle/1950/6251 (accessed on 10 June 2021).
32. *CDT Medición del Consumo Nacional de Leña y Otros Combustibles Sólidos Derivados de la Madera*; Ministerio de Energía: Santiago, Chile, 2015.
33. Instituto para la Diversificación y Ahorro de la Energía IDEA. Available online: http://www.idae.es/ (accessed on 30 July 2021).
34. Autodesk Autodesk. Available online: https://www.autodesk.com/ (accessed on 30 July 2021).
35. ISO ISO 52017-1:2017–Energy Performance of Buildings–Sensible and Latent Heat Loads and Internal Temperatures–Part 1: Generic Calculation Procedures. Available online: https://www.iso.org/standard/65698.html (accessed on 30 June 2017).
36. ISO ISO 13789:2017–Thermal Performance of Buildings–Transmission and Ventilation Heat Transfer Coefficients–Calculation Method. Available online: https://www.iso.org/standard/65713.html (accessed on 30 June 2017).
37. Albatici, R.; Passerini, F. Building Shape and Heating Requirements: A Parametric Approach Italian Climatic Conditions. In Proceedings of the CESB–Central Europe towards Sustainable Building Conference, Prague, Czech Republic, 30 June–2 July 2010.
38. Vásquez, C.; Encinas, F.; D'Alençon, R. Edificios de oficinas en Santiago: ¿Qué estamos haciendo desde el punto de vista del consumo energético? *Arq* **2015**, *89*, 50–61. [CrossRef]

 *sustainability*

*Article*

# Review of White Roofing Materials and Emerging Economies with Focus on Energy Performance Cost-Benefit, Maintenance, and Consumer Indifference

**Fadye Al Fayad [1,*], Wahid Maref [2] and Mohamed M. Awad [3]**

[1] Business Administration Department, Jubail University College, Royal Commission for Jubail, Al Jubail 35716, Saudi Arabia

[2] Construction Engineering Department, École de Technologie Supérieur (ÉTS), University of Québec, 1100 Rue Notre-Dame Ouest, Montreal, QC H3C 1K3, Canada; Wahid.Maref@etsmtl.ca

[3] Mechanical Power Engineering Department, Faculty of Engineering, Mansoura University, Mansoura 35516, Egypt; m_m_awad@mans.edu.eg

* Correspondence: Fayadf@ucj.edu.sa; Tel.: +966-500038138

**Abstract:** This article performed a comprehensive review of the different state-of-the-art of roofing technologies and roofing materials and their impact on the urban heat island (UHI) and energy consumption of buildings. The building roofs are the main sources of undesirable heat for buildings, especially in warm climates. This paper discusses the use and application of white roofing material in emerging economies. The use of white roofing material is a suggestion because of its cooling, evaporative and efficiency characteristics compared to traditional black roofing materials. Many research studies have shown that the darker roofing surfaces that are prevalent in many urban areas actually can increase temperature by 1 to 3 degrees Celsius to the environment surrounding these urban areas. Additionally, improved temperature control and heat reflection also work to reduce the energy requirements for the interior spaces of the structures that have white roofing surfaces. The white or lighter colored roofs tend to reflect a part of the solar radiation that strikes the roof's surface. Consequently, one might believe that white roofing material would be commonplace and especially so within emerging economies. Yet, this is hardly the case at all. This paper examines the issue of white roofing materials in emerging economies from a dual perspective. The dual perspective includes the technical details of white roofing material and its impact on lowering the interior temperature of the affected structures, which consequently reduces hours of indoor thermal discomfort and use of air conditioners in indoor spaces. The other element in this study, however, involves the marketing aspect of white roofing material. This includes its adoption, acceptance and cost-benefit in emerging economies.

**Keywords:** white roofs; cool roofs; reflective material; cost-benefit; energy savings; urban heat island

**Citation:** Fayad, F.A.; Maref, W.; Awad, M.M. Review of White Roofing Materials and Emerging Economies with Focus on Energy Performance Cost-Benefit, Maintenance, and Consumer Indifference. *Sustainability* **2021**, *13*, 9967. https://doi.org/10.3390/su13179967

Academic Editor: Alberto-Jesus Perea-Moreno

Received: 22 June 2021
Accepted: 30 August 2021
Published: 6 September 2021

**Publisher's Note:** MDPI stays neutral with regard to jurisdictional claims in published maps and institutional affiliations.



## 1. Introduction

Global warming occurs when $CO_2$ and other air pollutants absorb solar radiation that has bounced off the earth's surface resulting in an increase in the air temperature near the surface of the earth. As well, urban heat island effects contribute to global warming [1]. More recent research found that urban heat island effects contributed to climate warming by about 30% of all other issues that contribute to climate warming [2]. The study by [3] has shown increases in the severity of the effect of heat islands with the progress of climate change. Unlike vegetation and other natural ground cover, urban surfaces absorb and store more solar radiation, which leads to increasing the surrounding temperature [4,5]. It is important to point out that the higher surrounding temperature would cause a greater energy demand for air conditioning (A/C) systems [6]. Currently, there are five strategies that can be applied in order to mitigate the effects of urban heat islands [7–13]). These strategies include increasing vegetation and trees in urban areas [8], installing reflective pavements on streets, sidewalks and parking lots [9,10], utilizing smart growth practices that help protect the natural environment [11], installing green roofs by growing vegetative layers on the rooftops [12], and installing cool/white/reflective roofing systems [6,13–17]. Throughout this paper, unless otherwise specified, cool roofs, white roofs and reflective roofs have the same meaning.

According to [18], surfaces in urban cities that are exposed to solar radiation are primarily rooftop surfaces. For residential buildings, the intense and prolonged solar radiation heats up the roof assembly more than any other building envelope component as the rooftop surface has the highest exposure to solar radiation [19]. Heat gain through residential roofs is primarily in the form of radiation [20]) and accounts for about 50% to 70% of the total heat gain into the indoor spaces below the ceiling board [21].

Because of the urban expansion phenomena, several materials especially artificial have been introduced into the market in order to replace the natural vegetation such as asphalt, limes, etc., which have affected drastically the environment and its temperature, thus energy consumption of buildings. To overcome to this phenomenon, many technology solutions have been investigated such as the use of vegetation [22,23], phase change materials (PCM) [22,24] and reflective coatings for claddings and roofing components [25] especially the use of the roof coating with high near-infrared reflectance (NIR) as to be an effective solution to mitigate the UHI. Therefore, 52% of absorbed heat is due to the near-infrared component of solar radiation. Ref. [26] has demonstrated in his paper that cool coating is one of the most effective solutions to mitigate the UHI on both facades and roofing systems. He uses cool coating that contains color pigments, which do not absorb the infrared portion of the solar spectrum.

Ref. [27] has studied the influence of traditional and solar reflective coatings on the heat transfer of building roofs in four cities with warm climates in Mexico. He has proven by simulation and experiments that uninsulated and insulated concrete slab white reflective roofs to have a daily heat gain between 37 and 56% compared to uninsulated and insulated traditional slab roofs (gray roof).

Ref. [28] have focused in their research study on the modern residential roofs in Malaysia that employed mainly red and brown roof tiles due to aesthetic factors. The authors reveal the findings of their research study on the effect of roof tile colour on heat conduction transfer through roof tiles and ceiling boards, rooftop surface temperature and cooling load. They demonstrated that the selection of white roof tiles significantly reduces the peaks of heat conduction transfer and rooftop surface temperature as well as the values of heat conduction transfer and rooftop surface temperature throughout diurnal profiles, which consequently reduces hours of indoor thermal discomfort and use of air-conditioners in indoor spaces. A decline in peak rooftop surface temperature of up to 16.00 °C that results in annual energy savings of up to 13.14% can be achieved when the roof tile colour is changed from red to white.

Briefly, green roofs are mainly constructed by planting the rooftops of residential buildings [8,10–12]. In hot climates, shading the outer surface of the building envelope with green roofs has been shown to be more effective than increasing the amount of insulation. These roofs bring many benefits to the public, private, economic, social sectors, and the local and global environments. Both installation and the thermal performance of a green roof vary by the type and design, region, climate and building type. A list of benefits as a result of installing green roofs include:

- Green roofs reduce stormwater runoff where retention and/or delay of runoff eases stress on stormwater infrastructure and sewers.
- Green roofs enhance the energy performance of buildings due to reducing the heat flux through the roof. This results in fewer energy requirements for cooling during summertime and thus can lead to significant cost savings.
- Green roof's plant leaves trap dust particles from the air, and evapotranspiration cools ambient temperatures. This contributes to reducing global warming.

- Green roofs cover the waterproofing membrane and thus protect it from UV rays and extreme daily temperature fluctuations. This protection extends the lifespan of the waterproofing membranes under green roofs in relation to traditional roofs.
- Green roofs can be designed to enhance urban food security through rooftop gardening and food production.

Generally, the external surfaces of different types of roof systems (traditional, green and cool) are exposed to several environmental factors that include dust/dirt, cloud coverage, sunlight, rain, snow, wind, outdoor temperature and relative humidity. The performance of the roofs depends mainly on these factors and the roof specifications [6,14–17,29]. The absorbed solar radiation on the external surfaces of the roofing systems causes an increase in the surface temperature of the roof, thereby increasing the cooling load in summer and reducing the heating load in winter [30]. Unlike traditional roofing systems, both green roofs and cool roofs are designed to reduce the amount of absorbed solar radiation. The cool roofs, which are the focus of this paper, use reflective materials or coatings that have high short-wave solar reflectivity (i.e., low short-wave absorption coefficient) to reflect a substantial portion of the incident solar radiation.

The influences of dirt and/or particles such as dust accumulations on surfaces are important for the solar photovoltaic (PV) panel and cool roof applications. For PV panels, the dirt/dust accumulations on the panels obstruct or distract light energy from reaching the solar cells, resulting in reduction in PV performance (e.g., see [31,32]) for more details). As provided in [6,16,17] for cool roof applications, dust and dirt accumulations on the reflective materials or coatings installed on the external surfaces of roofs can decrease the short-wave solar reflectivity of these surfaces. This results in increasing the solar heat gains. Additionally, a number of studies reported on the change in the properties of roofing surfaces due to weathering factors and dirt accumulation, and also developed cleaning processes/procedures so as to minimize the loss in the solar reflecting of the rooftops [33–35].

The short-wave solar reflectivity is one of the most important properties for the reflective materials that have a great effect on the amount of energy savings [14,15,36]. Most recently, experimental and numerical studies were conducted to investigate the potential use of reflective roofing technology in hot, humid, dusty and polluted climates such as that in Kuwait and Saudi Arabia [6,16,17]. In these studies, the dust concentration on the surface of a Reflective Coating Material (RCM) that is currently available in the markets was measured in terms of the turbidity when the RCM was subjected to the natural weathering conductions of Jubail Industrial City (JIC). Additionally, due to the quite high pollution level in JIC [37,38], black carbon, inorganic carbon and some isolated dark spots of biomass can be seen on the surface of RCM as a thin layer between the coating and the dust. Beside dust, this thin layer of the contaminants can contribute to reducing the short-wave solar reflectivity of RCM. Furthermore, technical guide and cleaning processes were developed to increase solar reflectivity of the RCM that would result in high energy savings (see [6,16,17] for more details).

For cool roofing systems subjected to hot climates such as that in Gulf Cooperation Council (GCC) countries, a simple and user-friendly design tool was recently developed [14,15]. This tool can easily be used by building engineers and architects for determining all pairs of the insulation thickness and the corresponding solar reflectivity of the reflective roofing materials/coatings that resulted in the same levels of the energy performance as those for the black roofing systems of thicker insulation thickness. As well, this design tool can be used to upgrade the building codes in order to allow using less insulation in the roofs if cool/white/reflective roofing systems are installed [17].

This section has introduced the concept of the different roofing systems that exist and made a difference between green/white/reflective/cool roofs, etc. and identified the advantages of these different roofing systems.

## 2. Technical Background and Cost-Benefit

The use of white roofing materials is known to have thermal performance that makes it desirable in some climatic conditions and undesirable in others. The existing research indicates that solar energy striking roof structures have a significant effect on the thermal characteristics of the underlying structure. The use of white roofing material that is classified as a reflective surface material, relies on brightness as a means to reflect much of the short-wave radiation emitted by sunlight and striking the rooftop [36,39]. The result is, of course, a lower surface temperature of the roof material and therefore of the underlying structure. As such, the white roofing materials have affected significantly the interior heating and cooling energy loads of the structure itself.

On the technical side of this paper, some work on the specific materials that can be used in white roofs must be discussed. The feature of white roofs is much more involved than simply painting a roof with a reflective coating. Rather, reflective roofing material can be silicone-based and be painted or sprayed onto a structure or, alternatively, it can be a membrane material that is attached to the roof via adhesive [40]. Regardless of the specific material, the effectiveness of reflective roofing material is typically measured by determining the material's Solar Reflectance Index (SRI) rating. The SRI rating can see some materials that can reflect as much as 90% or more of the solar radiation striking the roof [41]. Regardless of the specific SRI rating, reflective roofing materials can take a number of different forms and applications and these should be discussed in some depth.

The costs of the reflective roofing materials along with the cost-benefit analysis can be a shared activity. The research process into costs of the materials itself is not technical but the ideal reflective roofing material for emerging markets does require some technical details such as its thermal properties, durability, etc. This is why costing and cost-benefit analysis benefits from a shared input format in this particular proposed article. The marketing element in the cost-benefit analysis can be integrated into the technical discussion either with a separate heading or as part of the overall discussion.

### 2.1. Application and Maintenance

The application process of white roofing material requires technical knowledge as does the maintenance processes involved. Yet, the maintenance processes involved in white roofing materials must also be known from a marketing perspective. This is because emerging economies typically have limited resources available. At a granular level, however, entities that utilize white roofing materials as a means to reduce the cooling energy load of a structure, reduce environmental heat build-up and as well seek to control costs have to consider the post-application phase of the material. As mentioned, the technical application of the white roof depends upon the form of white roofing material selected. Likewise, the maintenance is also dependent upon what material is selected. In effect, cost factors must also be considered with respect to ongoing maintenance of white roofs in developing markets.

### 2.2. Link between Reflective Roofing Material and Emerging Economies

This section introduces the concept of emerging economies and identifies the relation that exists between the reflective roofing materials and emerging economies.

Emerging markets are the centrepiece of global economic growth. As such, it is important to identify the link between reflective roofing material and emerging economies. In order to fully explore this link, one must first characterize what is meant by an emerging market. Emerging markets are those markets that are in the process of transitioning from a developing market to a developed market. They are characterized by some degree of market volatility, high growth potential for investors, and tend to have low to middle per capita income rates across their general populations. The link between energy demand, energy consumption, and climate change through the production of greenhouse gas emissions is irrefutable. While developed countries typically have the resources necessary to both develop new construction designs and technologies to reduce energy consumption and emissions, emergent economies lack such resources. Consequently, without an understanding of how to improve building design, efficiency, and construction in emergent economies, reducing global greenhouse gas emissions and especially carbon dioxide ($CO_2$) will be largely impossible. Research has demonstrated that as much as 75% of all energy demand involves industrial activities and/or buildings located in emerging economies globally [42]). Thus, even if developed markets are fully successful in reducing overall energy consumption, emissions, and gaining sustainability, climate change will continue unabated due to inefficient activities in emerging economies.

The use of passive energy efficiency designs is critical in any economy as a means to reduce energy consumption, achieve sustainable energy usage patterns and to reduce the human impact on global warming. Passive energy design-build techniques by definition do not require ongoing energy commitments in the same way that active or non-passive design-build solutions do. However, in emerging economies, such passive building designs are extremely important. This is because emerging economies tend to lack the resources necessary to reverse integrate more energy-efficient solutions with respect to energy-efficient materials used in building construction. This importance with respect to emerging economies can be seen in the degree to which energy consumption increases. For example, on average, overall energy consumption globally among all countries is expected to increase by more than 100% by the year 2050 but in emerging economies, this figure is believed to be more than 300% on average [43]. Hence, if emergent economies can deploy energy-efficient building designs and integrate energy-efficient building envelope materials at the outset, they can reduce this dramatic gap in expected energy usage growth moving forward. Specifically, in the design phase of new construction, passive energy-efficient solutions greatly reduce forward energy demands per structure.

In fact, the design-build phase of construction in emergent economies can address issues that have significant positive effects on energy consumption throughout the life of the structure. Even more pertinent to emergent economies is the fact that such solutions during the design-build process often do not come with excessive upfront costs if addressed during this phase. This is as opposed to addressing them later on during a deep restoration later in the structure's lifespan. Such solutions during this design-build phase include the simple use of increased insulating material, improved fenestration that reduces solar loading, the use of reflective surfaces including roofing material on all structure envelopes, improved sealing techniques, and designing out what are referred to as thermal bridges beneath the roof structure that conduct heat into the interior of the building [44]. All of these solutions reduce the energy use of the structure throughout its lifespan and work in conjunction with reflective roofing surfaces. The use of supplemental design-build solutions that address long-term energy use and demand work to amplify the positive effect of reflective roof materials in the operating costs of the structure over its entire lifespan. The result is structured so that it not only reduces the use of energy required to operate them on a long-term basis but also reduces the cost or even the necessity for major restorations in order to achieve long-term sustainability in the future. Such major restorations are themselves very costly and resources dependent which presents additional problems for emergent economies.

One aspect of energy-efficient building construction and materials such as the use of reflective roofing material that is often overlooked involves the financial element. This particular element is especially important if it is considering an emerging economy. The observation is that most if not all emerging economies implemented more energy-efficient design-build solutions if they were able to manage such solutions financially. Sustainability in energy use, usage growth, and zero-carbon emissions may be able to be achieved without significant cost increases but such solutions often require sophisticated financial structures during the funding process of a structure. Thus, emerging economies can benefit greatly from implementing sustainable finance methods as a precursor to new building design and construction. These are financial structures that include instruments such as blended financing, social responsibility factors, and green finance structures that are introduced as a criterion to foreign direct investment or FDI [42]. These types of financial structures are focused on building design and construction techniques that are inherently more energy-efficient than would otherwise be the case. This reduces the inclination within emerging economies to award construction contracts based purely on upfront costs. In essence, design-build solutions that are based primarily on upfront costs typically result in dramatically higher operating costs over the lifecycle of a structure. In essence, such long-term lifecycle costs ultimately end up costing more than the estimates for more energy-efficient design-build solutions. That is, energy-efficient financial structures incentivize energy-efficient design-build solutions for both local and outside economic participants.

Much of the need for both financial and technological energy-efficient solutions in emergent economies relates to the specific idiosyncrasies of emerging economies themselves. In effect, while developed markets do tend to have a high percentage of established infrastructure in place, emerging economies are still in the process of building out much of their internal infrastructure. Thus, where developed markets are both implementing new energy-efficient building construction techniques and directing enormous amounts of capital towards deep restorations of existing structures, emergent economies are focused primarily on building new infrastructure. This dynamic is expressed in certain ways with an emphasis on the type of energy that is utilized to fuel economic activity itself. For instance, countries such as Vietnam still rely heavily on dirty energy sources such as coal to support economic activity and Ethiopia which still needs to electrify some 50% of its market [45]. The net result is such that almost all of the downstream activity in these and similar emergent economies such as new building construction, employ solutions that themselves are inherently inefficient.

Still, in regards to emergent economies, the overall emphasis on sustainability in building design and construction must remain on the integration of efficient building techniques. Therefore, the use of reflective material in emerging markets is primarily the same as it is in developed markets. Yet, the scope and scale of reflective building design and construction techniques should be broadened for emerging markets simply because of the lack of resources compared to developed markets. The resource has demonstrated that the use of reflective roofing material as well as reflective material on other envelope structures is quite simply the most effective way to reduce the energy requirements necessary to cool/heat a structure. Yet, one other consideration that emerging markets can easily integrate into their development plans during the design-build process for new structures is to consider the reduction of urban heat islands in which heavily urbanized developments themselves become heat sinks [46]. Such heat sinks contribute to the energy requirements for the structures that exist within them. Hence, designing the development pattern of economic centres can contribute to the reduction of heat loading of a given structure. Additionally, emerging economies can further improve the effectiveness of reflective roofing solutions but augmenting them with solutions such as radiant barriers that block the transmission of any solar energy that does infiltrate the structure [47]. Regardless, emerging economies have a range of alternatives within the design-build phase of new construction that they can avail themselves of. Combined, the overall effect is the reduction not only of energy requirements but of future energy requirement growth.

Emerging economies experience certain economic and construction issues that developed markets do not. These issues are primarily resource-related as well as technology-related with respect to cost, affordability, and even availability. Yet, as this section has indicated, emergent markets tend to also experience sustainability issues involving energy efficiency due to a lack of financial sophistication in the funding process for new structures. These particular issues can prevent a new structure from being able to maximize energy efficiency at the outset resulting in the need for a deep restoration earlier in a structure's lifecycle than might otherwise be the case. Other solutions for emerging economies involve design-build techniques such as site selection and radiant barriers that reduce the long-term growth in energy demand itself.

## 3. White Roofing Systems Cost-Benefits

Many previous studies [4,48–57] including this current study have shown that green and reflective roofs are not only protection from solar radiation and rain but can save energy, mitigate urban heat islands, decrease greenhouse gas emissions, and reduce local air pollution while increasing the thermal comfort level (i.e., indoor air quality). The reflective roofs are widely known in the markets as white roofs and cool roofs. Ref. [58] have used the DOE-2 energy simulation tool to investigate if the reflective roofs have an effect on the heating and cooling energy use for buildings in the USA. That study showed that the estimated net saving of about \$750 M in annual energy payments for an annual electricity savings around 10 TWh and the peak electrical power reduction was about 7 GW.

Ref. [59] have shown in their study on a single-family one-storey building that by increasing the reflectivity of the roof from 20% to 60% is equivalent to a 50% reduction of the roof insulation thickness in hot climates. This comes to support the introduction of the reflective roof materials in the proposed ASHRAE SSPC 90.2. Four prototypes of commercial buildings in 236 US cities have been simulated by [60] in order to determine their annual cooling and heating energy loads. In that study, a short-wave solar reflectivity of 55% has replaced the cool roof and with short-wave solar reflectivity of 20% by a conventional grey roof. The results showed that the cool roofs showed an annual energy saving per unit conditioned roof area fluctuating from \$1.14/m$^2$ in Arizona to \$0.126/m$^2$ in West Virginia (\$0.356/m$^2$ the rest of the USA).

The study by [61] has shown that energy saving is highly dependent on roof type, climate and quantity of insulation used for cool and green roofs. As an example, for a typical one-storey building in Boston with a modified-bitumen roof and a thermal resistance RSI of 2.7 m$^2$ K/W (*R*-value of 15.3 ft$^2$ F hr/BTU), if you double the insulation thickness, you can save 13% in cooling and heating energy. However, when you install a green roof instead, you can save 12% energy. However, for the same building in Lisbon (Portugal), and by adding twice the quantity of the insulation, the results showed that there is almost no energy saving, whereas by installing a green roof, the result shows 26% reduction in energy use [61].

## 4. Energy Performance and Energy Saving in Buildings and Its Effect on the Environment in Hot and Cold Climate Zones

The continuous increase for example of greenhouse gas (GHG) emissions, the precipitation intensity and the atmospheric temperatures were the direct consequences of global warming. A high percentage of the local and global effects of climate change have been impacted by buildings and other infrastructures. In hot climate zones such as Saudi Arabia, most of the energy is used for cooling, which impacts directly global warming.

Ref. [57] have shown that 30–40% of the total energy demand is due to the high energy consumption in buildings. This can be different from one geographical zone to another, such as in Europe, the buildings are responsible for about 40–50% of energy use. A large portion of this energy is used for heating (European Commission 2010). Figure 1a,b and ref. [62,63] show that in Canada, the third important sector of energy use and GHG emissions is the building sector, just after industry and transportation, which is about 27% and 23% respectively.

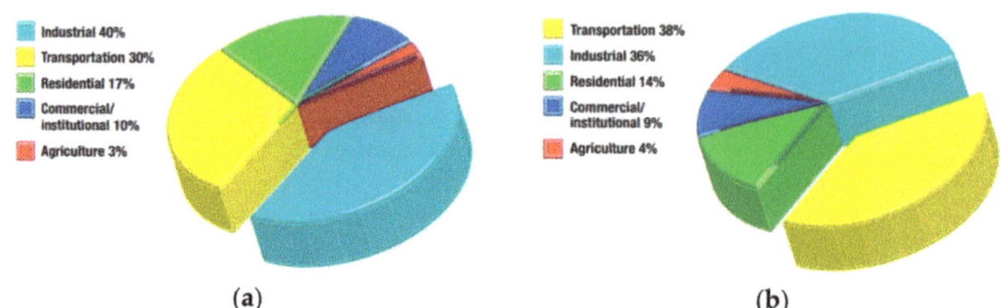


**Figure 1.** (**a**) Secondary energy use by sector [62,63]. (**b**) GHG Emission by sector [62,63].

The secondary energy use (Figures 1a,b and 2a,b) is the energy used by final consumers in various sectors of the economy, GHG emissions by sector and distribution of residential and commercial energy use respectively. This includes, for example, the energy used by vehicles in the transportation sector. Secondary energy use also encompasses energy required to heat and cool homes or businesses in the residential and commercial/institutional sectors. In addition, it comprises energy required to run machinery in the industrial and agricultural sectors. Energy is used in all five sectors of the economy: residential, commercial/institutional, industrial, transportation, and agriculture. The industrial sector accounted for the largest share of energy, followed by transportation, residential, commercial/institutional, and agriculture.

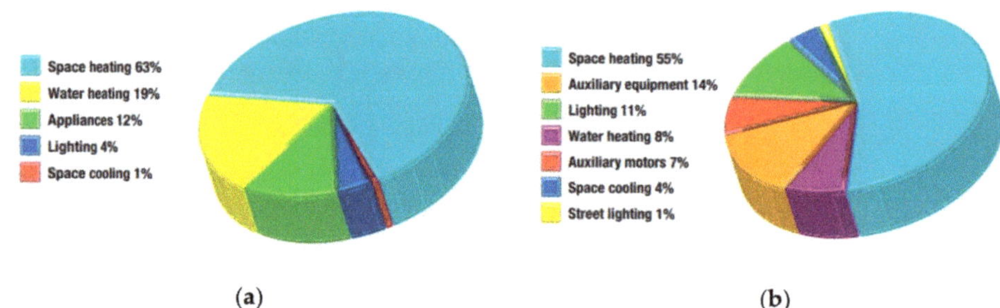


**Figure 2.** (**a**) Distribution of residential energy use by end use [62,63]. (**b**) Commercial/Institutional energy use by end use [62,63].

On the contrary, in hot climate zones such as in GCC countries, a large portion of the energy is used for cooling. Ref. [51] have shown in their study that in Kuwait, 90% of the electricity consumption is due to buildings, where 70% of the electricity peak demand in 2004 is due to the residential sector. The authors reported as well that in Canada (Figure 1a,b), the residential sector used about 17% of the energy where about 10% for the commercial and institutional sector are responsible for 14% and 9%, respectively, for the GHG emissions. Figure 2a,b shows the distribution of secondary energy use in residential and commercial/Institutional energy use by end-use. Figure 2a shows that the largest share in the distribution of energy use by end-use in residential buildings is the space heating which represents about 55–63% in commercial buildings (see Figure 2b).

In order to save energy in buildings, the building envelope design (i.e., walls, roofs and fenestration systems) plays a pivotal role. The energy conservation efforts will mainly not only reduce the energy consumption but also the GHG emissions by 50% (Figure 3) (see [64] for more details). As such, having a holistic approach is the best for energy efficiency. The first step is to improve the building envelope performance even before improving the energy performance of the mechanical systems.

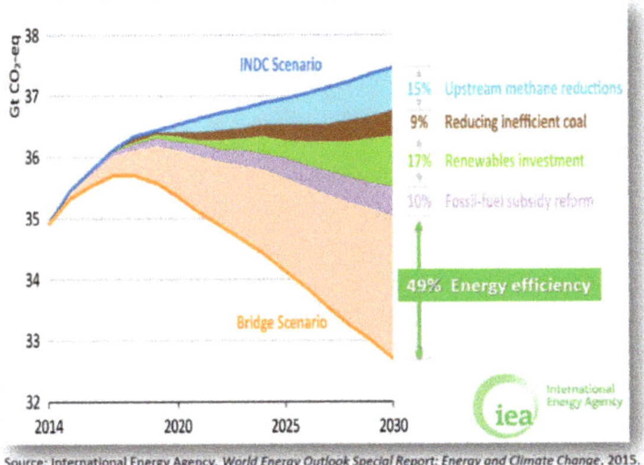



Source: International Energy Agency. *World Energy Outlook Special Report: Energy and Climate Change,* 2015.

**Figure 3.** Energy Consumption and GHG Emissions [64].

This research study focuses only on one part of building envelope, which is the cool roof system. Using an urban canyon model coupled to a global climate model, [4] have studied the impacts of globally installing cool roofs. The results showed that a reduction by 33% in the annual heat island can lead to reducing the daily maximum and minimum temperatures by 0.6 °C and 0.3 °C, respectively. In order to attenuate the effect of global warming a reduction of the energy demand for building operation will help and which imply a reduction of operating cost. This can be achieved by designing a high-performance building envelope having roofing systems with high potential of energy savings at no risk of condensation and its related problems such as mold growth [52,65–67].

Roofing systems have evolved in the construction industry. A "green" roofing has evolved for many decades now, and it contributes more and more in helping the environment by reducing GHG emissions and fighting against global warming in general. In order to enhance the roof's contribution to environmental requirements, "green" roofing has evolved into one of two types: highly reflective or "cool" roofing systems, and vegetative roofing systems. Ref. [67] presented a practical decision-making procedure for designing and selecting a sustainable green roofing system. In that study, the author considered several roofing assemblies available in the market, compared the disadvantages and advantages of each, and illustrated which assemblies meet cool roofing and vegetative roofing guidelines. That study also offered a good understanding of "green" roof options that are viable for different types of buildings [67].

Short-wave solar radiation consists of ultraviolet, visible light, and near-infrared radiation from the sun that reaches the Earth. The latter is one of the key drivers of urban heat islands. Urban surfaces reflect less solar radiation back to the environment. Instead, more solar radiation is stored and absorbs resulting in an increase in the temperatures of the surrounding [4,5,13,68]. Using cool pavements for example (either reflective or permeable) on parking areas, sidewalks and streets to reflect more solar energy back to the atmosphere. The temperatures of the surfaces of the pavement and the surrounding air are lower than traditional pavements because of the use of cool pavements [11]. To improve the urban climate during the heat waves, [21] used smart wetting of building materials and [69] evaluated the effect of different cool pavement strategies on the heat island mitigation. For example, in a city of 1 million inhabitants or higher, the annual mean air temperature can be higher than its close environment by 1 °C and 3 °C. In the nighttime, however, the difference can reach 12 °C or higher because the built environment radiates the heat absorbed during the daytime [5].

When the surrounding temperature increases, it means not only air conditioning systems demand will increase but also their coefficient of performance will decrease, which results in more energy consumption to operate these systems. Additionally, urban heat islands are not only uncomfortably warm, but are also smoggier. This is due to smog composition from the photochemical reactions of air contaminants can result in increasing air temperatures. For example, [68] showed that in some cities, the incidence of smog increases 3% for each degree higher than 21 °C.

According to the U.S. EPA [5], five approaches that can attenuate the impact of the urban heat island were recommended. These approaches include:

(1) Growing more trees and green areas in places with a potential of the urban heat island. This helps lower surface and air temperatures by providing shade and cooling through evapotranspiration. Note that plants absorb water through their roots and emit it through their leaves. This movement of water is called transpiration. As well, evaporation also occurs from vegetation's surfaces and the surrounding medium. The transpiration and evaporation together are called evapotranspiration. Consequently, evapotranspiration cools the air by absorbing the heat from the surrounding air to evaporate water [13].

(2) Installing green roofs by increasing a vegetative layer on the rooftops. The vegetative layer provides shade. This aids a decrease in the temperatures of the roof surface and the surrounding air. Another benefit of installing green roofs is that they have high thermal mass that attenuates the variations of the temperature on buildings during the day, which can help improve building comfort and as well reduce peak energy demands. Note that thermal mass is the ability of a material to store heat. During peak temperature hours, a material with high thermal mass absorbs heat rather than transfer it to the living space. This keeps the interior of the home comfortable during peak temperature hours. At nighttime, the absorbed heat is released, helping the home to stay warm. Correct use of thermal mass can delay heat flow through the building envelope by as much as 10 to 12 h. This produces a warmer house at nighttime and during winter months and a cooler house during the daytime and summer months. As of July 2012, there were nearly 59,000 completed green home projects in the United States, most of which were single-family homes with detached garages [70].

(3) Installing cool/reflective/white roofs, which is the focus of this paper. These roofs use roofing materials (e.g., Membranes and coatings) with high short-wave solar reflectivity. Good roofing systems can insulate, reduce heat transference and help save electrical energy. This eventually reduces the carbon footprint due to the reduction of the dependency on fossil fuels. In the United States, different types of reflective roofs have been used for more than 20 years [13]. Several studies on reflective roofs have shown a reduction in roof surface temperatures and energy demand that is needed for buildings [14,15,36,39,55,71–73].

(4) Applying innovative practices. These cover a variety of development and preservation approaches to protect the environment [5]. Because of the importance of reflective and green technology for saving energy in buildings, a workshop [74] entitled "Green Technologies and Energy Efficiency (GTEE 2017)", was held on 26 April 2017, at the King Faisal University in Saudi Arabia. Among many goals of this workshop, two of these goals were to: (a) gather experts in green technologies and energy efficiency and initiate communication and cooperation channels within the framework of the Kingdom of Saudi Arabia (KSA) 2030 Vision, and (b) initiate green technologies and energy efficiency-related research activities by institutions and companies.

This section explored how the energy demand has increased drastically in the last 10 to 20 years and what are the main factors that affect energy consumption negatively. It clarified how roofing technologies can be part of the solution in energy savings in hot climate zones, for example.

## 5. Marketing Considerations and Cost Benefit

The marketing discussion in this article addresses factors involving cost-benefit, maintenance and how the technology is adopted. In terms of the cost-benefit, it has to be understood that emerging economies often have limited capital and material resources. These factors have to be accounted for when introducing a new technology such as white roofing materials. If the cost to adopt the technology outweighs the long-term operational savings of the white roof material, then there is little rationale for the technology to be supported by either government or private enterprise. Among some of the cost factors involved in the cost-benefit are those such as the upfront cost of the material, the maintenance cost associated with specific technology deployed, the long-term cost savings associated with a particular white roof application and the durability of the specific material selected [40]. If not all of these factors are considered, then the targeted emerging economy is less likely to consider any such technology adoption.

Still, as is often the case in emerging markets, the general population has certain reservations about any technology uptake that is accompanied by increased costs. Furthermore, the type of roof structures that predominates in a given market also affects the decision to deploy the technology as well. If the constituents in the emerging economy believe that the white roofing material selected is less durable and thus more costly in terms of replacement frequency, no amount of marketing convinces them to adopt it. The fact is that how this technology is marketed to the constituents in the emerging economy is ultimately the factor that leads to its adoption or denial. One aspect of the marketing message with respect to white roofing materials that must be emphasized is durability since temperature is a main factor in durability where black roofs typically experience higher temperature. Additionally, the associated benefits of cooler interior temperatures, lower cooling costs and, in general, improved visual appearance of white roof materials should form the nucleus of the marketing message. Increasingly, the most relevant themes then in white roofing technology relate to global climate change.

The two unique features of white roofing materials that make its marketing and advertising much more feasible in the open market involve their inherent energy efficiencies and enhancing the indoor air quality (i.e., increasing the comfort level). White roofing materials and cool roofs in general have two unique features that differentiate them from other roofing materials vis-à-vis energy-related factors. The two unique features that have been discussed from a technical perspective in other sections of this study include: (1) white roofing materials have a higher solar reflectance factor than do traditional roofing materials, and (2) white roofing materials or cool roofs have a greater capacity to release heat that has been absorbed into the material and is usually given as a ratio somewhere in between 0 to 1 [75]. From a technical perspective, these two points are extremely important as they essentially determine the overall energy efficiency of the roofing system. However, from a marketing perspective, these two points offer enormous leverage in the marketing collateral developed to sell and market these materials to the public, contractors, and civil governments and agencies.

It should be noted, of course, that not all cool roofs are comprised of white roofing materials. However, white roofing materials are typically the easiest method to capture these benefits of cool roofing technologies because they are inherently more solar reflective than other colors and usually has a higher level of thermal emittance (i.e., higher long-wave thermal emissivity), which helps radiate more of the absorbed heat by the rooftop back to the environment. Any type of product or service that is sold and marketed to the public, corporate entities and government agencies benefit from marketing messaging that accurately communicates what exact properties about the product or service that differentiate them from others in their respective markets or industries [76]. Given this, cool roof technologies and white roofing materials would both benefit from marketing messaging that accurately captures what exactly differentiates them from traditional roofing technologies. Thus, a marketing message that succinctly informs the targeted consumer that white roofing materials and cool roofs save an enormous amount of energy and operate much more efficiently due to their thermal emittance are two themes that immediately capture the consumers' attention due to global factors such as climate change, peak oil and environmental decay [77]. Hence, consumers, whether public or private, government or corporate, all have some level of interest in cost savings. In addition, when such cost-savings can simultaneously benefit the earth's environment, the consumer benefits through goodwill, public relations and corporate social responsibility-related factors.

Yet, developing an effective marketing plan along with associated marketing messages might be difficult when speaking of esoteric technologies such as roofing materials. The marketing message that is developed for the target markets has to make the rationale for adopting such roofing solutions palatable from a cost-benefit perspective but also from a public relations/social equity perspective. In this regard, the marketing collateral developed to sell and market white roofing materials and cool roof technologies should address some or all of the following elements in one fashion or another [77]:

1. Characterize the lifecycle expectations associated with the roofing technology selected
2. Determine for the consumer how well the roofing technology will survive over time
3. Inform the consumer how the material resists the phenomena such as mold, mildew and discoloration
4. The cost of repairing any damage to the roofing technology
5. The routine cost of maintenance and upkeep of the roofing technology
6. Identifying what the installation process requires in terms of time and cost
7. Who is installing the roofing technology and are the installers skilled in the selected materials
8. The flexibility of the roofing technology selected during extreme fluctuations of temperature.

It is important to spend the necessary effort to fully explore each of the eight factors above prior to the development of a marketing platform for white roofing materials and cool roof technologies in order to ensure that the resulting marketing plan comprehensively responds to consumer needs. As well, some of these marketing-related initiatives must be tailored to those markets where white roofing and cool roofing technologies are more inclined to be installed.

## 6. Cool Roofs Pros and Cons and Its Relationship with the Weather

In the design stage of roofing systems, the energy savings and risk of condensation are important. The moisture-related problems can lead to deterioration of roofing materials and affect negatively the thermal performance by means of reduction in overall thermal resistance and service life of the roofing systems. This can lead as well to mold growth in those systems and affect the indoor air quality (IAQ) and the occupants' health and comfort [14,15,76–82].

The roof surface reflectivity can affect the quantity of the absorbed short-wave energy of the roof. Since the short-wave solar radiation has the capacity during the summer and daytime to dry out the roof, the solar reflectivity becomes an important parameter for the selection of roofing materials [6,14,15,36,39,83]. The characteristic of the cool roofs is to maintain surface temperatures lower than those from the black roof (dark) because of its low short-wave solar absorption coefficient. This may lead to moisture-related problems in cold climate zones [84] as such observed for both black and cool roofs [14,15]. Several studies have investigated the moisture-related problems due to the colour of the roofing membrane in commercial buildings with a low slope such as in [85] research. Their study was done for cold climate with the focus on single-ply roofing systems with highly reflective materials, white TPO/PVC and black membranes, where they are attached mechanically on low slope roofs ($\leq$2:12). In this research study, hygrothermal simulations were conducted for a one-year period on those roofing systems for the following cities in the USA: Boston, Albany, Chicago, Cleveland and Detroit. Two cases were studied: (i) 10% short-wave solar reflectivity for black roofs, and (ii) 70% short-wave solar reflectivity for white roofs. Since those systems are mechanically attached to the metal deck and to account for the leakage due to the attachment, the vapour permeance of metal was taken at 0.75 perms. The results showed that even there is condensation below the membrane TPO/PVC in the winter, all roofs have dried out in the summer.

Hygrothermal simulations were conducted [86] for a 5-year period. They explored the moisture accumulation in different roofing systems (i.e., white with 80% short-wave solar reflectivity and black roofs with 12%) for different cities in the USA and Germany (Phoenix, AZ, Chicago, IL, Anchorage, AK, and Holzkirchen, Germany). The results of this study showed that the white roofs have more moisture accumulation than the black ones. Three roofing systems cases, dark, bright and shaded flat roofs, have been studied by [87] for their hygrothermal performance with initial construction moisture. In the summer, the results showed that the bright roofs have the lowest surface temperatures with a smaller drying potential than the two other roofing systems. The highest surface temperature and humidity fluctuations were shown in the dark roofs with high heat fluxes. The roof with shaded surfaces has shown as well a low temperature and drying potential.

Hygrothermal simulations on white and black Modified-Bitumen (MOD-BIT) roofing systems have been conducted by [14,15]) to evaluate their energy and moisture accumulation for different climate zones in North America based on their Heating-Degree-Days (HDD) such as Toronto (ON), Montreal (QC), St. John's (NL), Saskatoon (SK), Seattle (WA), Wilmington (AZ), and Phoenix (AZ). In the cities of St. John's and Saskatoon, the white roofs showed the highest moisture accumulation over time than the black ones which could lead to moisture damage. On the contrary, there is no risk for the black roofs. The simulation results for Toronto, Montreal, Seattle, Wilmington, and Phoenix, showed that the white roofs have a low risk of experiencing moisture damage. The yearly heating loads of the white roof were slightly higher than that of the black roof. Conversely, the yearly cooling loads of the black roof were much important than the white roof. Thus, buildings with white roofs in these locations are predicted to result in net yearly energy savings compared to buildings with black roofs.

Most recently, several researchers [14–16,72–88] conducted several studies to investigate the performance of cool and black roofs in terms of energy and moisture when they were subjected to different hot and humid climates in GCC countries. These studies covered several thicknesses of roof insulation material and surface solar reflectivity. These roofing systems showed no risk of condensation, thus, no risk of mold growth and roofs with high solar reflectivity showed significant energy savings. For the Eastern Province of Saudi Arabia and Kuwait City, respectively, the highest hourly surface temperatures of the black roofs were found to be 93.2 °C and 84.0 °C compared to 65.4 °C and 61.5 °C for a white roof at no cleaning condition, and 52.7 °C and 52.6 °C for white roofs at weekly cleaning conditions. The full results of that study that include material characterizations, installing guidelines, cleaning procedures, test results, three-dimensional numerical results, etc., are available in [69,70].

## 7. Application and Maintenance of Cool Roofs

The application process of white roofing material requires technical knowledge as does the maintenance processes involved. Yet, the maintenance processes involved in white roofing materials must also be known from a marketing perspective. This is because emerging economies typically have limited resources available. At a granular level, however, entities that utilize white roofing materials as a means to reduce the cooling energy load of a structure, reduce environmental heat build-up and as well seek to control costs have to consider the post-application phase of the material. As mentioned, the technical application of the white roof depends upon the form of white roofing material selected. Likewise, maintenance is also dependent upon what material is selected. In effect, cost factors must also be considered with respect to ongoing maintenance of white roofs in developing markets.

The exterior surface of the roofing system is exposed to dust/dirt, rain, snow, wind, cloud index, exterior temperature and relative humidity, etc. All these external conditions as well as the roofing systems specifications (components, dimensions, etc.) could affect the roofing system's thermal and hygrothermal performance. As shown in Figure 4, when solar radiation hits the surface of a roofing system, a portion of this energy is reflected and the other portion is absorbed. Due to this energy absorption, the roof's surface temperature increased, thus in the summer, the cooling energy load increased and in the winter, the heating energy load decreased. On the other hand, white roofing systems use surfaces with low short-wave solar absorption coefficient to show a significant portion of the incident short-wave solar radiation, and therefore, lowers the cooling energy load and as well the roof's surface temperature compared to black roofing systems.

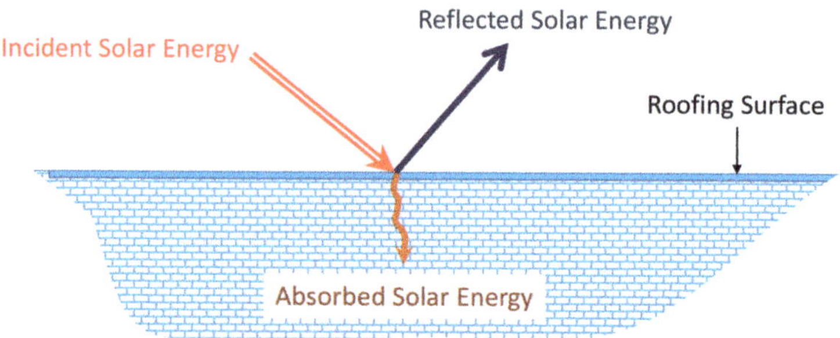


**Figure 4.** Schematic of roofing surface subjected to incident solar energy.

The surface solar reflectivity can be affected negatively by the accumulation of dust or dirt on those surfaces, which can result in increasing solar heat gains. Several researchers [29,30,33–35] have investigated the negative effect of the dust or dirt accumulation and weathering factors on the roofs' thermal properties. Additionally, several cleaning processes of dust or dirt on roofing systems' surfaces have been established to bring back roofing surfaces' solar reflectivity to its original value such as in Levinson et al. [34] studies on light-coloured roofing membranes' solar heat gain.

In the study [34], several roofing membranes with white or light gray polyvinyl chloride collected from different locations in the USA were tested. Black and inorganic carbon were found on the surface of the sample. These contaminants reduce the solar reflectivity of these membranes. To analyze the influence of several cleaning processes on the solar reflectivity values, the sample surfaces were firstly wiped to mimic the action of the wind action, then rinsed to mimic rain effect, and they were washed in the third step to simulate a homemade cleaning process using a phosphate-free dishwashing detergent. As a final step, all the surfaces were treated with sodium hydroxide and a mixture of sodium hypochlorite to mimic real cleaning processes. The outcomes of study [34] showed that after washing and rinsing processes, almost all the dirt deposited on the surface was removed except for thin layers of organic carbon and some isolated dark spots of biomass. Bleaching processes cleared these last two contaminants recovering the loss of solar reflectivity.

Akbari et al. [35] used an identical cleaning process established in [34] on unweathered (i.e., new materials) and weathered single ply roofing membranes from several North American sites. In this research study, 16 types of roofing materials were tested at Lawrence Berkeley National Laboratory (LBNL), following all the cleaning processes concerning the weathered samples surface treatment, and 25 other types were tested at the National Research Council of Canada, applying just wiping processes on the surface of the roofing material. The results showed that all types of cleaning recovered near 90% of their unweathered solar reflectivity and thus, showed their effectiveness.

To the best of our knowledge, most (if not all) previous studies related to characterizing the impact of dust or dirt accumulation on the rooftops and its impact on cool roofs' energy performance were conducted in non-dusty climates. As such, a new joint research study between the KSA's Jubail University College (JUC) and Kuwait Institute for Scientific Research (KISR), called "JUC-KISR project" was initiated to address this issue in a dusty climate. Several roofing materials with different emissivity are currently being tested and numerically modelled under the dusty and polluted climate of KSA's Jubail Industrial

City. The objectives of JUC-KISR project include: (i) motioning the short-wave solar reflectivity's variation of the tested materials, (ii) developing a technical guide for dust and dirt cleaning processes for the tested materials, and (iii) quantification the effect of dirt and dust accumulation of on the overall roofing systems' energy and moisture performance. The developed technical guide will be proposed later to International Organizations such as ASTM or ASHRAE to develop an international standard for the cleaning procedure of different reflective materials for use in residential and commercial buildings when they are subjected to dusty and non-dusty climates.

In the JUC-KISR project, one of the reflective coating materials was characterized and tested under the highly polluted and dusty weather of Jubail Industrial City. Thereafter, this reflective coating material was used in one of the most common roofing systems in GCC countries when it was exposed to Kuwait City and Eastern Province of Saudi Arabia's environmental conditions. For Saudi Eastern Province climates, Figure 5 shows that the yearly cooling load (EC, Y) of the conventional/black roof (1077 Wd/m2) decreased to 706 Wd/m$^2$ (i.e., a reduction in EC, Y by 53%) and 563 Wd/m$^2$ (i.e., a reduction in EC, Y by 91%) as a result of installing reflective coating material at no cleaning condition and weekly cleaning condition, respectively. Furthermore, as shown in Figure 6 for Kuwait City climates, the yearly cooling load of the black roof (1091 Wd/m$^2$) decreased to 730 Wd/m$^2$ (i.e., a reduction in EC, Y by 49%) and 592 Wd/m$^2$ (i.e., a reduction in EC, Y by 84%) as a result of installing reflective cooling material at no cleaning condition and weekly cleaning condition, respectively.

For the Eastern Province of Saudi Arabia and Kuwait City's environmental conditions, respectively, the highest hourly black roofs' surface temperatures were 93.2 °C and 84.0 °C compared to 65.4 °C and 61.5 °C for white roof at no cleaning condition, and 52.7 °C and 52.6 °C for white roofs at weekly cleaning conditions. The full results of the JUC-KISR project that include material characterizations, installing guidelines, cleaning procedures, test results, three-dimensional numerical results, etc., are available in [71,72].

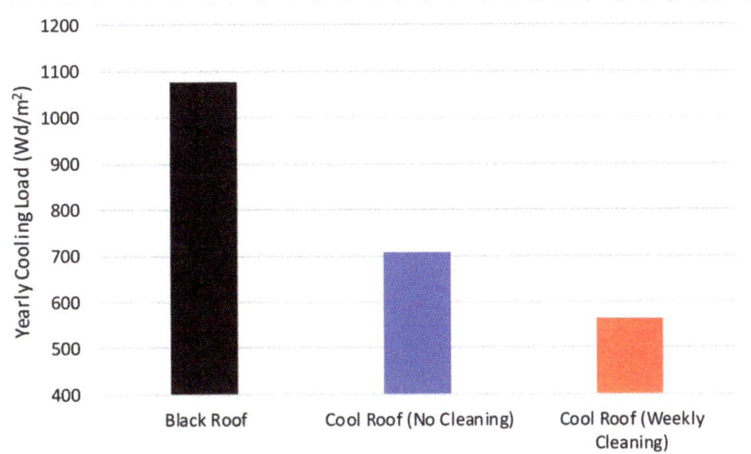


**Figure 5.** Comparison of the yearly cooling energy load of conventional/black roof and white roof based on no cleaning condition and weekly cleaning condition and subjected to Saudi Eastern Province climates [71,72].

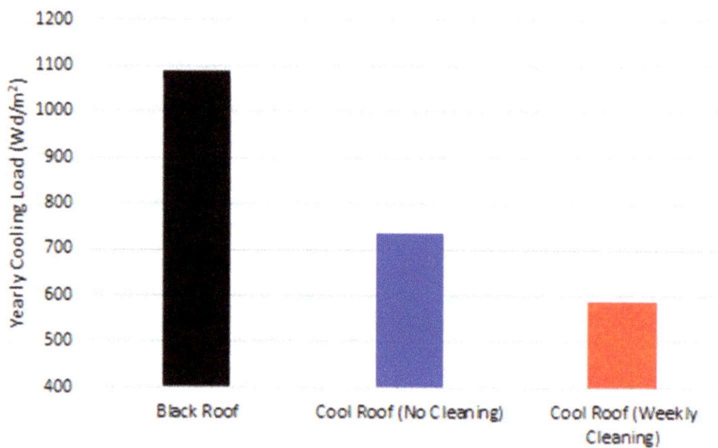


**Figure 6.** Comparison of the yearly cooling energy load of conventional/black roof and white roof based on no cleaning condition and weekly cleaning condition and subjected to Kuwait City climates [71,72].

In summary, a logical step towards achieving energy-efficient buildings in cities with extreme climatic conditions is to design roofing systems with simultaneous energy savings, no condensation and less risk of moisture problems. The main parameters affecting the energy and moisture performance of roofing systems are a type of roof, environmental conditions, amount of insulation and the solar reflectivity of the roofing surfaces. Several cleaning processes have been used to increase the solar reflectivity to reach approximately its initial value in case there is dust/dirt accumulation on the roofing surface.

## 8. Conclusions

This paper discussed in some depth the issue of white roofing materials and cool roofing technologies with a view to identifying how these technologies affect technology uptake and adoption. The technical aspects of what actually comprise white roofing solutions and cool roof technologies are explored in some depth. This paper reviewed most of the concepts of white roofing materials with the focus on their energy performance, cost-benefit, maintenance process and their impact on the emergent countries. This paper gave a retrospective as well on the concept of different roofing systems that exist and identified differences among them such as Green/White/Reflective/Cool roofs. The emerging economies concept was explored and a tight relation that exists between the reflective roofing materials and these emerging economies were established. The link between energy demand, energy consumption, and climate change through the production of greenhouse gas emissions is irrefutable. While developed countries typically have the resources necessary to both develop new construction designs and technologies to reduce energy consumption and emissions, emergent economies lack such resources. On average, overall energy consumption globally among all countries is expected to increase by more than 100% by the year 2050 but in emerging economies, this figure is believed to be more than 300% on average. Hence, the emergent economy has to deploy energy-efficient building designs and integrate energy-efficient building envelope materials at the outset, thus, they can reduce this drastic gap in expected energy usage growth moving forward. Specifically, in the design phase of new construction, passive energy-efficient solutions greatly reduce forward energy demands per structure. The design-build phase of construction in emergent economies can address issues that have significant positive effects on energy consumption throughout the life of the structure. This paper discussed as well in some depth the issue of white roofing materials and cool roofing technologies with a view to identify how these technologies affect technology uptake and adoption. The technical aspects of what actually comprise white roofing solutions and cool roof technologies are explored in some depth. The study focused on cool roofing systems subjected to hot climates such as that in Gulf Cooperation Council (GCC) countries. Factors such as a cool roof's solar reflectance and its overall thermal emittance were reviewed. These two factors in particular are extremely critical from a technical perspective as well as a marketing-related perspective. This is because these two factors (solar reflectance and thermal emittance) generally determine the overall energy efficiency of the roofing technology selected.

Yet, these two factors also work to determine the character of a cool roof technology's marketing messaging, platform and purchase rationale because they directly relate to global issues such as climate change, energy savings and environmental responsibility, and so forth. In essence, while the technological characteristics of white roofing material and cool roof technologies are critical from performance-based perspectives, and a technology adoption perspective, these factors must be related to practical outcomes from a consumer-related vantage point.

The adoption of white roof tiles or other non-white roof colours with high solar reflectance in countries subjected to tropical climates can be increased by spreading aware-ness among house owners, through various stakeholders, that emphasizes the cost savings that the house owners can enjoy, which can potentially, from the stance of the house owners, outweigh all of the other disadvantages. Thus, further research on the development of solar-reflective paint or coating products that can significantly increase the solar reflectance values of non-white roof tiles is essential to overcome issues related to maintenance diffi-culties and lack of preference among house buyers towards white roof tiles.

This review study has shown the importance in achieving energy-efficient building in cities with extreme climatic conditions by designing roofing systems that considered simultaneously the energy savings with less risk of moisture problems. The main parame-ters affecting the energy and moisture performance of roofing systems are types of roof, environmental conditions, amount of insulation and the solar reflectivity of the roofing surfaces. Several cleaning processes have been identified to increase the solar reflectivity to reach approximately its initial value in case there is dust/dirt accumulation on the roofing surface.

**Author Contributions:** Conceptualization, F.A.F. and W.M., Methodology, F.A.F. and W.M., Analy-sis, F.A.F., W.M. and M.M.A. investigation, F.A.F. and W.M., Resources, F.A.F., W.M. and M.M.A., Writing—original draft preparation, F.A.F. and W.M., Writing—Review and editing, F.A.F., W.M. and M.M.A. Project administration, F.A.F. and W.M. All authors have read and agreed to the pub-lished version of the manuscript.

**Funding:** This research received no external funding.

**Institutional Review Board Statement:** Not applicable.

**Informed Consent Statement:** Not applicable.

**Data Availability Statement:** Not applicable.

**Conflicts of Interest:** The authors declare no conflict of interest.

## References

1. Peterson, T.C.; Gallo, K.P.; Lawrimore, J.; Owen, T.W.; Huang, A.; McKittrick, D.A. Global rural temperature trends. *Geophys. Res. Lett.* **1999**, *26*, 329–332. [CrossRef]
2. Huang, Q.; Lu, Y. The effect of urban heat island on climate warming in the Yangtze river delta urban agglomeration in China. *Int. J. Environ. Res. Public Health* **2015**, *12*, 8773–8789. [CrossRef]
3. Sachindra, D.A.; Ng, A.W.M.; Muthukumaran, S.; Perera, B.J.C. Impact of climate change on urban heat island effect and extreme temperatures: A case study. *Q. J. R. Meteorol. Soc.* **2016**, *142*, 172–186. [CrossRef]
4. Oleson, K.W.; Bonan, G.B.; Feddema, J. Effects of white roofs on urban temperature in a global climate model. *Geophys. Res. Lett.* **2010**, *37*, 3. [CrossRef]

5.  U.S. Environmental Protection Agency (EPA). Keeping Your Cool: How Communities Can Reduce the Heat Island Effect. Available online: https://www.epa.gov/sites/production/files/2016-09/documents/heat_island_4-page_brochure_508_120413.pdf (accessed on 1 June 2019).

6.  Saber, H.H.; Hajiah, A.E.; Alshehri, S.; Hussain, H.J. Investigating the Effect of Dust Accumulation on Solar Reflectivity of Coating Materials for Cool Roof Applications. *Energies* **2021**, *14*, 445. [CrossRef]

7.  Aflaki, A.; Mirnezhad, M.; Ghaffarianhoseini, A.; Ghaffarianhoseini, A.; Omrany, H.; Wang, Z.-H.; Akbari, H. Urban Heat Island Mitigation Strategies: A State-of-the-Art Review on Kuala Lumpur, Singapore and Hong Kong. *Cities* **2017**, *62*, 131–145. [CrossRef]

8.  Huang, M.; Cui, P.; He, X. Study of the cooling effects of urban green space in Harbin in terms of reducing the heat island effect. *Sustainability* **2018**, *10*, 1101. [CrossRef]

9.  Nuruzzaman, M. Urban heat island: Causes, effects and mitigation measures—A review. *Int. J. Environ. Monit. Anal.* **2015**, *3*, 67–73. [CrossRef]

10. Sebastiani, A.; Marando, F.; Manes, F. Mismatch of regulating ecosystem services for sustainable urban planning: PM10 removal and urban heat island effect mitigation in the municipality of Rome (Italy). *Urban For. Urban Green.* **2021**, *57*, 126938. [CrossRef]

11. Mohajerani, A.; Bakaric, J.; Jeffrey-Bailey, T. The urban heat island effect, its causes, and mitigation, with reference to the thermal properties of asphalt concrete. *J. Environ. Manag.* **2017**, *197*, 522–538. [CrossRef]

12. Lee, J.S.; Kim, J.T.; Lee, M.G. Mitigation of urban heat island effect and greenroofs. *Indoor Built Environ.* **2014**, *23*, 62–69. [CrossRef]

13. U.S. Environmental Protection Agency. Cool Roofs. Reducing Urban Heat Islands: Compendium of Strategies. 2008. Available online: https://www.epa.gov/heat-islands/heat-island-compendium (accessed on 1 January 2019).

14. Saber, H.H.; Maref, W.; Hajiah, A.E. Hygrothermal Performance of Cool Roofs Subjected to Saudi Climates. *J. Front. Energy Res.* **2019**, *7*, 39. [CrossRef]

15. Saber, H.H.; Maref, W. Energy Performance of Cool Roofs Followed by Development of Practical Design Tool. *J. Front. Energy Res.* **2019**, *7*, 122. [CrossRef]

16. Saber, H.H. Hygrothermal Performance of Cool Roofs with Reflective Coating Material Subjected to Hot, Humid and Dusty Climate. *J. Build. Phys.* **2021**. [CrossRef]

17. Saber, H.H. Experimental Characterization of Reflective Coating Material for Cool Roofs in Hot, Humid and Dusty Climate. *Energy Build.* **2021**, *242*, 110993. [CrossRef]

18. Al-Obaidi, K.M.; Ismail, M.; Abdul Rahman, A.M. Investigation of passive design techniques for pitched roof systems in the tropical region. *Mod. Appl. Sci.* **2014**, *8*, 182–191. [CrossRef]

19. Miranville, F.; Boyer, H.; Mara, T.; Garde, F. On the thermal behaviour of roof-mounted radiant barriers under tropical and humid climatic conditions: Modelling and empirical validation. *Energy Build.* **2003**, *35*, 997–1008. [CrossRef]

20. Lee, S.W.; Lim, C.H.; Chan, S.A.; Von, K.L. Techno-economic evaluation of roof thermal insulation for a hypermarket in equatorial climate: Malaysia. *Sustain. Cities Soc.* **2017**, *35*, 209–223. [CrossRef]

21. Vijaykumar, K.C.K.; Srinivasan, P.S.S.; Dhandapani, S. A performance of hollow tiles clay (HTC) laid reinforced cement concrete (RCC) roof for tropical summer climates. *Energy Build.* **2007**, *39*, 886–892. [CrossRef]

22. Ferrari, A.; Kubilay, A.; Derome, D.; Carmeliet, J. Design of smart wetting of building materials as evaporative cooling measure for improving the urban climate during heat Waves. *E3S Web Conf.* **2020**, 03001. [CrossRef]

23. Berardi, U.; Ghaffarian Hoseini, A.H. State-of-the-art analysis of the environmental benefits of green roofs. *Appl. Energy* **2014**, *115*, 411–428. [CrossRef]

24. Berardi, U.; Gallardo, A.A. Properties of concretes enhanced with phase change materials for building applications. *Energy Build.* **2019**, *199*, 402–414. [CrossRef]

25. Hernández-Pérez, I.; Xam´an, J.; Macías-Melo, E.V.; Aguilar-Castro, K.M.; Zavala- Guill'en, I.; Hernandez-L'opez, I.; Sim'a, E. Experimental thermal evaluation of building roofs with conventional and reflective coatings. *Energy Build.* **2018**, *158*, 569–579. [CrossRef]

26. Pisello, A.L. State of the art on the development of cool coatings for buildings and cities. *Sol. Energy* **2017**, *144*, 660–680. [CrossRef]

27. Hernández-Pérez, I. Influence of Traditional and Solar Reflective Coatings on the Heat Transfer of Building Roofs in Mexico. *Appl. Sci.* **2021**, *11*, 3263. [CrossRef]

28. Farhan, S.A.; Ismail, F.I.; Kiwan, O.; Shafiq, N.; Zain-Ahmed, A.; Husna, N.; Hamid, A.I.A. Effect of Roof Tile Colour on Heat Conduction Transfer, Roof-Top Surface Temperature and Cooling Load in Modern Residential Buildings under the Tropical Climate of Malaysia. *Sustainability* **2021**, *13*, 4665. [CrossRef]

29. Berdahl, P.; Akbari, H.; Levinson, R.; Miller, W.A. Weathering of roofing materials—An overview. *Constr. Build. Mater.* **2008**, *22*, 423–433. [CrossRef]

30. Suehrcke, H.; Peterson, E.L.; Selby, N. Effect of roof solar reflectance on the building heat gain in a hot climate. *Energy Build.* **2008**, *40*, 2224–2235. [CrossRef]

31. Mani, M.; Pillai, R. Impact of dust on solar photovoltaic (PV) performance: Research status, challenges and recommendations. *Renew. Sustain. Energy Rev.* **2010**, *14*, 3124–3131. [CrossRef]

32. Sulaimana, S.A.; Singhb, A.K.; Mokhtara, M.M.M.; Bou-Rabeec, M.A. Influence of dirt accumulation on performance of pv panels. *Energy Procedia* **2014**, *50*, 50–56. [CrossRef]

33. Algarni, S.; Nutter, D. Influence of dust accumulation on building roof thermal performance and radiant heat gain in hot-dry climates. *Energy Build.* **2015**, *104*, 181–190. [CrossRef]
34. Levinson, R.; Berdahl, P.; Asefawberhe, A.; Akbari, H. Effects of soiling and cleaning on the reflectance and solar heat gain of a light-colored roofing membrane. *Atmos. Environ.* **2005**, *39*, 7807–7824. [CrossRef]
35. Akbari, H.; Berhe, A.; Levinson, R.; Graveline, S.; Foley, K.; Delgado, A.H.; Paroli, R.M. Aging and Weathering of Cool Roofing Membranes. In Proceedings of the First International Conference on Passive and Low Energy Cooling for the Built Environment, Athens, Greece, 17 May 2005.
36. Saber, H.H.; Swinton, M.C.; Kalinger, P.; Paroli, R.M. Long-term hygrothermal performance of white and black roofs in North American climates. *Build. Environ.* **2012**, *50*, 141–154. [CrossRef]
37. Alkhalifa, A.H.; Al-Homaidan, A.A.; Shehata, A.I.; Al-Khamis, H.H.; Al-Ghanayem, A.A.; Ibrahim, A.S.S. Brown macroalgae as bio-indicators for heavy metals pollution of Al-Jubail coastal area of Saudi Arabia. *Afr. J. Biotechnol.* **2012**, *11*, 15888–15895.
38. Al-A'ama, M.S.; Nakhla, G.F. Wastewater reuse in Jubail, Saudi Arabia. *Water Res.* **1995**, *29*, 1579–1584. [CrossRef]
39. Saber, H.H.; Swinton, M.C.; Kalinger, P.; Paroli, R.M. Hygrothermal simulations of cool reflective and conventional roofs. In Proceedings of the 2011 NRCA International Roofing Symposium, Emerging Technologies and Roof System Performance, Washington, DC, USA, 7–9 September 2011.
40. Alchapar, N.L.; Correa, E.N. Aging of roof coatings. Solar reflectance stability according to their morphological characteristics. *Constr. Build. Mater.* **2016**, *102*, 297–305. [CrossRef]
41. Muscio, A. The Solar Reflectance Index as a Tool to Forecast the Heat Released to the Urban Environment: Potentiality and Assessment Issues. *Climate* **2018**, *6*, 12. [CrossRef]
42. Ziolo, M.; Jednak, S.; Savic, G.; Kragulj, D. Link between energy efficiency and sustainable economic and financial development in OECD countries. *Energies* **2020**, *13*, 5898. [CrossRef]
43. *Technology Roadmap: Energy Efficient Building Envelopes*; International Energy Agency: Paris, France, 2013; Volume 7, pp. 1–68.
44. Ochedi, E.T.; Taki, A.; Painter, B. Low cost approach to energy efficient buildings in Nigeria: A review of passive Design Options. In Proceedings of the 21st Century Human Habitat: Issues, Sustainability and Development, Akure, Nigeria, 21–24 March 2016; Available online: https://dora.dmu.ac.uk/handle/2086/11876 (accessed on 1 June 2021).
45. Tsafos, N. *Energy Transitions in Emerging Economies: What Success Looks like and How to Replicate It*; CSIS: Jakarta, Indonesia, 2020; Volume 12, pp. 1–9.
46. Santamouris, M. *Minimizing Energy Consumption, Energy Poverty and Global and Local Climate Change in the Built Environment: Innovating to Zero: Causalities and Impacts in a Zero Concept World*; Elsevier: Amsterdam, The Netherlands, 2018.
47. Feng, W.; Zhang, Q.; Ji, H.; Wang, R.; Zhou, N.; Ye, Q.; Lau, S.S.Y. A review of net zero energy buildings in hot and humid climates: Experience learned from 34 case study buildings. *Renew. Sustain. Energy Rev.* **2019**, *114*, 109303. [CrossRef]
48. Al-Homoud, D.M.S. Performance characteristics and practical applications of common building thermal insulation materials. *Build. Environ.* **2005**, *40*, 353–366. [CrossRef]
49. Solecki, W.D.; Rosenzweig, C.; Parshall, L.; Pope, G.; Clark, M.; Cox, J.; Wiencke, M. Mitigation of the heat island effect in urban New Jersey. *Glob. Environ. Chang. Part B Environ. Hazards* **2005**, *6*, 39–49. [CrossRef]
50. Green Public Procurement Thermal Insulation Technical Background Report. 2010. Available online: https://ec.europa.eu/environment/gpp/pdf/thermal_insulation_GPP_%20background_report.pdf (accessed on 1 June 2021).
51. Krarti, M.; Hajiah, A. Analysis of impact of daylight time savings on energy use of buildings in Kuwait. *Energy Policy* **2011**, *39*, 2319–2329. [CrossRef]
52. Zirkelbach, D.; Schafaczek, B.; Künzel, H. Long-term hygrothermal performance of green roofs. In Proceedings of the Eleventh International Conference on Thermal Performance of the Exterior Envelopes of Whole Buildings XI, Clearwater, FL, USA, 4–8 December 2016.
53. Ismail, A.; Samad, M.H.A.; Rahman, A.M.A. The investigation of green roof and white roof cooling potential on single storey residential building in the Malaysian climate. *Proc. World Acad. Sci. Eng. Technol.* **2011**, *76*, 129–137.
54. Jo, J.H.; Carlson, J.; Golden, J.S.; Bryan, H. Sustainable urban energy: Development of a mesoscale assessment model for solar reflective roof technologies. *Energy Policy* **2010**, *38*, 7951–7959. [CrossRef]
55. Levinson, R.; Akbari, H.; Berdahl, P.; Wood, K.; Skilton, W.; Petersheim, J. A novel technique for the production of cool colored concrete tile and asphalt shingle roofing products. *Sol. Energy Mater. Sol. Cells* **2010**, *94*, 946–954. [CrossRef]
56. Xu, T.; Sathaye, J.; Akbari, H.; Garg, V.; Tetali, S. Quantifying the direct benefits of cool roofs in an urban setting: Reduced cooling energy use and lowered greenhouse gas emissions. *Build. Environ.* **2012**, *48*, 1–6. [CrossRef]
57. Vrachopoulos, M.; Koukou, M.; Stavlas, D.; Stamatopoulos, V.; Gonidis, A.; Kravvaritis, E. Testing reflective insulation for improvement of buildings energy efficiency. *Open Eng.* **2012**, *2*, 83–90. [CrossRef]
58. Akbari, H.; Konopacki, S.; Pomerantz, M. Cooling energy savings potential of reflective roofs for residential and commercial buildings in the United States. *Energy* **1999**, *24*, 391–407. [CrossRef]
59. Akbari, H.; Konopacki, S.; Parker, D. Updates on revision to ASHRAE standard 90.2: Including roof reflectivity for residential buildings. *Proc. ACEEE Summer Study Energy Effic. Build.* **2000**, *1*, 11–111.
60. Levinson, R.; Akbari, H. Potential benefits of cool roofs on commercial buildings: Conserving energy, saving money, and reducing emission of greenhouse gases and air pollutants. *Energy Effic.* **2010**, *3*, 53–109. [CrossRef]

61. Ray, S.; Glicksman, L. Potential energy savings of various roof technologies. In Proceedings of the Eleventh International Conference on Thermal Performance of the Exterior Envelopes of Whole Buildings XI, Clearwater, FL, USA, 4–8 December 2016.

62. Maref, W. For Better Performance of the Building Envelope: Improvement and Complexity, ASHRAE Virtual Conference, Seminar 67—Back to the Future on High Efficiency Design and Operation. In Proceedings of the ASHRAE Annual Meeting, Austin, TX, USA, 30 June 2020.

63. Natural Resources Canada (NRCan). *Energy Efficiency Trend in Canada de 1990 à 2013*; Cat. No. M141-1E-PDF; Natural Resources Canada: Ottawa, ON, Canada, 2016. Available online: https://www.nrcan.gc.ca/energy/publications/19030 (accessed on 1 June 2019).

64. International Energy Agency (IEA). *World Energy Outlook Special Report: Energy and Climate Change*; International Energy Agency: Paris, France, 2015.

65. Mchugh, B.; Petrick, R. Chicago's green and garden roofing codes and technology. In Proceedings of the 2011 international Roofing Symposium, Washington, DC, USA, 7–9 September 2011.

66. Durhman, A.; Collins, M.; Mcgillis, W.R. Utilizing green technology and research to assess green roofing benefits. In Proceedings of the 2011 International Roofing Symposium, Washington, DC, USA, 7–9 September 2011.

67. Bentz, S.P. Decision-making process for green options in reroofing. In Proceedings of the 2011 International Roofing Symposium, Washington, DC, USA, 7–9 September 2011.

68. Peltier, R.; Lee, S.; Hennigan, C.; Arhami, E. Urban Meteorology. 2018. Available online: https://slideplayer.com/slide/7880769/ (accessed on 1 June 2021).

69. Li, H. *Evaluation of Cool Pavement Strategies for Heat Island Mitigation [Improving Outdoor Thermal Environment in Hot Climates through Cool Pavement Design Strategies]*; Final Research Report [D05-4]; Institute of transportation studies, Department of Civil Environmental Engineeing, Unversity of California: Davis, CA, USA, 2012; Available online: https://escholarship.org/content/qt6mr4k9t1/qt6mr4k9t1_noSplash_9ab61b14762e84b7d61994b7aa5e0d1b.pdf?t=pyhuvm (accessed on 1 June 2021).

70. Green Roofs. Reducing Urban Heat Islands: Compendium of Strategies: Heat IslandReduction Activities. 2018. Available online: https://healthyplacesindex.org/wp-content/uploads/2018/01/epa_heat_island_reduction_activities.pdf. (accessed on 1 June 2021).

71. Saber, H.H.; Alshehri, S.A.; Alnofaie, M.; Alghamdi, A.; Alraghi, S.; Hajiah, A.E. Energy Savings Due to Using Reflective Roofing Materials in Buildings of Kuwait and Saudi Arabia. Client Report—Part II. Unpublished work. 2020.

72. Saber, H.H.; Hajiah, A.E. 3D Numerical Modeling for Assessing the Energy Performance of Single- and Two-Zone Buildings with and without Phase Change Materials. In Proceedings of the Gulf Conference on Sustainable Built Environment, Kuwait City, Kuwait, 10–13 March 2019; pp. 10–13.

73. Hajiah, A.E.; Saber, H.H. Long-Term Energy and Moisture Performance of Reflective and Non-Reflective Roofing Systems with and without Phase Change Materials under Kuwaiti Climates. In Proceedings of the Gulf Conference on Sustainable Built Environment, Kuwait City, Kuwait, 10–13 March 2019; pp. 10–13.

74. Green Technologies and Energy Efficiency (GTEE 2017) workshop at the King Faisal University in Saudi Arabia, 26 April 2017. Available online: https://img0cf.b8cdn.com/images/course/35/68573835_1594764313.pdf (accessed on 1 June 2021).

75. Pisello, A.L.; Castaldo, V.L.; Pignatta, G.; Cotana, F.; Santamouris, M. Experimental in-lab and in-field analysis of waterproof membranes for cool roof application and urban heat island mitigation. *Energy Build.* **2016**, *114*, 180–190. [CrossRef]

76. Miller, W.; Crompton, G.; Bell, J. Analysis of Cool Roof Coatings for Residential Demand Side Management in Tropical Australia. *Energies* **2015**, *8*, 5303–5318. [CrossRef]

77. William, R.; Goodwell, A.; Richardson, M.; Le, P.V.; Kumar, P.; Stillwell, A.S. An environmental cost-benefit analysis of alternative green roofing strategies. *Ecol. Eng.* **2016**, *95*, 1–9. [CrossRef]

78. Saber, H.H.; Maref, W.; Abdulghani, K. *Properties and Position of Materials in the Building Envelope for Housing and Small Buildings*; National Research Council of Canada: Ottawa, ON, Canada, 2014.

79. Saber, H.H.; Maref, W. Risk of condensation and mould growth in wood-frame wall systems with different exterior insulations. In Proceedings of the Building Enclosure Science & Technology Conference (BEST4 Conference), Kansas City, MO, USA, 16 April 2015.

80. Lacasse, M.A.; Saber, H.H.; Maref, W.; Ganapathy, G.; Plescia, S.; Parekh, A. Field Evaluation of Thermal and Moisture Response of Highly Insulated Wood-Frame Walls. In Proceedings of the 13th International Conference on Thermal Performance of the Exterior Envelopes of Whole, Buildings XIII, Clearwater, FL, USA, 4–8 December 2016.

81. Saber, H.H.; Lacasse, M.A.; Ganapathy, G.; Plescia, S.; Parekh, A. Risk of condensation and mould growth in highly insulated wood-frame walls. In Proceedings of the 13th International Conference on Thermal Performance of the Exterior Envelopes of Whole, Buildings XIII, Clearwater, FL, USA, 4–8 December 2016.

82. Saber, H.H.; Lacasse, M.A.; Moore, T.V. Hygrothermal performance assessment of stucco-clad wood frame walls having vented and ventilated drainage cavities. In *Advances in Hygrothermal Performance of Building Envelopes: Materials, Systems and Simulations, ASTM STP1599*; ASTM International: West Conshohocken, PA, USA, 2017; pp. 198–231.

83. Brehob, E.; Desjarlais, A.; Atchley, J. Effectiveness of cool roof coatings with ceramic particles. In Proceedings of the 2011 International Roofing Symposium, Washington, DC, USA, 7–9 September 2011.

84. Urban, B.; Roth., K. Guidelines for Selecting Cool Roofs. Building Technologies Program; 2010. Available online: https://www.energy.gov/sites/prod/files/2013/10/f3/coolroofguide.pdf (accessed on 1 June 2021).

85.  Ennis, M.; Kehrer, M. The effects of roof membrane color on moisture accumulation in low-slope commercial roof systems. In Proceedings of the 2011 international roofing symposium, Oak Ridge, TN, USA, 1 January 2011.
86.  Bludau, C.; Zirkelbach, D.; Kuenzel, H.M. Condensation problems in cool roofs. *Interface J. RCI* **2009**, *XXVII*, 11–16.
87.  Bludau, C.; Künzel, H.M.; Zirkelbach, D. Hygrothermal performance of flat roofs with construction moisture. In Proceedings of the Eleventh International Conference on Thermal Performance of the Exterior Envelopes of Whole Buildings, Clearwater, FL, USA, 4–8 December 2016.
88.  Saber, H.H.; Alshehri, S.A.; Alnofaie, M.; Alghamdi, A.; Alraghi, S.; Hajiah, A.E. *Energy Savings Due to Using Reflective Roofing Materials in Buildings of Kuwait and Saudi Arabia; Internal and Client Report (Confidenti—Part I)*; Kuwait Institute for Scientific Research (KISR): Kuwait City, Kuwait, 2019; Volume 67.

 *sustainability*

*Article*

# Investigation on Subjects' Seasonal Perception and Adaptive Actions in Naturally Ventilated Hostel Dormitories in the Composite Climate Zone of India

Sanjay Kumar [1], Manoj Kumar Singh [2,3,*], Nedhal Al-Tamimi [4], Badr S. Alotaibi [4] and Mohammed Awad Abuhussain [4]

[1] Mechanical Engineering Department, Dr. B R Ambedkar National Institute of Technology, Jalandhar 144011, Punjab, India; sanjay@nitj.ac.in
[2] National Institute of Technology Arunachal Pradesh, Jote, Papum Pare 791123, Arunachal Pradesh, India
[3] Faculty of Civil and Geodetic Engineering, University of Ljubljana, Jamova Cesta 2, 1000 Ljubljana, Slovenia
[4] Architectural Engineering Department, College of Engineering, Najran University, Najran 66462, Saudi Arabia; naaltamimi@nu.edu.sa (N.A.-T.); bsalotaibi@nu.edu.sa (B.S.A.); maabuhussain@nu.edu.sa (M.A.A.)
* Correspondence: mksinghtu@gmail.com

**Abstract:** A seasonal adaptive thermal comfort study was done on university students in naturally ventilated dormitories in the composite climate zone of India. A total of 1462 responses were collected from the students during the field study spread over the autumn, winter, spring, and summer seasons of the academic year for 2018 and 2019. A "Right Here Right Now" type of surveying method was adopted, and the indoor thermal parameters were recorded simultaneously using high-grade instruments. The subjects' mean thermal sensation (TS) was skewed towards a slightly cool feeling for the combined data. Most occupants preferred a cooler thermal environment during the summer season, while hostel residents desired a warmer temperature during autumn, winter, and spring seasons. During the summer season, the PMV−PPD model overestimated the subjects' actual thermal sensation, while it underestimated the their thermal sensation in the winter season. The mean comfort temperature $T_{comf}$ was observed to be close to 27.1 ($\pm 4.65\ ^{\circ}$C) for the pooled data. Mean clo values of about 0.57 ($\pm 0.25$), 0.98 ($\pm 0.12$), 0.45 ($\pm 0.27$), and 0.36 ($\pm 0.11$) were recorded during the autumn, winter, spring, and summer seasons, respectively. Furthermore, switching on ceiling fans and opening doors and windows improved occupants' thermal satisfaction during different seasons. The study results show the effective use of environmental controls and the role of thermal adaptation in enhancing the subjects/overall thermal satisfaction in the composite climate of India.

**Keywords:** field surveys; thermal perceptions; adaptive actions; hostel dormitories; composite climate of India

**Citation:** Kumar, S.; Singh, M.K.; Al-Tamimi, N.; Alotaibi, B.S.; Abuhussain, M.A. Investigation on Subjects' Seasonal Perception and Adaptive Actions in Naturally Ventilated Hostel Dormitories in the Composite Climate Zone of India. *Sustainability* **2022**, *14*, 4997. https://doi.org/10.3390/su14094997

Academic Editor: Dino Musmarra

Received: 13 March 2022
Accepted: 18 April 2022
Published: 21 April 2022

**Publisher's Note:** MDPI stays neutral with regard to jurisdictional claims in published maps and institutional affiliations.



## 1. Introduction

In the last two decades, an increasing trend in the building sector's energy consumption has been observed worldwide, contributing to more than 40% of the total global greenhouse emissions. With the ever-increasing expectation of the indoor thermal environment, this is projected to increase further in the near future [1]. In India, the building sector is the second-largest contributor to greenhouse emissions and overall energy consumption [2]. In the context of India, many educational buildings exist and are emerging as the leading sector in overall building energy consumption [3]. A large chunk of the energy is consumed by buildings to restore thermal comfort. According to ASHRAE Standard 55 and ISO 7730, thermal comfort is the condition of the subject's mind that expresses thermal satisfaction with the thermal environment surrounding them. A questionnaire-based subjective evaluation methodology is generally adopted to evaluate the thermal comfort conditions in different building types [4,5]. International standards such as ASHRAE Standard 55 [4] and

ISO 7730 [5] are widely used to assess thermal comfort conditions in built environments. Primarily, ASHRAE Standard 55 [4] and ISO 7730 [5] are based on Fanger's heat balance model (PMV/PPD) [6]. Over the last two and half decades, numerous field studies carried out by researchers have shown that the PMV/PPD approach fails to capture the entire spectrum of parameters associated with the psychological, physiological, and socio-cultural aspects. These parameters play a vital role in the adaptation mechanism the occupants of different built environments go through, thus impacting their thermal comfort perception and expectations [7–12]. The above-mentioned parameters of thermal comfort become part of the study when the adaptive thermal comfort principle is applied to record the thermal response of the subjects in the field surveys. Adaptive thermal comfort through various field studies has shown massive potential in evaluating the indoor thermal environment and minimizing buildings' energy consumption without compromising occupant comfort [13].

Recently, researchers have reported several comprehensive reviews [14–17] helping to understand the causes of individual differences and thermal adaptation mechanisms of human subjects in different buildings, climates, ventilation strategies, and other contextual factors of thermal comfort. With the availability of the most extensive thermal comfort data in the ASHRAE Global Thermal comfort database, researchers have pointed out that thermal adaptation may significantly define the thermal acceptability ranges in different buildings and climates [18–20]. In India, research on thermal comfort gained popularity after the pioneering work of Nicol [21], and Sharma and Ali [22] in the late 1980s. After this, many researchers started evaluating the thermal adaptation mechanism of Indian inhabitants, considering the country's diverse climatic and geographical diversity. For instance, Indraganti [23] conducted a series of field monitoring in multi-story residential apartments in Southern India and stated that regional static comfort limits were not applicable to define the thermal comfort needs of residents in the hot and humid climate of Southern India. The author also analyzed adaptive strategies to achieve thermal comfort, such as windows, balconies, the use of external doors, fans, and clothing adaptations [24]. Singh et al. [9,25] also reported results related to the thermal adaptation of residents in vernacular houses in the north-east part of India. The authors argued that an adaptive approach to thermal comfort is more suitable for analyzing the thermal adaptation of people under different climatic zones of this region. The National Building Code of India [26] has adopted the India Model for Adaptive Comfort (IMAC) [27] to define the 80% and 90% thermal acceptability ranges of thermal comfort for naturally ventilated and mixed-mode buildings in different climatic zones of India. Since then, adopting similar approaches and methods, researchers in India have conducted field studies considering different building types, i.e., classrooms [28,29], university buildings [30], offices [27,31,32], residential buildings [23,25], hostel dormitories [33–35], and special metabolic activity spaces [36]. Researchers have concluded that thermal comfort is a complex phenomenon and depends on the different adaptation mechanisms and contextual factors inculcated in the adaptive approach to thermal comfort.

Educational buildings and the associated built environment play a significant role in students' learning and wellbeing [3,17]. Students in university generally fall in the age group of 18–26 years old and spend a lot of time in hostel dormitories for their undergraduate or postgraduate studies. Thus, emphasis should be placed on designing and constructing hostel dormitories so that they provide a conducive and quality thermal environment to stimulate the learning process [29,30], without compromising students' needs of comfort and health [37]. Moreover, the indoor thermal environment and air quality in hostels are very different from other building types because of significant differences in the age groups, occupancy patterns, behaviours, and activities carried out by students [34,37]. Considering the importance of a quality built environment in students' learning process, many researchers have carried out field studies to investigate the thermal performance of educational buildings in different climates and their relation to occupants' overall comfort requirements. For instance, Dalhan et al. [37] conducted a research study on thermal comfort in three high-rise hostel buildings in Malaysia's hot and humid climate.

They found that the mean neutral temperature of hostel buildings was close to 28.8 °C. Lai [38] used gap theory for a post-occupancy evaluation in order to explore the role of six parameters, namely visual comfort, acoustic comfort, fire safety, hygiene, and information and communication technology, in students' expectation and satisfaction. In India, Dhaka et al. [33] carried out a field study in a hostel building in a composite climate during the peak summer season in Jaipur City. The study found a higher comfort bandwidth of approximately 7.9 °C and a mean comfort temperature of 30.2 °C. Kumar et al. [34,35] carried out a questionnaire-based field study in naturally ventilated hostel buildings in Jaipur and Jalandhar City during the autumn and winter seasons, and compared the results. The study found that students living in the hostel had different comfort expectations that those in office or residential buildings. Mean Griffiths comfort temperatures ($T_c$) of 30.4 °C and 29.7 °C were observed for Jaipur and Jalandhar City, respectively. The data analysis also showed the extension of comfort boundaries by 1.8 °C at a high airspeed (ceiling fans).

University buildings consist of different built environments such as offices, residential buildings, classrooms, lecture theatres, and hostel dormitories. In India, researchers have carried out thermal comfort studies and found the expectation and preferences of occupants in offices, residential buildings, and classrooms during the summer and winter seasons for different climatic zones [9,22,27]. However, a literature review carried out here by the authors showed that very few studies have been done in hostel dormitories that have highlighted the subject's behavioral adaptation, use of controls, and thermal adaptation in different seasons of the composite climate of India. Therefore, the present study systematically investigates the thermal preferences and sensations of residents during different seasons in hostel dormitories under the composite climate of India. Furthermore, the study reports the behavioral adaptations and environmental controls of the occupants for their thermal comfort requirements.

## 2. Methodology

### 2.1. Location and the Selected Hostel Buildings

A questionnaire-based field study was done in naturally ventilated dormitory buildings at the National Institute of Technology premises, Jalandhar (latitude = 31.3° N, longitude = 75.58° E, mean sea level = +228 m). Jalandhar City is in the state of Punjab and is in the composite climate of India. The composite climate zone has a large geographical spread, so it has more climatic diversity than other climatic zones of India. It has four distinct seasons, i.e., winter (November–February), spring (March), summer (April–September), and autumn (October) [31]. The summer season is spread over six months. It is characterized by scorching and dry weather conditions and a maximum temperature exceeding 45 °C, while, during the winter season, the outdoor temperature dips below 2 °C. Figure 1 shows the recorded outdoor temperature and relative humidity profile in different months of the year at the study location. It can be seen clearly that during the summer season, the air temperature peaks start from March (mean temperature = 29 °C) and attain a maximum temperature during May (mean temperature = 35 °C) and June (mean temperature = 30.3 °C) at the study location. The relative humidity is generally very low during the summer season, and the months are mostly dry. Following June, July and August are considered rainy months, characterized by a high relative humidity and low air temperature. Autumn and spring have generally moderate ambient conditions with a mean air temperature not exceeding 25 °C. The winter season consists of December and January months under the composite climate of India. During the winter season, the minimum air temperature falls below 2 °C, with an average temperature range between 15–22 °C with moderate relative humidity conditions.

A naturally ventilated dormitory environment with students in typical clothing on a typical survey day is shown in Figure 2. The investigated hostel dormitories were multi-story buildings and were constructed using high thermal capacity construction materials, i.e., concrete mixture, brick burned, plaster, etc. The roofs of the dormitories were made up of reinforced concrete cement (RCC) with a thickness of ~0.15 m, with ~0.015 m thick gypsum plaster on both sides. The external walls of the hostel buildings had a thickness of about 0.20−0.23 m. The window assemblies consisted of a single piece of clear glass of ~0.003 m thickness with a U-value of ~5.7 W/m²K.

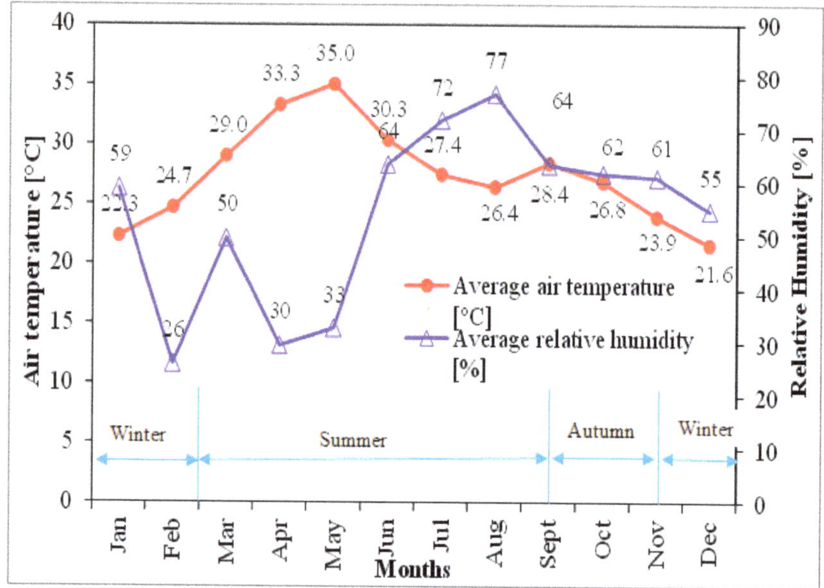


**Figure 1.** Ambient climatic parameters for different months at the location.

### 2.2. Sample Size Description

A total of 1462 questionnaire-based responses were returned during the field study. The subjects were undergraduate and postgraduate students with a mean age of about 20 years and were healthy individuals. The field study was spread over different seasons of the academic year for 2018 and 2019. Therefore, the number of subjects who participated in the survey varied in each season, as shown in Table 1. Furthermore, all of the subjects voluntarily participated in the field study.

**Table 1.** Information of the selected location, sample size, and gender (N = 1462).

| City | Location | Season | No of Samples |
|---|---|---|---|
| Jalandhar, India | Latitude—31.33° N, Longitude—75.58° E, Altitude—228 m | Autumn | 135 |
| | | Winter | 181 |
| | | Spring | 248 |
| | | Summer | 898 |

### 2.3. Field Study and Survey Protocols

The "Right Here Right Now"-based questionnaire was employed to record the students' thermal sensation votes and preferences. The questionnaire used in the field study is provided as "Appendix A". The ASHRAE seven-point thermal sensation scale and Nicol's [39] five-point thermal preference scale were used to record the subjects' thermal sensations and thermal preferences in the indoor environment (Table 2). Laboratory-grade industry calibrated instruments, with a high precision and accuracy, used in the field study are shown in Figure 2d. The make, range, and accuracy of the instruments used in the field study are presented in Table 3. During the interaction with the subjects, the indoor thermal parameters were recorded using the instruments placed close to the students and at a height of 1.1 m [4]. The clothing values of each student were estimated using the insulation values of each clothing ensemble provided in ISO-7730 standard and ASHRAE standard 55-2020 [4,5]. The subject's metabolic rates were calculated according to the checklist provided in the ISO-7730 standard and ASHRAE standard 55-2020.

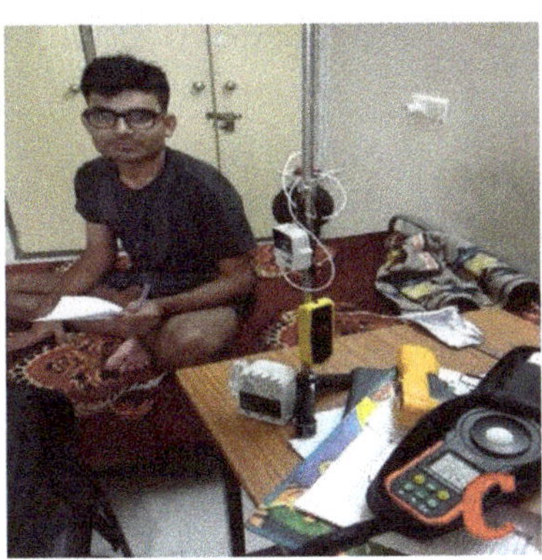
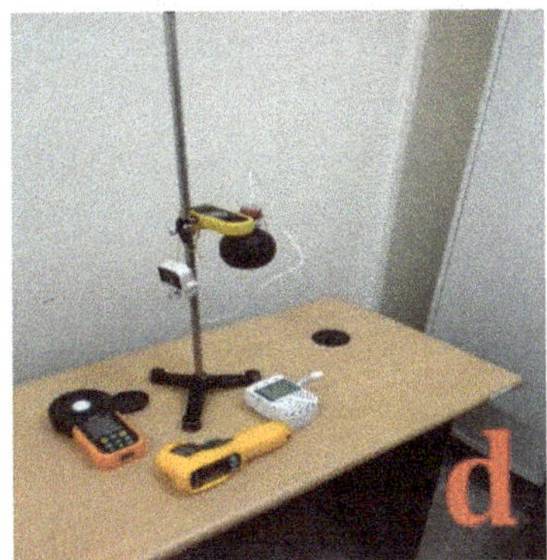

**Figure 2.** (**a,b**) Pictures of studied dormitories, (**c**) typical survey environment, and (**d**) instruments used to record the thermal parameters during the field study.

**Table 2.** Sensation and preference scales used in the present study.

| Scale Values | Thermal Sensation | Thermal Preference | Overall Comfort |
|---|---|---|---|
| +3 | Hot | | |
| +2 | Warm | | |
| +1 | Slightly warm | Cooler | Uncomfortable |
| 0 | Neutral | No change | Comfortable |
| −1 | Slightly cool | Warmer | |
| −2 | Cool | | |
| −3 | Cold | | |

**Table 3.** Make, range, and accuracy of instruments used in the field study.

| Description | Make of Instruments | Parameter Used | Range | Accuracy |
|---|---|---|---|---|
| Thermo-hygro $CO_2$ meter | TR—76Ui | Air temperature | 0–55 °C | ±0.5 °C |
| | | Relative humidity | 10–95% RH | ±5% RH |
| | | $CO_2$ level | 0–9999 ppm | ±50 ppm ± 5% |
| Globe thermometer | Tr-52i, globe (dia. 75 mm) | Globe temperature | −60–155 °C | ±0.3 °C |
| Infrared thermometer | Fluke 61 | Surface temperature | −18–275 °C | ±2 °C |
| Thermal anemometer | Testo-405 | Air velocity | 0.01–10.00 m/s | 0.01 m/s |
| | | Air temperature | −20–50 °C | ±0.1 °C |

## 3. Results and Discussion

### 3.1. Indoor and Outdoor Thermal Environmental Conditions

Table 4 presents the descriptive statistical summary of measured indoor thermal environment parameters during the field study. The mean indoor air temperature varied between 17.3 °C to 30.8 °C, and the mean indoor relative humidity varied between 30–78% from winter to summer at the study location. The average air was recorded to. Be higher during the autumn and summer seasons than in spring and winter. The mean airspeed was about 0.71 m/s for the combined dataset, and this value is well within the limits of no paper blowing conditions as defined in ASHRAE Standard 55-2020 [4].

**Table 4.** Statistics of the indoor thermal parameters measured in different seasons.

| Parameters | Autumn | | Winter | | Spring | | Summer | | All Season Data | |
|---|---|---|---|---|---|---|---|---|---|---|
| | Mean | sd | Mean | sd | Mean | sd | Mean | sd | Mean | sd |
| $T_a$ | 25.6 | 2.52 | 17.3 | 1.8 | 21.5 | 3.5 | 30.8 | 2.9 | 27.3 | 5.6 |
| $T_g$ | 24.6 | 2.69 | 16.6 | 1.6 | 20.8 | 3.5 | 30.3 | 2.5 | 26.7 | 5.7 |
| $T_{out}$ | 28.7 | 3.45 | 14.5 | 2.8 | 24.5 | 2.8 | 35.4 | 3.7 | 29.4 | 6.4 |
| $Rh_i$ | 48.9 | 6.5 | 61.5 | 8.8 | 54.1 | 10.7 | 61.1 | 16.0 | 59.1 | 15.4 |
| AS | 0.29 | 0.47 | 0.13 | 0.28 | 0.18 | 0.30 | 1.06 | 0.72 | 0.71 | 0.69 |

N: no. of samples; $T_a$: indoor air temperature (°C); $T_g$: indoor globe temperature (°C); $RH_i$: indoor relative humidity (%); AS: airspeed (m/s).

### 3.2. Analysis of Seasonal Thermal Sensation Votes and Preference Votes

As the field study was carried out across the different seasons of the year, a significant variation in the measured indoor and outdoor thermal parameters was observed. The thermal sensation voting patterns of the surveyed subjects during different seasons and from the pooled data are shown in Figure 3a,b. From the figure, a proportionally higher number of subjects voted "slightly cool", "cool"," neutral", and "slightly warm" during the autumn, winter, spring, and summer seasons, respectively (Table 5). The subject's mean thermal sensation skewed towards "slightly cool" (mean TS = −0.15; sd = ±1.37) in the pooled dataset. From Figure 3, it can be concluded that the students perceived the existing thermal environment as being "slightly cool" rather than "neutral" in the surveyed dormitories. About 71.2% of subjects voted in three central categories on the thermal sensation scale, i.e., ±1, and can be assumed to be comfortable. Furthermore, 32%, 59%, 23%, and 5% of the subjects voted for the cooler side of the TS scale (TS ≤ −1) during the autumn, winter, spring, and summer seasons. Conversely, only 18.1% of subjects voted for the warmer side (TS ≥ +1) during the summer season.

The mean thermal preference was observed to be +0.98, +0.87, +0.01, and −0.72 during the autumn, winter, spring, and summer seasons, respectively. The positive sign indicates that the subjects preferred to be warmer, and the negative sign indicates that the subjects preferred to be cooler. It can be seen that subjects preferred a cooler thermal environment in the summer season, while a warmer thermal environment was desired by hostel students in the autumn, winter, and spring seasons. A total of 39.5%, 19.9%, 25.5%, and 40.5% of subjects voted for "no change", i.e., "neutral" for the existing thermal environment during the autumn, winter, spring, and summer season at the location. However, 39.2%, 65%, and 31.2% of students preferred warm thermal environments during the autumn, winter, and spring seasons, respectively, whereas only 13% of occupants preferred a cooler thermal environment during the entire study period.

We also recorded the overall thermal comfort of the students in the prevailing thermal environment on a binary scale. The subjects voted on a binary scale, i.e., 1 indicating uncomfortable and 0 indicating comfortable, corresponding to the prevailing indoor thermal environment. Figure 4 shows the subjects' voting patterns regarding overall comfort in different seasons and in the pooled dataset. About 88.8%, 76.2%, 92.3%, and 75.5% of subjects voted "comfortable" in autumn, winter, spring, and summer, respectively. In the pooled dataset, about 87% of subjects indicated their immediate thermal environment as being comfortable, whereas about 13% of students found their thermal environment uncomfortable.

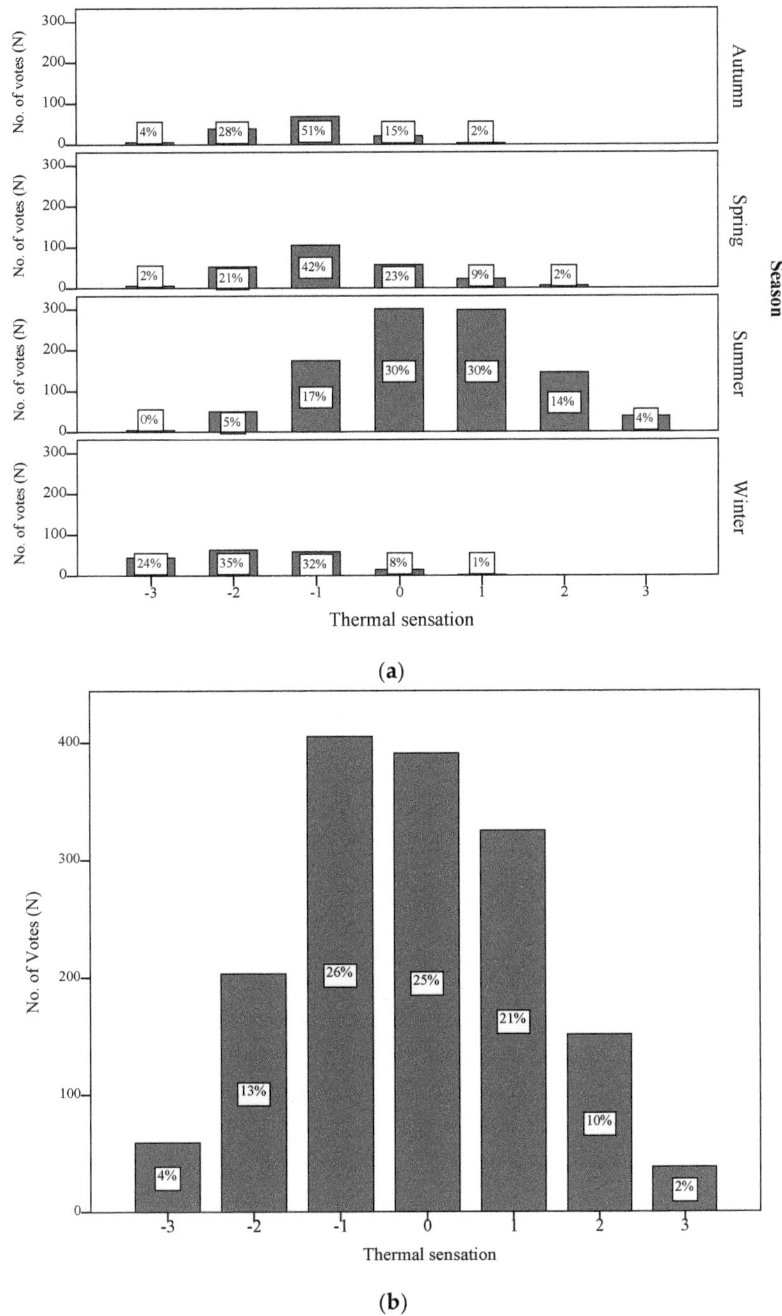

**Figure 3.** Thermal sensation votes distribution (**a**) for different seasons and (**b**) on combined database.

**Table 5.** Statistical summary of subjective and objective comfort parameters in different seasons.

| Season | Thermal Sensation (TS) | | Thermal Preference (TP) | | Overall Comfort (°C) | |
|---|---|---|---|---|---|---|
| | Mean | sd | Mean | sd | Mean | sd |
| Autumn | −1.16 | 0.81 | 0.98 | 0.78 | 0.11 | 0.32 |
| Winter | −1.73 | 0.95 | 0.87 | 0.72 | 0.24 | 0.43 |
| Spring | −0.78 | 1.03 | 0.01 | 0.87 | 0.08 | 0.17 |
| Summer | 0.51 | 1.20 | −0.72 | 0.75 | 0.24 | 0.41 |
| All seasons combined | −0.15 | 1.37 | −0.38 | 0.91 | 0.20 | 0.40 |

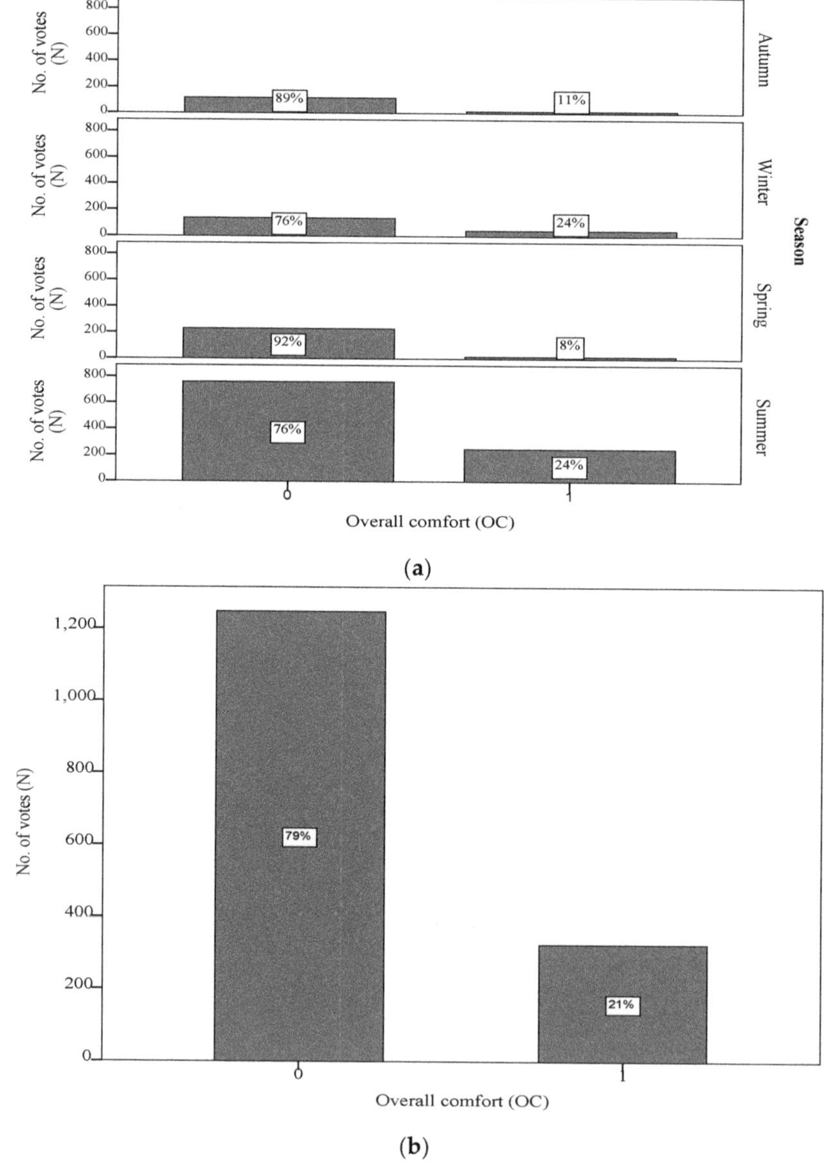

**Figure 4.** Overall comfort votes recorded in (**a**) different seasons and (**b**) in the pooled dataset.

## 4. Characteristics of Seasonal Comfort

### 4.1. Linear Regression and PMV−PPD Model Analysis

The adaptive thermal comfort principle assumes that people in built environments are not only the recipients, but actively participate and take actions to adapt themselves to the existing indoor environmental conditions through physiological, psychological, and behavioral adaptation in different seasons and climates across the world [7–10]. Therefore, in the current study, the seasonal comfort temperatures of the surveyed students were estimated using the PMV−PPD model, linear regression, and Griffiths approach. The procedure defined in standard ASHRAE 55-2020 calculated the PMV and PPD values [4]. The standard suggests that 80% of occupants will be comfortable within a PMV bandwidth of ±0.5 [4]. It can be seen in Figure 5 and Table 6 that there is a discrepancy between PMV with TSV for different seasons and pooled dataset. It can be concluded from Figure 5 that the PMV values overestimated the actual thermal sensation of subjects during the summer season. In the winter season, the PMV values underestimated the thermal sensation of the subjects (Figure 5).

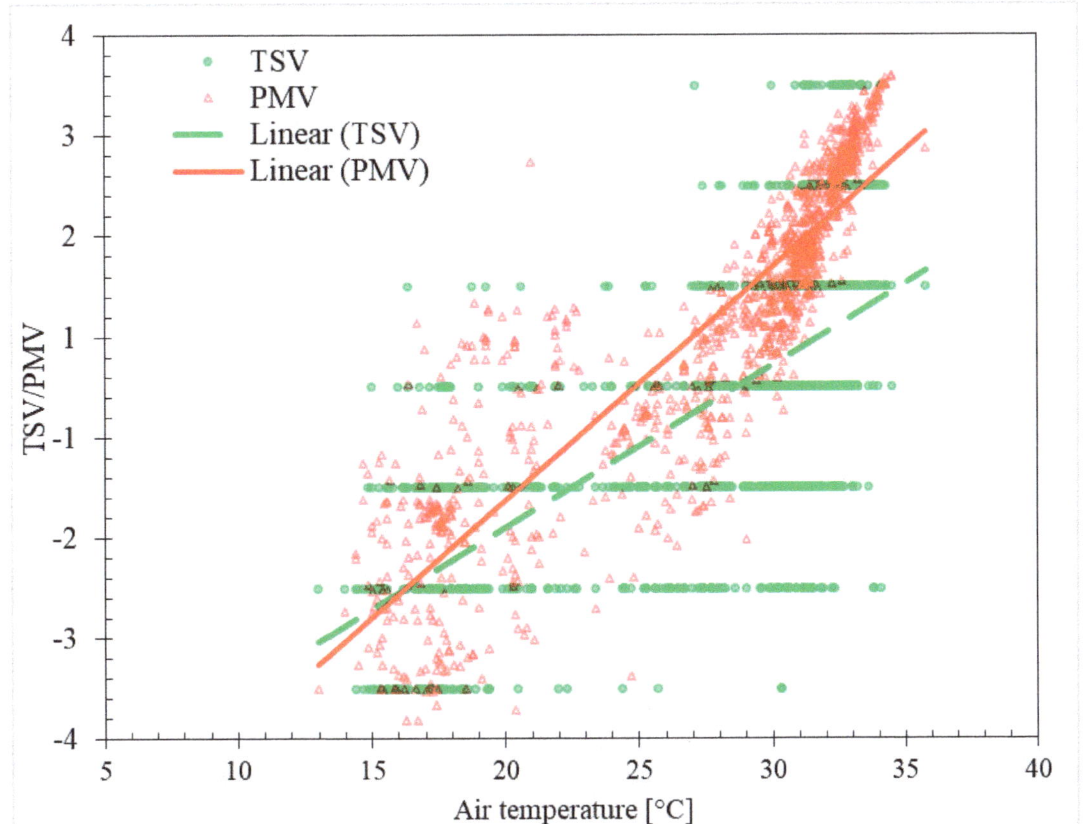


**Figure 5.** Linear regression analyses of PMV votes and TSV.

**Table 6.** Statistics of the linear regression analysis of TSV and PMV on a seasonal basis.

| Case | N | Regression Models * | $R^2$ | $T_n$ (°C) | Mean $T_{comf} \pm$ sd (°C) |
|---|---|---|---|---|---|
| All Season data | 1462 | TSV = 0.12 $T_a$ − 3.12 | 0.11 | 26 | 27.1 0 ± 4.6 |
| | | PMV = 0.12 $T_a$ − 3.24 | 0.21 | 27 | |
| Autumn | 135 | TSV = 0.08 $T_a$ − 3.18 | 0.06 | 39.7 | 26.9 ± 2.68 |
| | | PMV = 0.07 $T_a$ − 3.12 | 0.09 | 44.5 | |
| Winter | 181 | TSV = 0.13 $T_a$ − 4.01 | 0.06 | 30.8 | 19.9 ± 2.11 |
| | | PMV = 0.14 $T_a$ − 3.96 | 0.13 | 28.3 | |
| Spring | 248 | TSV = 0.12 $T_a$ − 3.41 | 0.17 | 28.4 | 22.4 ± 3.2 |
| | | PMV = 0.12 $T_a$ − 3.38 | 0.22 | 28.2 | |
| Summer | 898 | TSV = 0.12 $T_a$ − 3.22 | 0.05 | 26.8 | 29.5 ± 2.6 |
| | | PMV = 0.12 $T_a$ − 3.54 | 0.11 | 29.5 | |

N = sample size; TSV = thermal sensation vote; $T_a$ = indoor air temperature; $T_n$ = regression neutral temperature; $T_{comf}$ = Griffiths comfort temperature (°C) with 0.50 as a coefficient. * The regression models are all significant at ($p < 0.001$).

The mean indoor air temperature is considered a neutral temperature at which an average subject will vote neutral "0" on the TSV scale [23,25]. Researchers extensively use the linear regression method to estimate the thermal neutrality of surveyed subjects for different building types and in other climates. We used the linear regression approach to calculate the seasonal neutral temperature in hostel dormitories, as depicted in Figure 6. Behavioral adaptation is evident from the low regression coefficient values ($R^2$) [25]. Analyzing the data, it was found that the mean neutral temperature was 26 °C for the pooled data. A higher comfort bandwidth was also noticed, which varied by more than 16 °C (17.6–33.3 °C) from winter to summer for the hostel residents, showing the wider thermal adaptability corresponding to the climatic variations. Interestingly, the findings of the present study are supported by the studies done by Mishra and Ramgopal [30], Dhaka et al. [33], and Dahlan et al. [37].

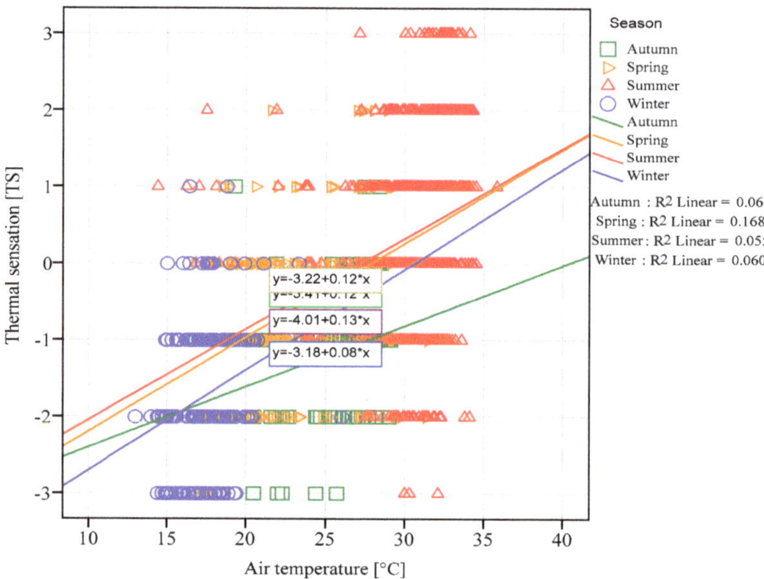


**Figure 6.** Linear regression analysis of TSV for different seasons.

### 4.2. Mean Comfort Temperature ($T_{comf}$): Griffiths Approach

Some researchers have challenged the applicability of the linear regression approach in field studies due to the effects of the adaptive behavior [23,27,40]. Therefore, the mean comfort temperature for each season was estimated using the Griffiths method [41]. The Griffiths equation can be written as follows:

$$T_{comf} = T_a + \frac{[0 - TS]}{GC} \tag{1}$$

where $T_{comf}$ = Griffiths comfort temperature, $T_a$ = air temperature, TS = thermal sensation votes, and GC = Griffiths constant.

Previous studies carried out in the composite climate of India have suggested the use of the Griffiths coefficient of 0.50/°C for the calculation of a neutral temperature [20,27]. Hence, a 2 °C perturbation, i.e., 0.50/°C, was considered for the analysis of the mean comfort temperature ($T_{comf}$) in the present study. The mean comfort temperature ($T_{comf}$) was found to be 27.1 ± 4.65 °C in the pooled data, and is shown in Figure 7. In addition, the mean $T_{comf}$ was about 26.9 ± 2.68 °C, 19.9 ± 2.11 °C, 22.4 ± 3.2 °C, and 29.5 ± 2.6 °C for the autumn, winter, spring, and summer seasons, respectively.

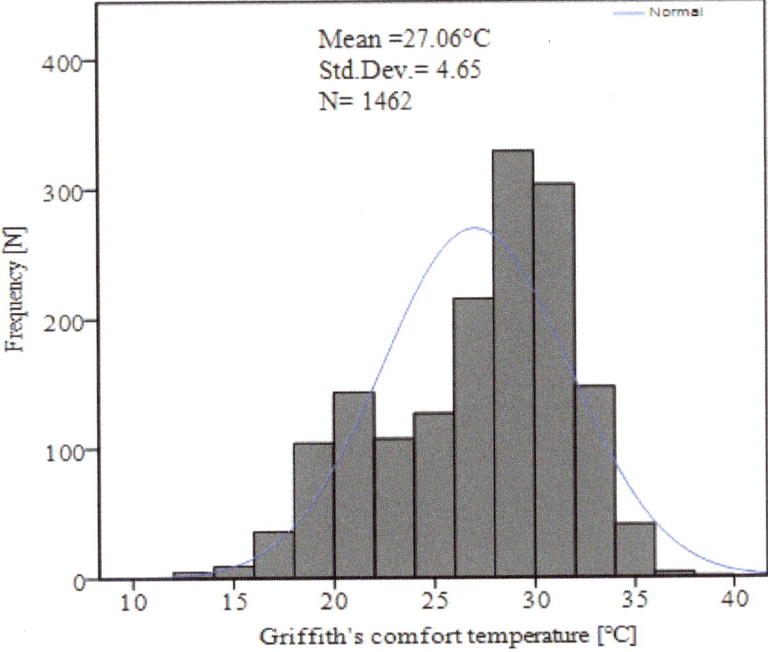


**Figure 7.** Distribution of the calculated mean comfort temperature using the Griffiths method for the pooled data.

## 5. Investigation of Thermal Adaptation Behavior of Residents

The adaptive thermal comfort principle considers that subjects in a built environment are active agents and can exercise various thermal and behavioral adaptations to restore their comfort or make themselves thermally comfortable [7,39]. Therefore, the adaptive behavior of subjects in university hostel dormitories was analyzed in the context of clothing adjustments; the application of environmental controls, i.e., opening and closing of windows and doors, and the use of ceiling fans in different seasons; and in the pooled dataset.

### 5.1. Clothing Adjustments

Students mostly wore clothing ensembles such as shirts/t-shirts and trousers/jeans during the daytime, and nightwear, i.e., pajamas and half sleeve t-shirts, during the holidays and late evening hours. The clothing ensembles ranged between 0.32–1.84 clo for the hot summer season to the cold winter months at the study location. Mean clo values of about 0.57 (±0.25) clo, 0.98 (±0.12) clo, 0.45 (±0.27) clo, and 0.36 (±0.11) clo were recorded during the autumn, winter, spring, and summer seasons, respectively. An average clothing value of about 0.49 ((±0.31) clo was recorded in the pooled database, matching closely with the ASHRAE Standard 55 recommendation for the summer season.

To analyze the characteristic of the adaptation behavior related to the clothing, linear and quadratic regression analyses for indoor air temperature were carried out to observe the inflection points [7,9,31]. Figure 8 shows the linear and quadratic regression fit for predicting the adaptive behavior of students regarding clothing corresponding to the change in indoor air temperature during different seasons at the study location. The inflection points were observed at 18 °C and 31 °C. A sudden change in clothing value was observed at these points, showing adaptation. It can be seen that the correlation coefficient was reasonably strong, which suggests that subjects adaptively use clothing adjustments to restore their comfort. The authors also found similar observations in studies carried out in different climates and building types [29,30,35].

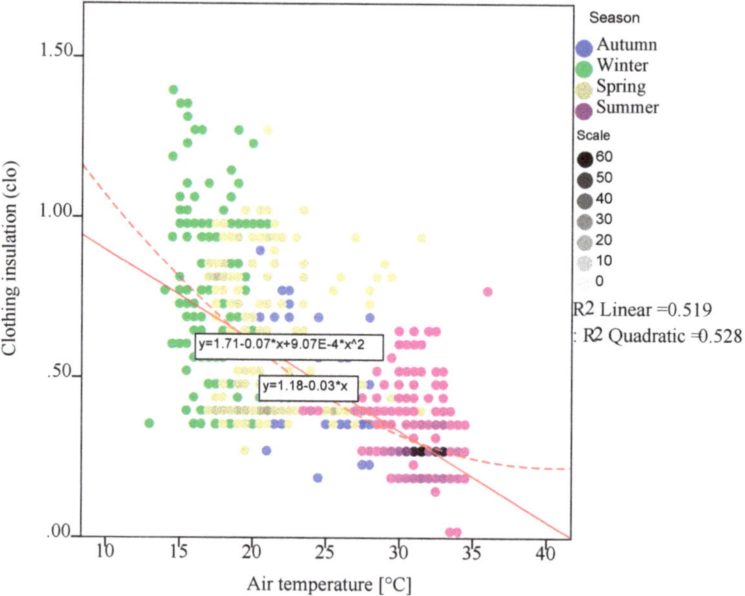


**Figure 8.** Seasonal variations of clothing insulation with indoor air temperature.

### 5.2. Impact of Controls and Exercised Controls on Comfort

Analyzing the use of controls plays an essential role in adaptive thermal comfort studies. The effective use of controls in built environments by subjects enhances thermal comfort and extends the comfort boundaries [16,24,34]. In this context, the authors recorded the use of environmental controls by the students in the dormitory, i.e., windows and ceiling fans, in binary variables (i.e., window open: 1; window closed: 0; fans on: 1; fans off: 0) during the field surveys. Figure 9 shows the percentage of subjects voting feeling comfortable (corresponding to three central categories of the TS scale) when available controls were used. It can be seen that when subjects used the general controls, such as opening windows and doors and switching on ceiling fans, the occupants' thermal comfort during different seasons improved significantly. In addition, during the summer season, it was observed that more than 80% of students voted feeling comfortable when ceiling fans were operating. In contrast, only 60% of students voted feeling comfortable when windows were open at the time survey.

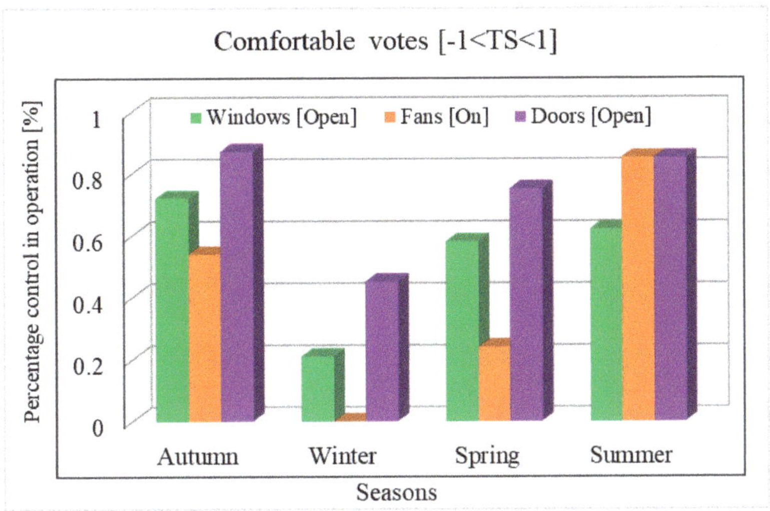


**Figure 9.** Percentage of occupants voting feeling comfortable when environmental controls are in operation during different seasons.

The study results have close resemblances with the finding of other studies conducted under similar climatic conditions but at different locations. Rijal et al. [41] found that about 81% of fans were in use when the indoor air temperature exceeded 28.5 °C in offices at Pakistan. Indraganti et al. [42] noted about 80% fans were in operation at 30 °C in office buildings of India. Manu et al. [43] analyzed the windows and fan use behavior of office occupants based on the field data collected for different climatic zones of India, and concluded that maximum fans as well windows were used under hot and dry, and hot and humid climates of India compared to other climatic zones of India. Similarly, Kumar et al. [31,34] observed that about 50% of windows and 80% of fans were used when indoor air temperature peaked at 28 °C in university buildings situated under the composite climate of India. Singh et al. [44] also predicted a similar observation in office buildings located in the north-east part of India.

To gain more insight, we further plotted the comfortable [±1 TS votes] on the ASHRAE Standard 55-2020 comfort zone when windows and ceiling fans were in operation. ASHRAE Standard 55-2020 [4] graphically defines thermal comfort boundaries on a typical psychrometric chart describing the operative temperature and humidity range for occupants corresponding to the sedentary activity level (1−1.3 met) and clo value in the range of 0.5−1 clo. Furthermore, ASHRAE Standard 55-2020 recommends a maximum indoor airspeed of 0.80 m/s to avoid paper blowing conditions in office buildings. Figure 10a,b shows the plotting of comfortable votes when windows were "open" or ceiling fans were "on" during the field surveys. It was observed that when windows were open, subjects felt comfortable at a high relative humidity and high indoor air temperatures. In addition, the maximum airspeed was recorded at close to 2 m/s during the summer. The subjects voted feeling comfortable even when the indoor air temperature was about 34 °C and the relative humidity was more than 70%. These results are supported by the authors' previous findings under similar climatic conditions for office buildings [8,32,33].

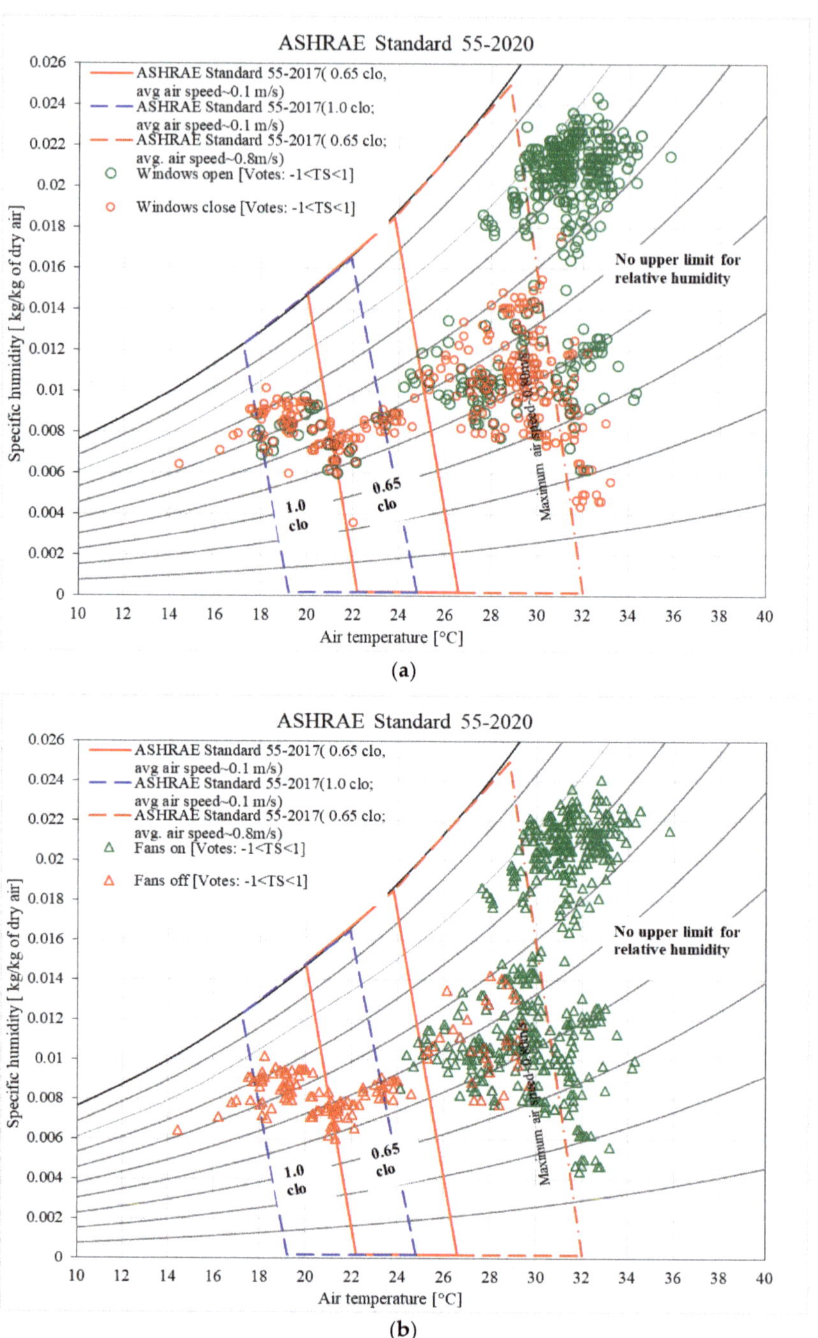



**Figure 10.** Plotting comfortable votes on the ASHRAE Standard 55 – 2020 comfort zone when (**a**) windows are open/closed and (**b**) fans are on/off.

## 6. Summary of Work and Conclusions

In this study, a seasonal comfort study was carried out in naturally ventilated hostel dormitories under the composite climate of India. One of the prime objectives of the built environment is to provide the desired thermal comfort to the occupants. If this is disregarded, occupants make use of mechanical and electrical devices to achieve the desired thermal comfort. This involves costs at various levels and impacts sustainability of the building sector. To improve the sustainability of the building sector, presently, the entire world is working on various issues to reduce the energy consumption. Precise evaluation of the comfort parameters of different built environments and occupants' behaviour characteristics are key for the reduction and optimal use of energy in buildings. To achieve comfort in the built environment, it is of the utmost importance that the comfort parameters of different built environments and occupants' behaviour characteristics must be known by building designers and architectures so that they can design buildings that will provide optimum comfort to the occupants and consume less energy so as to provide the necessary comfort. In this study, university students participated as subjects, under the composite climatic of India, considering ASHRAE Class II protocols. The following is a summary of the findings from the analysis of the collected data:

1. The mean thermal sensations for the students in the dormitory were recorded as "slightly cool", "cold", "slightly cool", and "slightly warm" during the autumn, winter, spring, and summer seasons. The subject's mean thermal sensation was skewed towards "slightly cool" (mean TS = $-0.15$; sd = $\pm1.37$) for the combined dataset.
2. A total of 39.5%, 19.9%, 25.5%, and 40.5% of subjects in the hostel dormitories voted for "no change" in the persisting indoor thermal environment during the autumn, winter, spring, and summer seasons. However, 39.2%, 65%, and 31.2% of subjects preferred a warm thermal environment in the autumn, winter, and spring seasons, respectively. In comparison, about 13% of students preferred a cooler thermal environment in the combined dataset.
3. The PMV$-$PPD model overestimated and underestimated the actual thermal sensations in the summer and winter seasons.
4. The mean $T_{comf}$ was about $26.9 \pm 2.68$ °C, $19.9 \pm 2.11$ °C, $22.4 \pm 3.2$ °C, and $29.5 \pm 2.6$ °C for the autumn, winter, spring, and summer seasons, respectively.
5. Mean clo values of 0.57 ($\pm0.25$) clo, 0.98 ($\pm0.12$) clo, 0.45 ($\pm0.27$) clo, and 0.36 ($\pm0.11$) were recorded in the autumn, winter, spring and summer seasons, respectively. An average clothing value of about 0.49 (($\pm0.31$) clo was recorded for the pooled dataset, closely matching with the ASHRAE Standard 55 recommended clo value for the summer season.
6. More than 80% of subjects responded that they were comfortable when ceiling fans were operating. In contrast, only 60% of the subjects voted being comfortable when the windows were open at the survey time.

The study put forth the idea of future studies involving subjects of university dormitories regarding their comfort expectations and the use of environmental controls in different climates and geographical locations. An effective quantification of their thermal adaptation behavior and its impact on the comfort parameters will be advantageous for improving the students' thermal comfort and overall indoor thermal environment. It is also anticipated that the findings of this study will help building designers, architects, and engineers in designing energy-efficient and comfortable university hostel dormitories in the near future.

**Author Contributions:** Data curation, S.K., M.K.S., B.S.A., M.A.A. and N.A.-T.; Formal analysis, S.K., M.K.S. and B.S.A.; Methodology, S.K., M.K.S. and M.A.A.; Supervision, M.K.S.; Visualization, S.K., M.K.S. and N.A.-T.; Writing—original draft, S.K.; Writing—review & editing, M.K.S., B.S.A., M.A.A. and N.A.-T. All authors have read and agreed to the published version of the manuscript.

**Funding:** This research did receive any funding from any sources.

**Institutional Review Board Statement:** This is certified that the field surveying to study the thermal comfort expectation of subjects residing in hostel dormitories, located at Dr. B R Ambedkar National Institute of Technology, Jalandhar (Punjab) 144011, India, was carried out during the academic year 2018–2019. The field monitoring and subjective study doesn't violate any ethics and was fully in accordance with the data collection procedure laid down by the institute as well as International standard i.e. ASHRAE Standard 55-2013.

**Informed Consent Statement:** A brief introduction and discussion were provided to subjects by surveyor regarding the objective of the present study. A brief discussion with the occupant's also included the information about the type of data to be collected. However, no data has been collected regarding their personal information. The consent was also taken from each subject for the use of collected data for academic and academic research only. Same statement is also made by the authors in the acknowledgement section of the manuscript.

**Data Availability Statement:** The data collected in the study is available with the authors. Since authors are working on another project and present data will be a part of grand data base, authors would not like to make the data public at present. But if any researcher finds thae study interesting and requires some specific information and data for academic research then he/she can contact the corresponding author at the given e-mail address and the authors will be happy to help the researcher by providing the data and requested information.

**Acknowledgments:** Despite their class schedule, the author thanks all the subjects for participating in the field surveys. The authors also acknowledge the support received from authorities of Dr. B R Ambedkar National Institute of Technology, Jalandhar, towards the successful completion of the study. The authors would like to mention that this study did not receive any funding from any sources.

**Conflicts of Interest:** Corresponding author has informed all co-authors about this manuscript and took their permission before uploading it to the Sustainability journal. Corresponding author also like to state that there is no conflict of interest.

## Appendix A

**Thermal comfort Questionnaire used in study (Part-A)**

Date… … … …

| Name: | | Height: |
|---|---|---|
| Age: | Gender (M/F): | Weight: |
| Permanent Address: | | |
| Timing: | | |

**1.Feeling: At Present I feel temperature as:**

| | | |
|---|---|---|
| Cold | -3 | |
| Cool | -2 | |
| Slightly Cool | -1 | |
| Neutral | 0 | |
| Slightly Warm | +1 | |
| Warm | +2 | |
| Hot | +3 | |

**2.Prefrence:I would prefer to feel Temperature as:**

| | | |
|---|---|---|
| Much Cooler | -2 | |
| A bit Cooler | -1 | |
| No Change | 0 | |
| A bit Warmer | +1 | |
| Much Warmer | +2 | |

**3. Feeling: At Present I feel air movement as:**

| | | |
|---|---|---|
| Very Still | -3 | |
| Moderately Still | -2 | |
| Slightly still | -1 | |
| Neutral | 0 | |
| Slightly moving | +1 | |
| Moderately Moving | +2 | |
| Much Moving | +3 | |

**4.Prefrence:I would prefer to feel air movement as:**

| | | |
|---|---|---|
| Moderately Still | -2 | |
| Slightly still | -1 | |
| Acceptable | 0 | |
| Slightly Moving | +1 | |
| Moderately Moving | +2 | |

**5. Feeling: At Present I feel relative humidity level as:**

| | | |
|---|---|---|
| Very Dry | -3 | |
| Dry | -2 | |
| Slightly dry | -1 | |
| Neutral | 0 | |
| Slightly humid | +1 | |
| Humid | +2 | |
| Very Humid | +3 | |

**6.Prefrence: I would prefer Humidityas**

| | | |
|---|---|---|
| Moderately dry | -2 | |
| Slightly dry | -1 | |
| Acceptable | 0 | |
| Slightly Humid | +1 | |
| Moderately Humid | +2 | |

**7.Overall Comfort: Right Now I feel this *Environment***

| | | |
|---|---|---|
| Comfortable | 0 | |
| uncomfortable | 1 | |

**8.What would you prefer to do when you feel *uncomfortable***

| | |
|---|---|
| Open Windows | |
| On ceiling fan | |
| Open Door | |
| Clothing change | |
| Change posture | |
| Walking indoor /Outdoor | |
| Drink beverage(Coffee/Tea/Water) | |

**9.Activity: What are you doing in last 15 minutes-**

| | |
|---|---|
| Sleeping | |
| Sitting(Office/School work) | |
| Sitting(Passive work) | |
| Standing relaxed | |
| Standing Walking | |
| Walking indoor/ outdoor | |
| Eating | |

**Thermal comfort Questionnaire used in study (Part-B)**

Date… … … …

| Name: | Floor No: |
|---|---|
| | Building Name: |
| No.of student: | Orientation of Room: |

**Measurement(Environment Variables):**

| | |
|---|---|
| 1.   Room Temperature | |
| 2.   Globe Temperature | |
| 3.   Wall Temperature | |
| 4.   Air Velocity | |
| 5.   Relative Humidity | |
| 6.   Light Intensity (LUX) | |
| 7.   $CO_2$ Concentration (ppm) | |

**8.Controls : At present Location:**

| | |
|---|---|
| External Door | |
| Windows | |
| Blinds/Curtains | |
| Fan | |
| Lights | |

**9.Clothing Right Now:**

| | |
|---|---|
| T-Shirt | |
| Short  Sleeve Shirt | |
| Long Sleeve Shirt | |
| Lower | |
| Jeans | |
| Trouser | |
| Sandals/Slippers | |
| Socks & Shoes | |
| Payjamas | |
| Jacket/Woolen Jacket | |
| Thermal Inner | |
| Sweater/Pullover | |
| Scarf/Woolen cap | |
| Other(Mention) | |

**Figure A1.** Questionnaire used in the study.

## References


1. Abergel, T.; Dean, B.; Dulac, J. Towards Zero-Emission, Efficient and Resilient Buildings and Construction Sector: Global Status Report 2017. Paris: Global Alliance for Buildings and Construction, United Nations Environment Programme (UNEP) and International Energy Agency (IEA). Available online: https://www.globalabc.org/uploads/media/default/0001/01/ (accessed on 12 March 2022).
2. Shukla, Y.; Rawal, R.; Shnapp, S. *Residential Buildings in India: Energy Use Projections and Savings Potentials*; GBPN: Paris, France, 2014.
3. Singh, M.K.; Ooka, R.; Rijal, H.B.; Kumar, S.; Kumar, A.; Mahapatra, S. Progress in thermal comfort studies in classrooms over last 50 years and way forward. *Energy Build.* **2019**, *188–189*, 149–174. [CrossRef]
4. *Thermal Environmental Conditions for Human Occupancy*; ASHRAE 55-2020; American Society of Heating, Refrigerating and Air-conditioning Engineers Inc.: Atlanta, GA, USA, 2020.
5. *ISO 7730*; Ergonomics of the Thermal Environment—Analytical Determination and Interpretation of Thermal Comfort Using Calculation of the PMV and PPD Indices and Local Thermal Comfort Criteria. International Organization for Standardization: Geneva, Switzerland, 2005.
6. Fanger, P.O. Thermal comfort. In *Analysis and Application in Environment Engineering*; Danish Technology Press: Copenhagen, Denmark, 1970.
7. Brager, G.S.; de Dear, R. *Developing an Adaptive Model of Thermal Comfort and Preferences, Final Report, ASHRAE RP-884*; ASHRAE: Atlanta, GA, USA, 1998.
8. Kumar, S.; Mathur, J.; Mathur, S.; Singh, M.K.; Loftness, V. An adaptive approach to define thermal comfort zones on psychrometric chart for naturally ventilated buildings in composite climate of India. *Build. Environ.* **2016**, *109*, 135–153. [CrossRef]
9. Singh, M.K.; Mahapatra, S.; Atreya, S.K. Adaptive thermal comfort model for different climatic zones of North-East India. *Appl. Energy* **2011**, *88*, 2420–2428. [CrossRef]
10. Yao, R.; Li, B.; Liu, J. A theoretical adaptive model of thermal comfort—Adaptive Predicted Mean Vote (aPMV). *Build. Environ.* **2009**, *44*, 2089–2096. [CrossRef]
11. Al-Tamimi, N.A.; Fadzil, S.F.S. Energy-efficient envelope design for high-rise residential buildings in Malaysia. *Archit. Sci. Rev.* **2012**, *55*, 119–127. [CrossRef]
12. Al-Tamimi, N.A. Toward Sustainable Building Design: Improving Thermal Performance by Applying Natural Ventilation in Hot–Humid Climate. *Indian J. Sci. Technol.* **2015**, *8*, IPL0662. [CrossRef]
13. Hellwig, T.R.; Teli, D.; Schweiker, M.; Choi, J.; Lee, M.C.J.; Mora, R.; Rawal, R.; Wang, Z.; Al-Atrash, F. A framework for adopting adaptive thermal comfort principles, in design and operation of buildings. *Energy Build.* **2019**, *205*, 109476. [CrossRef]
14. Mishra, A.K.; Ramgopal, M. Field studies on human thermal comfort—An overview. *Build. Environ.* **2012**, *64*, 94–106. [CrossRef]
15. Halawa, E.; Van Hoof, J. The adaptive approach to thermal comfort: A critical overview. *Energy Build.* **2012**, *51*, 101–110. [CrossRef]
16. Rupp, R.F.; Vásquez, N.G.; Lamberts, R. A review of human thermal comfort in the built environment. *Energy Build.* **2015**, *105*, 178–205. [CrossRef]
17. Lamberti, G.; Salvadori, G.; Leccese, F.; Fantozzi, F.; Bluyssen, P.M. Advancement on Thermal Comfort in Educational Buildings: Current Issues and Way Forward. *Sustainability* **2021**, *13*, 10315. [CrossRef]
18. Ličina, F.; Cheung, T.; Zhang, H.; de Dear, R.; Parkinson, T.; Arens, E.; Chungyoon, C.; Schiavon, S.; Luo, M.; Brager, G.; et al. Development of the ASHRAE global thermal comfort database II. *Build. Environ.* **2018**, *142*, 502–512. [CrossRef]
19. Parkinson, T.R.; de Dear, G. Brager Nudging the adaptive thermal comfort model. *Energy Build.* **2020**, *260*, 109559. [CrossRef]
20. Kumar, S. Subject's thermal adaptation in different built environments: An analysis of updated metadata-base of thermal comfort data in India. *J. Build. Eng.* **2022**, *46*, 103844. [CrossRef]
21. Nicol, J.F. An analysis of some observations of thermal comfort in Roorkie, India, and Baghadad, Iraq. *Ann. Hum. Biol.* **1974**, *1*, 411–426. [CrossRef]
22. Sharma, M.R.; Ali, S. Tropical summer index—A study of thermal comfort of Indian subjects. *Build. Environ.* **1986**, *21*, 11–24. [CrossRef]
23. Indraganti, M. Using the adaptive model of thermal comfort for obtaining indoor neutral temperature: Findings from a field study in Hyderabad, India. *Build. Environ.* **2010**, *45*, 519–536. [CrossRef]
24. Indraganti, M. Adaptive use of natural ventilation for thermal comfort in Indian apartments. *Build. Environ.* **2010**, *45*, 1490–1507. [CrossRef]
25. Singh, M.K.; Mahapatra, S.; Atreya, S.K. Thermal performance study and comfort temperature in vernacular buildings of North–East India. *Build. Environ.* **2010**, *45*, 320–329. [CrossRef]
26. Bureau of Indian Standards (BIS). *National Building Code of India*; Bureau of Indian Standards: Delhi, India, 2016.
27. Manu, S.; Shukla, Y.; Rawal, R.; Thomas, L.E.; de Dear, R. Field studies of thermal comfort across multiple climate zones for the subcontinent: India model for adaptive comfort (IMAC). *Build. Environ.* **2016**, *98*, 55–70. [CrossRef]
28. Singh, M.K.; Kumar, S.; Ooka, R.; Rijal, H.B.; Gupta, G.; Kumar, A. Status of thermal comfort in naturally ventilated classrooms during the summer season in the composite climate of India. *Build. Environ.* **2018**, *128*, 287–304. [CrossRef]
29. Kumar, S.; Singh, M.K.; Mathur, A.; Mathur, J.; Mathur, S. Evaluation of comfort preferences and insights into behavioural adaptation of students in naturally conditioned classrooms in a tropical country, India. *Build. Environ.* **2018**, *143*, 532–547. [CrossRef]

30. Mishra, A.K.; Ramgopal, M. Thermal Comfort Field Study in Undergraduate Laboratories–An Analysis of Occupant Perceptions. *Build. Environ.* **2014**, *76*, 62–72. [CrossRef]
31. Kumar, S.; Singh, M.K.; Loftness, V.; Mathur, J.; Mathur, S. Thermal comfort assessment and characteristics of occupant's behavior in naturally ventilated buildings in composite climate of India. *Energy Sustain. Dev.* **2016**, *33*, 108–121. [CrossRef]
32. Tewari, P.; Mathur, S.; Mathur, J.; Kumar, S.; Loftness, V. Field study on indoor thermal comfort of office buildings using evaporative cooling in the composite climate of India. *Energy Build.* **2019**, *199*, 145–163. [CrossRef]
33. Dhaka, S.; Mathur, J.; Wagner, A.; Agarwal, G.; Garg, V. Evaluation of thermal environmental conditions and thermal perception at naturally ventilated hostels of undergraduate students in composite climate. *Build. Environ.* **2013**, *66*, 42–53. [CrossRef]
34. Kumar, S.; Singh, M.K.; Kukreja, R.; Chaurasiya, S.; Gupta, V. Comparative study of thermal comfort and adaptive actions for modern and traditional multi-storey naturally ventilated hostel buildings during monsoon season in India. *J. Build. Eng.* **2019**, *23*, 90–106. [CrossRef]
35. Kumar, S.; Singh, M.K. Field investigation on occupant's thermal comfort and preferences in naturally ventilated multi-storey hostel buildings over two seasons in India. *Build. Environ.* **2019**, *163*, 106309. [CrossRef]
36. Kumar, S.; Singh, M.K.; Mathur, A.; Kosir, M. Occupant's thermal comfort expectations in naturally ventilated engineering workshop building: A case study at high metabolic rates. *Energy Build.* **2020**, *217*, 109970. [CrossRef]
37. Dahlan, N.D.; Jone, P.J.; Alexander, D.K. Operative temperature and thermal sensation assessments in non-air-conditioned multi-storey hostels in Malaysia. *Build. Environ.* **2011**, *46*, 457–467. [CrossRef]
38. Lai, J.H.K. Gap theory-based analysis of user expectation and satisfaction: The case of a hostel building. *Build. Environ.* **2013**, *69*, 183–193. [CrossRef]
39. Rijal, H.B.; Tuohy, P.; Humphreys, M.A.; Nicol, J.F.; Samuel, A.; Clarke, J. Using results from field surveys to predict the effect of open windows on thermal comfort and energy use in buildings. *Energy Build.* **2007**, *39*, 823–836. [CrossRef]
40. Nicol, F.; Humphreys, M.; Roaf, S. *Adaptive Thermal Comfort Principles and Practice*; Routledge: London, UK, 2012.
41. Griffiths, I.D. *Thermal Comfort in Buildings with Passive Solar Features: Field Studies*; EN3S-090; Report to the Commission of the European Communities: London, UK, 1990.
42. Indraganti, M.; Ooka, R.; Rijal, H.B.; Brager, G.S. Occupant behavior and obstacles in operating the openings in offices in India. In Proceedings of the 8th Windsor Conference, Windsor, UK, 10–13 April 2014.
43. Sanyogita, M.; Shukla, Y.; Rawal, R.; Thomas, L.E.; de Dear, R.; Dave, M.; Vakharia, M. Assessment of air velocity preferences and satisfaction for naturally ventilated office buildings in India. In Proceedings of the Passive and Low Energy Architecture (PLEA) Annual International Conference, Geneva, Switzerland, 6–8 September 2014.
44. Singh, M.K.; Ooka, R.; Rijal, H.B.; Takasu, M. Adaptive thermal comfort in the offices of North-East India in autumn season. *Build. Environ.* **2017**, *124*, 14–30. [CrossRef]

 *sustainability*

*Article*

# Experimental Assessment of the Reflection of Solar Radiation from Façades of Tall Buildings to the Pedestrian Level

Alberto Speroni [1], Andrea Giovanni Mainini [1,*], Andrea Zani [1,2], Riccardo Paolini [3], Tommaso Pagnacco [1,4] and Tiziana Poli [1]

1   Architecture, Built Environment and Construction Engineering Department, Politecnico di Milano, Via Ponzio 31, 20133 Milano, Italy; alberto.speroni@polimi.it (A.S.); a.zani@permasteelisagroup.com (A.Z.); tpagnacco@bollinger-grohmann.it (T.P.); tiziana.poli@polimi.it (T.P.)
2   Permasteelisa North America, 1179 Centre Pointe Circle, Mendota Heights, MN 55120, USA
3   School of Built Environment, Faculty of Arts, Design & Architecture, University of New South Wales, Sydney, NSW 2052, Australia; r.paolini@unsw.edu.au
4   Bollinger + Grohmann Ingegneria, Via Garofalo 31, 20133 Milano, Italy
*   Correspondence: andreagiovanni.mainini@polimi.it; Tel.: +39-022-399-6015

**Abstract:** Urban climates are highly influenced by the ability of built surfaces to reflect solar radiation, and the use of high-albedo materials has been widely investigated as an effective option to mitigate urban overheating. While diffusely solar reflective walls have attracted concerns in the architectural and thermal comfort community, the potential of concave and polished surfaces, such as glass and metal panels, to cause extreme glare and localized thermal stress has been underinvestigated. Furthermore, there is the need for a systematic comparison of the solar concentration at the pedestrian level in front of tall buildings. Herein, we show the findings of an experimental campaign measuring the magnitude of the sunlight reflected by scale models reproducing archetypical tall buildings. Three 1:100 scaled prototypes with different shapes (classic vertical façade, 10% tilted façade, curved concave façade) and different finishing materials (representative of extremes in reflectance properties of building materials) were assessed. A specular surface was assumed as representative of a glazed façade under high-incidence solar angles, while selected light-diffusing materials were considered sufficient proxies for plaster finishing. With a diffusely reflective façade, the incident radiation at the pedestrian level in front of the building did not increase by more than 30% for any geometry. However, with a specular reflective (i.e., mirror-like) flat façade, the incident radiation at the pedestrian level increased by more than 100% and even by more than 300% with curved solar-concentrating geometries. In addition, a tool for the preliminary evaluation of the solar reflectance risk potential of a generic complex building shape is developed and presented. Our findings demonstrate that the solar concentration risk due to mirror-like surfaces in the built environment should be a primary concern in design and urban microclimatology.

**Keywords:** reflective materials; mitigation; urban heat island; outdoor comfort; visual comfort; heat stress; optimization; skyscrapers

**Citation:** Speroni, A.; Mainini, A.G.; Zani, A.; Paolini, R.; Pagnacco, T.; Poli, T. Experimental Assessment of the Reflection of Solar Radiation from Façades of Tall Buildings to the Pedestrian Level. *Sustainability* **2022**, *14*, 5781. https://doi.org/10.3390/su14105781

Academic Editor: Baojie He

Received: 17 March 2022
Accepted: 26 April 2022
Published: 10 May 2022

**Publisher's Note:** MDPI stays neutral with regard to jurisdictional claims in published maps and institutional affiliations.



## 1. Introduction and State of the Art

The accelerating city climate change in combination with local and global climate change heightens the need for decarbonization of the built environment through energy efficiency and mitigation of urban overheating [1]. In particular, solar reflective roofs and walls have been largely investigated to reduce the solar absorption by the urban envelope and thus reduce the release of turbulent sensible heat that increases the ambient temperature [2,3]. Cool surfaces have high reflectance and emissivity and are capable of reducing both solar gains and surface temperatures, positively affecting the energy use of the building and helping to mitigate heat island effects at the mesoscale and local level [4–8]. Heat mitigation technologies can reduce the ambient temperature by 2–2.5 °C when combined with building positive synergies, with the reduction of the solar gains being one of the main pathways to minimize urban overheating [9]. Cool walls in Los Angeles, for instance, may reduce the peak temperature by approximately 0.60 °C [2]. Furthermore, increasing the albedo of walls by 0.10 reduces the cooling energy needs of residential buildings in Mediterranean climates by 2.9 kWh/m$^2$ and reduces the indoor operative temperature of unconditioned buildings by 1.1 °C [10]. While the use of solar-reflective walls has been common in the Mediterranean and other vernacular architecture, there is an increasing concern among architects and urban designers about the potential increased solar reflection towards pedestrians [11,12]. While there are some limitations in current outdoor thermal comfort models, the need to limit the downward reflection of solar radiation towards the bottom of street canyons has received considerable attention in the literature [13–15]. This led to the identification and testing of retro-reflective surfaces as an option to minimize the shortwave radiation entrapment within the urban canopy layer [16]. Retro-reflective materials reflect the direct component of the solar radiation back towards the sun, thus upwards and not directed towards other urban surfaces. However, the proliferation in the use of specular reflective (i.e., mirror-like) materials in architecture has been underinvestigated with respect to their impact on outdoor visual and thermal comfort.

The research of eye-catching shapes for tall buildings, without considering its impact on the urban context [17,18], sometimes leads to increased incident solar radiation on other buildings and at street level due to unwanted reflections [19]. This phenomenon is due to the geometry being able to concentrate increasing solar radiation and the materials used in façade applications, especially high-reflectance glass or polish metal. Considering the transparent part of the building envelope, reflecting glazing systems with a reflectivity of more than 30–40% are chosen to reduce the cooling load of office and commercial buildings with large, exposed curtain walls. Reflectance properties of glazing are angular-dependent and influenced by the direction of the light source falling on their surface. The more the rays strike toward a direction parallel to the surface, the more the reflectance of the surface rapidly increases [20]. This is the case, for example, in temperate climates, particularly during the winter months, for surfaces facing south and during the early and late hours of the day, or in tropical climates for the same orientation and during the central hours of the day. High-reflectance surface treatment increases the possibility of external reflections for lower incidence angles as well.

Most unwanted reflections affect the vision and the visual comfort of pedestrians, users of concurrent buildings, car drivers, train conductors, and plane pilots [21]. The temporary visual disabilities resulting from these phenomena can also raise security issues by potentially causing an accident due to visual impairment, risking people's lives. Glare has also been reported as a critical issue in an urban environment, concerning angular reflective surfaces such as Photovoltaic panels [22] or concentrating solar collector plants with small and large highly reflective surfaces [23].

More disturbing effects have been reported when concentrated solar energy led to direct damage of properties, plants, and people caused by increased and focused solar radiation. The consequences of these effects can be temporary or repeated cyclically during the day/year, depending on the location, orientation, and urban context. Among the most significant cases are the "20 Fenchurch Street" building in London (Figure 1a, Table 1) and the "Vdara" building in Las Vegas (Table 1), which obliged the owners to either modify the façade or change the previously programmed use of the surrounding area in order to provide for costly subsequent mitigation and unplanned mitigation measures [24].

(a)          (b)

**Figure 1.** (**a**) "Walkie-Talkie" Building, London, UK [25]. (**b**) Walt Disney Concert Hall, Los Angeles, USA [25].

**Table 1.** Comparative analysis of the cited buildings.

| Name of the Building | Façade Material | Building Geometry | Effects on the Surroundings | Mitigation Strategies Applied |
|---|---|---|---|---|
| 20 Fenchurch Street-London, UK (Figure 1a) | Laminated glazing | Concave, single curvature | Reported focused solar radiation spot at pedestrian level six times higher than direct sunlight | External fins and shading systems |
| Vdara Building-LasVegas, USA | Reflective glazing | Concave, single curvature | Reported raised temperatures in the surrounding areas and sunburns on pedestrian bystanders | Application of nonreflective solar films on the façade |
| Walt Disney Concert Hall-Los Angeles, USA (Figure 1b) | Stainless steel panels | Multiple double curvature surfaces | Multiple disabling glare sources, melted asphalt pavements due to the concentrated sunlight | Diffuse and satin-finishing of surfaces |

A further representative example is the Walt Disney Concert Hall in Los Angeles (Figure 1b, Table 1), in which the freeform façade cladding in polished metal was responsible for glare and concentrated solar radiation phenomena. The latter caused the asphalt-covered pavement around the building to melt. A measurement campaign for temperature monitoring around the building recorded a 150 °C temperature over a piece of painted black foam core blackboard used as a reference absorber [19].

Concentrated irradiance is reported as a source of possible damage for all plastic or temperature-sensitive surfaces, which may experience localized melting or burns [26]. As a reference, we report that a minimum value of 8000 W/m$^2$ and 10 min of continuous exposure is needed to ignite common combustible materials, although autoignition is possible, depending on the material, only for values between 16,000 and 25,000 W/m$^2$ [27]. User comfort boundaries are included under lower values of the irradiance threshold. For short term exposures, but longer than the safe exposure time limit, which is 10 min, a radiation exposure of 1500 W/m$^2$ is considered a source of strong thermal discomfort. On the other hand, 2500 W/m$^2$ is considered the maximum value for people's safety [28] with a maximum exposure of 30 s [28]. The presence of clothing can contribute to mitigating this effect by allowing higher exposures times. A secondary effect related to the user's experience of the space, the risk of accidental damage due to direct contact with individual urban surfaces increases as their general or localized temperature increases.

Only a few cities have implemented measures that are prescriptive for the reflectance properties of the building surfaces. The city of Sydney applies a limit of 20% maximum reflectance for all the façade materials [29], and the same limit is implemented by the city of Hong Kong [30]. The planning strategies [31] mainly focus on reducing glazing areas or reshaping the texture of building surfaces to avoid any interference with the surroundings. In the literature, different approaches were tested to provide an adequate assessment of the effect of the solar radiation at the local and urban scale, but at present, there are no universally accepted criteria for the assessment of the maximum tolerance for reflected solar radiation affecting urban areas [28]. An experimental campaign aiming at testing the response of users exposed to glare found that users tend to be more tolerant to visually uncomfortable scenes while resting in an outdoor environment and performing no task or simple tasks, such as reading. Under these conditions, the subjects evaluated the glare conditions between perceptible and disturbing [32].

Reflected sunlight is in some ways unexpected due to its dependency on a scenario that is generally complex and the additional strict dependency on building geometry [33]. The definition of the right-angular optical properties of the involved surfaces [32] is also critical. The general approach leads to simulations created with dedicated software, but the results are unreliable when the accuracy of the surroundings, or of the building model itself, is not adequate.

Typically, simulations are performed during the early design stages, considering only the building masses and overlooking the presence of some façade details. Simplifications of the model are generally performed to retain simulation times within a limit of acceptability. How the building and its surroundings are simplified can strongly affect the results [34].

Raytracing methods can be used for caustics evaluation and identification of the Reflection Glare Area (RGA) [35]. The main limitation of the computational approach lies in the computing power and the level of detail requested for the model [36]. Custom-made tools are a solution to effectively include the geometry of the buildings and the optical reflectance properties of the material through the use of bidirectional reflectance distribution functions (BRDF) [22].

Experimental procedures refer to direct analysis of the scenario with High Dynamic Range (HDR) imaging of samples of exterior glare and a post-process digital analysis through a bespoke MATLAB tool [32] to identify glare sources within the context. Some other researchers have tested scale models of buildings with standardized geometries and surface materials able to redirect or concentrate the solar radiation. This is the case in [37], which tested cylindrical, concave, and triangular glass curtain walls, assessing the peak shift and the intensity of the solar radiation on their surroundings due to the building geometry.

The Boundary Reflection Area (BRA) was already proposed as a performance index for the reflection glare [38]. This approach neglects the reflectance of the building surfaces but identifies the type and the possible dimension of the region over which the reflected radiation impacts. This region, as an example of a standard test cubic building, presents a characteristic butterfly shape. All the reflections occur with the movement of the sun at the horizon. A forward-sloping façade between 10° and 20° can reduce the BRA, but determines an increase in the possible sun positions that can cause glare.

Other experiments were conducted with pure reflective surfaces resembling concave building geometries, trying to understand the effect of different parameters on the caustics. The variables considered in this study were building height, width, the radius of curvature, orientation, sun elevation, and azimuth angles [39]. The research aimed to test mathematical correlations, derived from optics, with simulations using precision software [40,41] and scale models [39].

However, as the critical review by Danks et al. evidenced [28], most of the literature focused on glare issues and general visual comfort [32,42,43], with limited investigation of the solar irradiance levels at the pedestrian level in front of tall buildings. While measurements have been performed for some case studies, these were usually carried out for a single building, without comparison between different design scenarios. A comprehensive raytracing modelling campaign by Wong assessed different buildings, but did not consider some of the geometries that lead to solar concentration [44].

Additionally, numerical modelling techniques are still in need of improvement [27], probably due to issues in the representation of the diffuse fraction of solar radiation in raytracing models, angular properties of materials, or a combination of these factors.

Yet, no systematic study compares the influence of building shape and material beyond some modelling attempts that require validation.

It is therefore important to provide the designers with a clear overview of the problem and its intensity in a way that prevents general errors during the early design stages and guides a detailed analysis that will reduce the risk of future local environmental problems that can lead to extra costs after the construction [45]. Therefore, the objectives of this research are to:

(i) quantify the solar concentration (expressed in units of sun or suns) at the pedestrian level in front of archetypes of tall buildings with diffuse and specular reflective facades;
(ii) identify the archetypes at risk of causing excessive solar concentration and harming pedestrians; and
(iii) devise a measurement protocol that can be used to quantify shortwave radiative impacts (and solar concentrations) in real buildings, to assist in the identification of the need for façade retrofits and dispute resolution.

## 2. Methodology

This paper presents the measurement process and the results obtained during the experimental campaign carried out in order to investigate the effects of reflections due to sunlight for three standardized types of skyscrapers' geometries with two different façade finishings. The cases studied have been defined based on the preliminary review carried out.

The three geometries identified as representative typologies are:

- Vertical planar façade;
- Planar façade with 10% of vertical tilt;
- Curved concave façade (with a curvature radius of 60 meters, rescaled then at model scale).

In addition to the geometries, two façade finishings were analyzed which represented two possible extreme (worst- and best-case scenario) behaviors (detailed description in Section 3.1): specular and scattering.

### 2.1. The On-Site Measurements

### 2.1.1. Experimental Scenario

The experimental tests were carried out in Milan, Italy (45°28′45.713″ North–9°13′47.937″ East, 121 m above mean sea level) on an unobstructed rooftop of a university building, equipped with a complete weather and radiometric station.

The experimental set-up consisted of a 6 m × 1.5 m (Figure 2) grey coated work plane placed at 1 m height (over the building roof). The dimension of the plane and optimal measurement grid was defined using preliminary hourly simulations (with Rhinoceros 5.0 [46] and Grasshopper [47]), considering forward raytracing algorithms and Fresnel geometrical reflections. The grey matte base (albedo = 0.36) was selected for the plane as a representative mean reflectance of urban albedo.

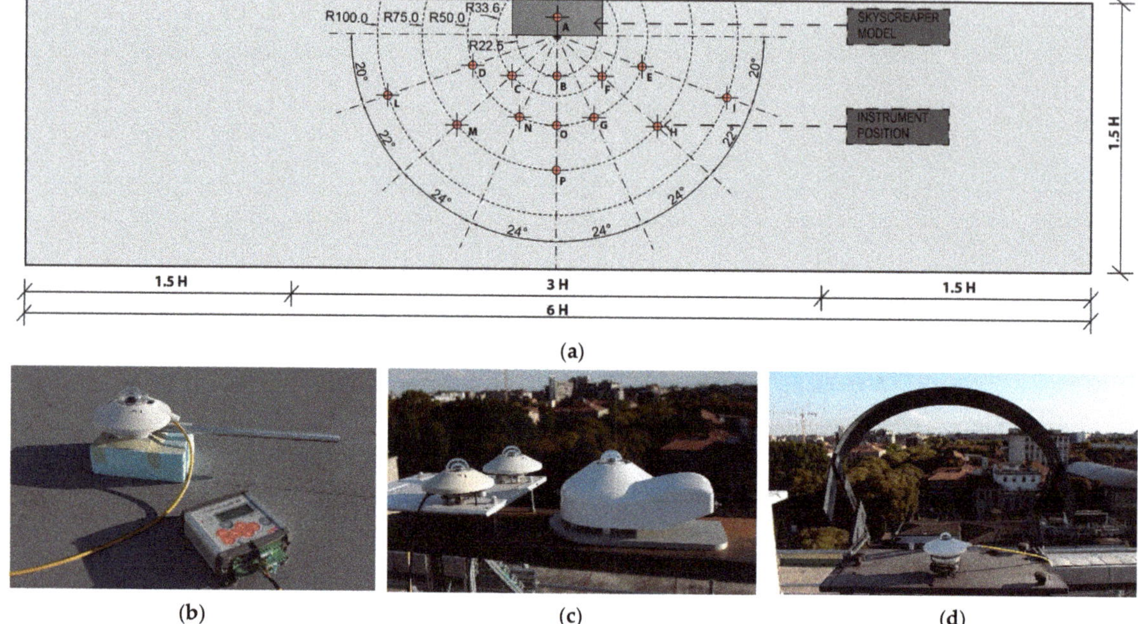

**Figure 2.** (**a**) 6 H × 1.5 H measurement plane (H is the height of the Skyscraper model that in the presented experiment is equal to 1 m) with measurement positions. (**b**) Moveable albedometer (CMA11) with Datalogger. (**c**) Pyranometer CM21 and CM22 are part of the weather station that provides undisturbed reference values. (**d**) Pyranometer CM6 with shadow-band that provides undisturbed reference values.

Based on the results obtained from the preliminary analysis, the positions of the measurement points were defined with a double construction: the points were placed in the intersection between a radial subdivision (relatively 20° and 22° in the external part and 24° in the central part) over a circumferential construction (relatively with a radius of 100 cm, 75 cm, 50 cm, 33.6 cm, 22.5 cm), as shown in Figure 2. This approach allowed the identification of the behavior of both diffuse and specular reflectors for the three geometries. Furthermore, this configuration allowed the evaluation of the impact of the façade on possible relevant context areas close to the building model, such as squares, streets, and adjacent buildings.

### 2.1.2. Scale Models

In order to assess the increment of solar radiation generated by tall building solar reflection, three skyscraper scale models were built based on a review of contemporary skyscraper dimensions and shapes [35,41,45,48].

The 1:100 scale models were 20 × 50 × 100 cm parallelepiped-shaped wood structures with white diffusive finishing, except for the front façade surface which could be replaced according to the required analysis. Other authors have proposed another similar reference model in a virtual scenario [35], where four buildings (with concave, convex, angular, and planar geometries) with a façade dimension of 100 × 40 m were considered representative (Figure 3).

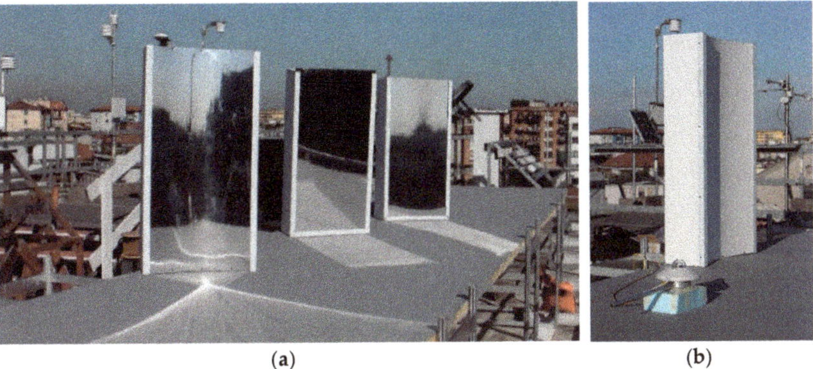


**Figure 3.** (**a**) Three skyscraper scale models with reflective film as a finishing surface. (**b**) Concave skyscraper scale model with white scattered film as a finishing surface.

Two finishing materials were selected for the front changeable façade: a white scattering diffusing surface and a specular reflective material, which are representative of the extreme cases in façade applications.

These case study geometries permitted the investigation of the reflection phenomenon, during a clear summer day, for three characteristic tall building shapes.

### 2.2. Experimental Sample Material Properties

To identify the adequate façade diffusive and specular materials for the tests, different white finishing paints and mirror films were measured to find their relative spectral reflectance values. Two of them with similar solar reflectance values were selected for the experiment to ensure the differences were due to variations in their optical angular behavior and not in their total reflectance. A Perkin Elmer Lambda 950 Spectrometer was used (for wavelengths between 250 and 2500 nm) according to UNI 14500 [49], and the values were post-processed following the ASTM E903 [50] procedure. Figure 4 shows the results of the measurement procedure for the selected materials used during the experimental campaign.

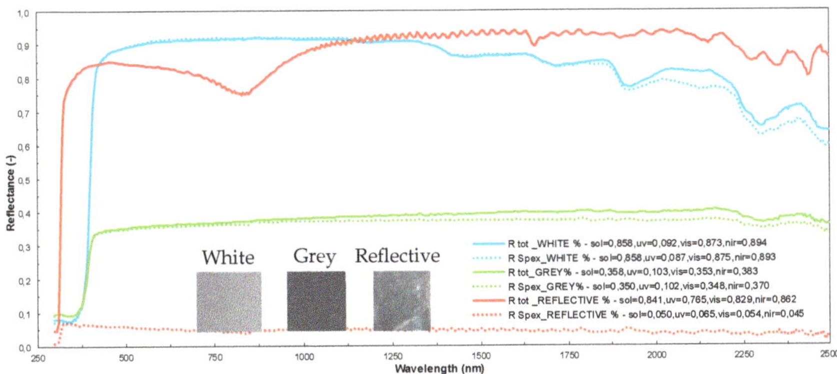


**Figure 4.** Spectral and computed solar (sol), UV, visible (vis), and near-infrared (nir) reflectance of the diffusive materials and specular film. The measurements are repeated for total (R tot) and specularity-excluded (R Spex) reflectance.

As a scattering diffuse surface (Figure 4, blue line), a typical white painting with a solar reflectance equal to 0.86 was chosen. It showed a typical spectral curve of light-diffusive materials. The specular surface selected was a metallized mirror-like (especially for high-reflectance angles) polyester film (Figure 4, red line) that had a mean of 0.85 as its solar reflectance value. From a preliminary qualitative analysis, not having the possibility to perform complete BRDF for the material under analysis through the comparison of the reflectance curves (total and specular excluded), we can say that the specular component remains predominant regardless of the angle of incidence, although we cannot state this with certainty.

With the aim of preliminarily evaluating if the mirror film is perfectly reflective, the specular excluded reflectance was measured. Analyzing the obtained results, it is possible to highlight that 95% of solar reflectance is due to a specular component. Both finishing materials presented almost the same integral value of solar reflectance, and the choice was intentional to compare the results of solar radiation insisting on the surroundings under different reflective behavior of the building models.

Figure 4 also shows both total and specular excluded reflectance of the background plane used. This grey surface was selected because it represents the standard urban surface with an albedo of 0.35 [51–53].

Instruments and Measurements Procedure

Three different sets of instruments (some of which were from the weather station placed next to the measurement site) were used:

- An albedometer (CMA11 by Kipp & Zonen), with data recorded by an M-Log logger (by LSI) placed over a specific plastic support, was used to measure solar irradiance in different positions over the test plane (Figure 2b). The CMA11 is a secondary standard albedometer with a maximum solar irradiance value equal to 4000 $W/m^2$ and 5 s of response time.
- CM21 and CM22 pyranometers (by Kipp Zonen) were used to measure the undisturbed solar irradiance and calibrate the albedometer.
- A CM6 pyranometer with a shadow-band was used to measure the diffuse component of solar radiation.
- A thermal infrared camera was used to verify the temperature increase on the plane due to reflections.

All the measurements were carried out in one week in order to have similar sun position and radiation values. The survey was performed during a typical Italian summer clear-sky day, from the 19 to the 25 of September, during the daylight hours from 9 a.m. to 6 p.m. with a maximum solar elevation of 46.04°. The measurements were taken under equal solar diffuse fraction, namely with the same diffuse/global ratio during the same hour.

The Unit of Sun (UoS) was defined as a normalized value describing the ratio of the reflected irradiance over the ambient solar irradiance measured on the horizontal surface on the top of the scaled building mock-up, as defined in [34].

A total of 54 scenarios were measured, combining the three building geometries with the two previously described alternatives for the façade finishing material.

During each measurement session, the solar irradiance values in the 14 points over the plane were measured, moving the albedometer every minute (so that the measurement readings could be considered as stable) and over each position. Data acquisition time depended on the instrument response time lag and sky conditions. Tmeasurement sequence, namede with with progressive letters; start from the acquisition of undisturbed solar radiation (above the skyscraper), passing through the points over the measurement plane and concluding with a vertical solar irradiance. All the geometries and materials have been analyzed with the same procedure, as previously explained. All the recorded values were compared with the ones gathered from a reference weather station located on the same floor of the measurement plane.

### 2.3. The Simulations Workflow

#### 2.3.1. The Façade Material Benchmark

The optical performance of glass is typically angular; namely, it depends on the angle of incidence of radiation that hits the surface. In general, and for incidence angles $\alpha < 60$–$70°$, the visible solar transmittance of a single or double glass is close to the value measured for a normal incidence ($\alpha = 0°$). For angles of incidence above this threshold, the transmittance value decreases, while the reflectance increases exponentially.

In order to understand how representative the choice of a highly reflective finishing was for our building model, the yearly high solar reflectance behavior of a typical glazing façade made with a Double-Glazed Unit (DGU) was evaluated.

Two types of glass were selected to describe two common DGUs that could be installed in a new skyscraper building in accordance with its optical and energy needs (Table 2). The two selected materials also allowed us to evaluate the extreme behaviors that could include all the possible causes related to intermediate properties of systems and components. The selected DGUs included one with high reflectance and solar control properties, penalizing light transmission properties, and a second DGU with good solar control values and a high selectivity index.

**Table 2.** Solar and visual properties of the Double-Glazed Units (DGUs) selected as a reference for our analysis, considering solar transmission ($\tau_s$), solar reflectance ($\rho_s$), and visual transmittance ($\tau_v$).

| Description | Code | $\tau_s$ [%] | $\rho_s$ [%] | $\tau_v$ [%] |
|---|---|---|---|---|
| High Reflective Sun Control_DGU | HR_SunC | 9.9 | 53.3 | 15.1 |
| Selective_DGU | SEL | 31.6 | 23.9 | 66.5 |

In Figure 5, the angular solar transmittance and reflectance properties are reported for the DGUs listed in Table 2. The values were computed using LBNL WINDOW 7.7 [54]. The normal reflectance value of the HR_SunC case was approximately twice that measured for the Selective (SEL) DGU. For high incidence angle of solar radiation (i.e., >70°), the percentage increases of the reflectance of the two analyzed DGU types are comparable, regardless of whether the two measured reflectance values at normal incidence ($\rho_s$) are very different.

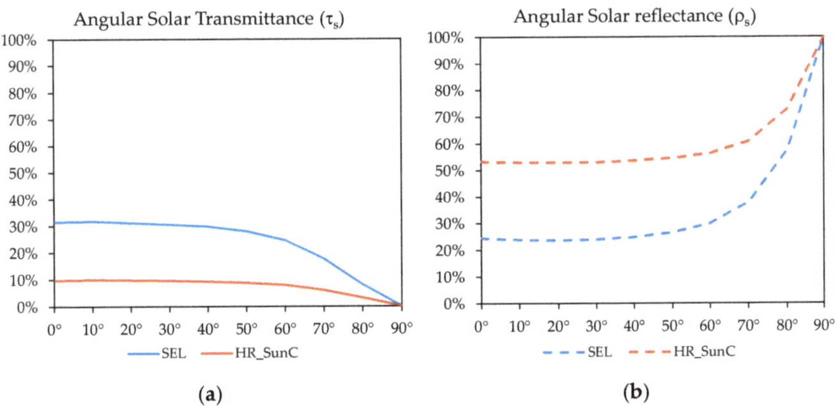


**Figure 5.** Angular solar transmittance—$\tau_s$ (a) and angular solar reflectance properties—$\rho_s$ (b) of a selective (SEL) and High Reflective Sun control (HR_SunC) double glass unit. On the x-axis, the incidence angle $\alpha$ is reported.

#### 2.3.2. A Parametric Analysis Script for Unwanted Reflections of a Glazed Façade

Regarding the above considerations (Section 2.3.1) on the variation of the optical performance of transparent systems, a parametric script was developed in Rhinoceros

5 [46] within the Grasshopper [47] environment to evaluate the probability of occurrences of unwanted solar reflection phenomena for a generic curtain wall surface. The script considered: the variability of the site (latitude/longitude); the orientation of the building masses; and the slope of the surfaces [55].

The script considered the solar rays as vectors to reduce calculation time, allowing a comprehension of the number of rays that hit the surface. The script was based on the following hypotheses:

- Every "solar ray" represents the sun's position in the middle of each sun hour of radiation;
- A quad mesh subdivision of the building mass surface was used in order to replicate a realistic building façade panelization, made through the use of discrete glazed elements.

The working flow proposed (Figure 6) lists the incident solar rays coupled with the related normal vector of every mesh tile. This approach allowed the script to correlate the initial vector list with the angular degree, excluding (by the use of filters) the portion of rays not required, based on the designer criteria.

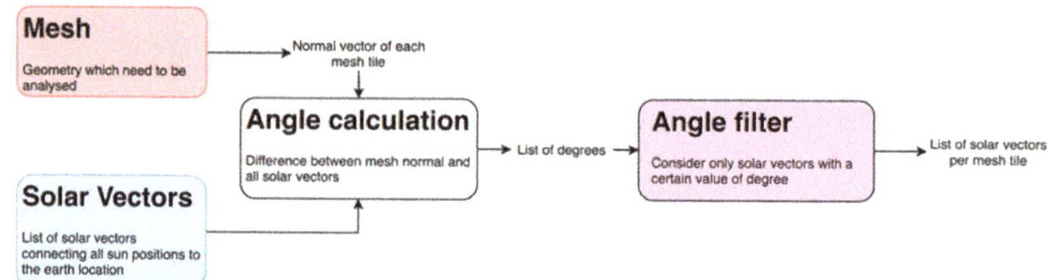


**Figure 6.** Grasshopper script workflow.

The parametric model defined is capable of parsing sunrays incidence angles based on their inclination for each surface normal vector. The system creates a virtual circular radiation cone (Figure 7a), with its vertex on the surface's central point and a perpendicular orientation to the face.

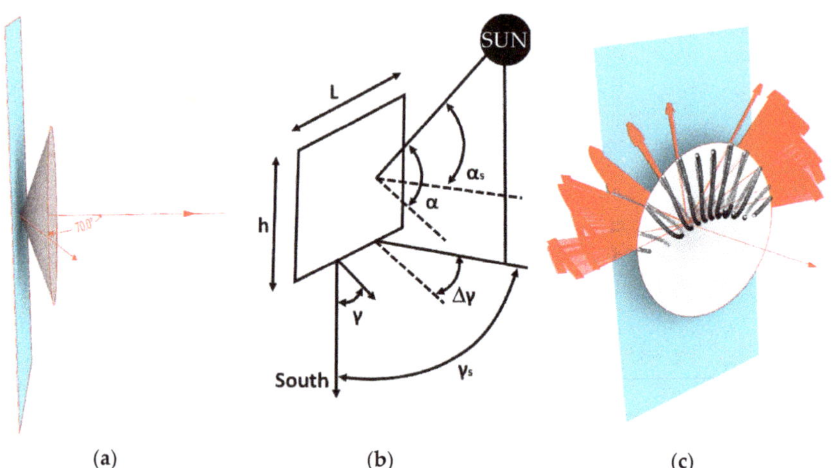


**Figure 7.** (**a**) The extent of the radiation cone that has a lower probability of creating unwanted solar reflections. (**b**) The model for evaluating the incidence solar radiation angle. (**c**) Representation with, highlighted in red, the hourly annual solar positions that have the maximum probability of creating unwanted reflections.

Such geometry allows the control of the cone angle by increasing or reducing the number of rays hitting the surface beyond a certain degree. As shown in Figure 5, an angle between 70° and 90° could generate reflected radiation on the surroundings. Considering this, the scripts can directly provide the number of hours over the entire year in which a mesh tile has this behavior (Figure 7c).

For every mesh tile, a virtual circular cone of a certain amplitude, based on the glass properties, was created in its center. The cone represents the solar rays filter; in this way, it is possible to visualize the number of vectors between 90° (the tile plane) and the angle $\alpha$ of the cone.

The generic incidence angle of the solar radiation over the surface was evaluated in accordance with [56], using the following equation (based on Figure 7b):

$$\tan(\alpha) = \frac{\tan(\alpha_s)}{\cos(\Delta\gamma)}$$

where:

- ($\alpha$) is the incidence angle;
- ($\alpha_s$) is the hourly solar altitude; and
- ($\Delta\gamma$) is the difference between the hourly solar azimuth ($\gamma_s$) and the azimuth of the surface normal ($\gamma$), both measured from the South.

Based on the geographical location and orientation of each mesh, the script can evaluate the entire spatial distribution of solar rays hitting the interested surfaces over a year. Complex façade geometries and double-curved envelopes can potentially create over-shadowing effects, hiding a façade portion from solar rays, because of the coverage of part of the sky vault.

The parametric script (Figure 8) provides a double filter level which excludes from the analysis all the rays screened by the obstructions and the ones out of the portion of the skydome seen by each analyzed surface. Once the filtering has been performed, it is possible to retrieve the amount of the reflected radiation by using the radiation cone.

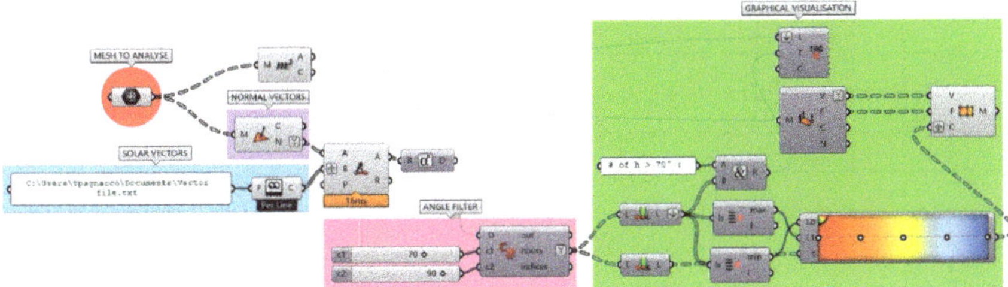

**Figure 8.** Example of the parametric script developed in grasshopper.

## 3. Results

### 3.1. Experimental Measurements

The experimental measurement presented in the following section is a part of the entire measurement campaign carried out and completely reported in Appendices A–F. In order to present the recorded value in a comparable way the undisturbed values recorded by the weather station will be taken as reference.

The 25th of September had almost a completely clear sky condition; during the other two days, some atmospheric turbidity was present in the central part of the day (Figure 9b).

Tables 3 and 4 and Appendices A–F include all the irradiance values recorded for the registered interval and for the points that were selected as representatives to describe the magnitude of the solar radiation over the surroundings of each model.

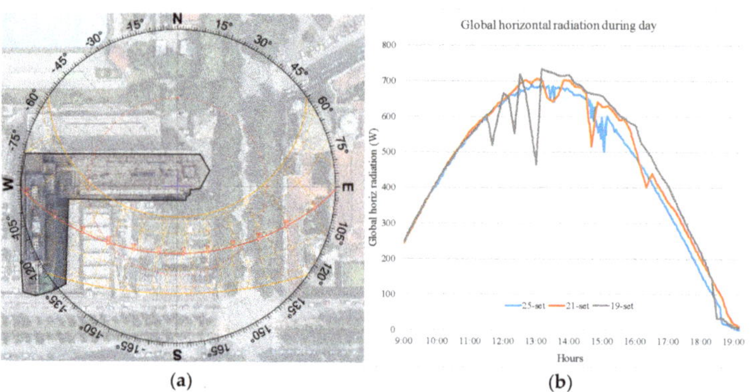


**Figure 9.** (**a**) Test site plan view (Politecnico di Milano, Nave Building) with September sun path. (**b**) Global horizontal radiation during test days (19, 21 and 25 September).

**Table 3.** Reflection for flat façade models: preliminary analysis and measurement results for 10:00, 13:00, and 16:00. For each point and in bold: the measured value (Meas.), the non-disturbed measure (N.d.Meas), the Unit of Sun (UoS), and the time (Time) of the measure. All the measurements are shown in Appendix B. In grey the measurements points within the solar reflection area.

| Flat Reflective Façade | | | | |
|---|---|---|---|---|
| **h. 10:00** | | **Meas.** [W/m²] | **N.d.Meas** [W/m²] | **UoS** [-] | **Time** [hh:mm] |
| | A | 406 | 410 | 0.99 | 10:00 |
| | B | 422 | 413 | 1.02 | 10:01 |
| | C | 754 | 416 | 1.81 | 10:02 |
| | D | 431 | 420 | 1.03 | 10:03 |
| | E | 467 | 422 | 1.11 | 10:04 |
| | F | 445 | 424 | 1.05 | 10:05 |
| | G | 444 | 424 | 1.05 | 10:05 |
| | H | 450 | 428 | 1.05 | 10:06 |
| | I | 458 | 428 | 1.07 | 10:06 |
| | L | 439 | 431 | 1.02 | 10:07 |
| | M | 870 | 434 | 2.00 | 10:09 |
| | N | 441 | 439 | 1.01 | 10:10 |
| | O | 438 | 442 | 0.99 | 10:11 |
| | P | 437 | 446 | 0.98 | 10:12 |
| **h. 13:00** | | **Meas.** [W/m²] | **N.d.Meas** [W/m²] | **UoS** [-] | **Time** [hh:mm] |
| | A | 693 | 686 | 1.01 | 13:06 |
| | B | 1321 | 687 | 1.92 | 13:07 |
| | C | 705 | 685 | 1.03 | 13:08 |
| | D | 698 | 685 | 1.02 | 13:09 |
| | E | 694 | 685 | 1.01 | 13:10 |
| | F | 708 | 684 | 1.03 | 13:11 |
| | G | 711 | 684 | 1.04 | 13:12 |
| | H | 691 | 685 | 1.01 | 13:13 |
| | I | 697 | 686 | 1.02 | 13:14 |
| | L | 686 | 681 | 1.01 | 13:15 |
| | M | 685 | 676 | 1.01 | 13:16 |
| | N | 704 | 676 | 1.04 | 13:17 |
| | O | 1327 | 683 | 1.94 | 13:18 |
| | P | 1374 | 684 | 2.01 | 13:19 |

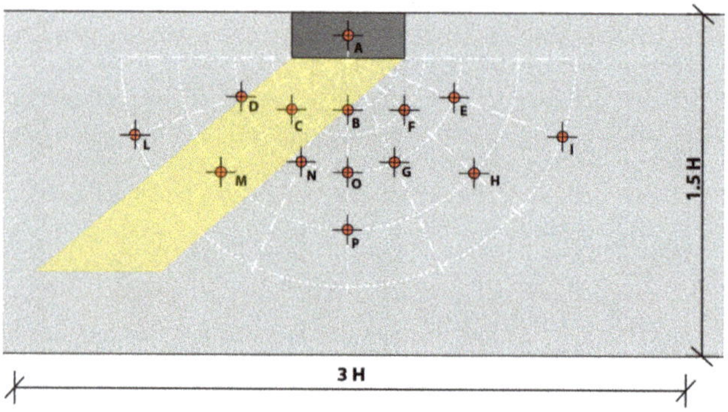

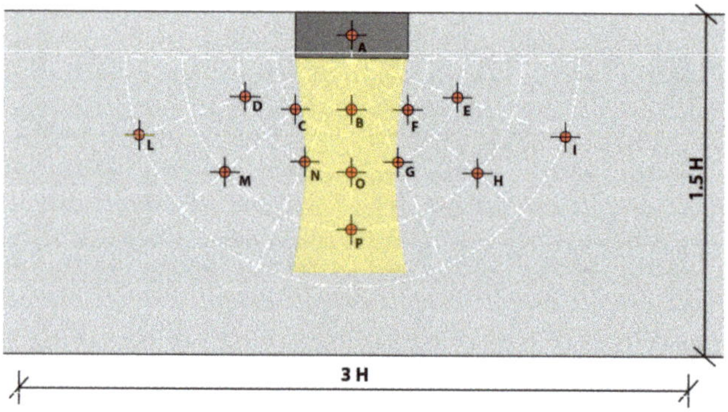

The plan view diagrams (h. 10:00 and h. 13:00) show measurement points A–P with dimensions 3 H (width) and 1.5 H (height).

**Table 3.** *Cont.*

| Flat Reflective Façade | | | | |
|---|---|---|---|---|
| **h. 16:00** | **Meas. [W/m²]** | **N.d.Meas [W/m²]** | **UoS [-]** | **Time [hh:mm]** |
| A | 462 | 456 | 1.01 | 16:14 |
| B | 481 | 452 | 1.06 | 16:15 |
| C | 468 | 452 | 1.04 | 16:15 |
| D | 486 | 449 | 1.08 | 16:16 |
| E | 455 | 449 | 1.01 | 16:16 |
| F | 782 | 446 | 1.75 | 16:17 |
| G | 428 | 441 | 0.97 | 16:19 |
| H | 635 | 441 | 1.44 | 16:19 |
| I | 453 | 439 | 1.03 | 16:20 |
| L | 445 | 438 | 1.02 | 16:21 |
| M | 442 | 434 | 1.02 | 16:22 |
| N | 440 | 428 | 1.03 | 16:23 |
| O | 440 | 428 | 1.03 | 16:23 |
| P | 435 | 428 | 1.02 | 16:24 |

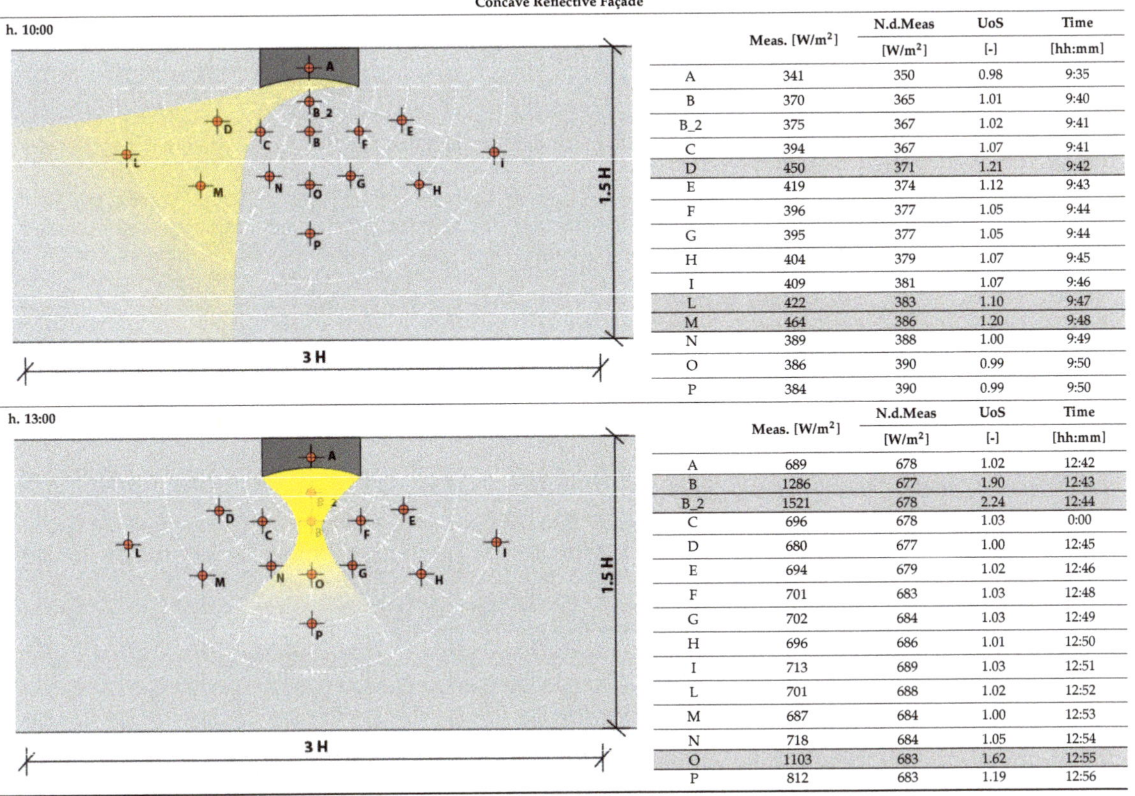

**Table 4.** Reflection for curved façade models: preliminary analysis and measurement results at 10:00, 13:00, and 16:00. For each point and in bold: the measured value (Meas.), the non-disturbed measure (N.d.Meas), the Unit of Sun (UoS), and the time (Time) of the measure. B_2 was introduced to better characterize the reflection next to the façade. All the values are shown in Appendix F. In grey the measurements points within the solar reflection area.

| Concave Reflective Façade | | | | |
|---|---|---|---|---|
| **h. 10:00** | **Meas. [W/m²]** | **N.d.Meas [W/m²]** | **UoS [-]** | **Time [hh:mm]** |
| A | 341 | 350 | 0.98 | 9:35 |
| B | 370 | 365 | 1.01 | 9:40 |
| B_2 | 375 | 367 | 1.02 | 9:41 |
| C | 394 | 367 | 1.07 | 9:41 |
| D | 450 | 371 | 1.21 | 9:42 |
| E | 419 | 374 | 1.12 | 9:43 |
| F | 396 | 377 | 1.05 | 9:44 |
| G | 395 | 377 | 1.05 | 9:44 |
| H | 404 | 379 | 1.07 | 9:45 |
| I | 409 | 381 | 1.07 | 9:46 |
| L | 422 | 383 | 1.10 | 9:47 |
| M | 464 | 386 | 1.20 | 9:48 |
| N | 389 | 388 | 1.00 | 9:49 |
| O | 386 | 390 | 0.99 | 9:50 |
| P | 384 | 390 | 0.99 | 9:50 |
| **h. 13:00** | **Meas. [W/m²]** | **N.d.Meas [W/m²]** | **UoS [-]** | **Time [hh:mm]** |
| A | 689 | 678 | 1.02 | 12:42 |
| B | 1286 | 677 | 1.90 | 12:43 |
| B_2 | 1521 | 678 | 2.24 | 12:44 |
| C | 696 | 678 | 1.03 | 0:00 |
| D | 680 | 677 | 1.00 | 12:45 |
| E | 694 | 679 | 1.02 | 12:46 |
| F | 701 | 683 | 1.03 | 12:48 |
| G | 702 | 684 | 1.03 | 12:49 |
| H | 696 | 686 | 1.01 | 12:50 |
| I | 713 | 689 | 1.03 | 12:51 |
| L | 701 | 688 | 1.02 | 12:52 |
| M | 687 | 684 | 1.00 | 12:53 |
| N | 718 | 684 | 1.05 | 12:54 |
| O | 1103 | 683 | 1.62 | 12:55 |
| P | 812 | 683 | 1.19 | 12:56 |

**Table 4.** *Cont.*

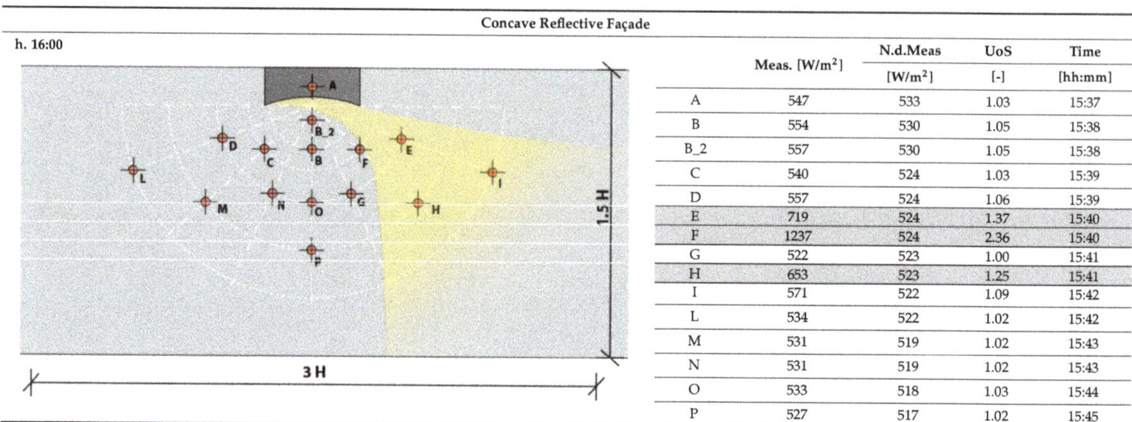



| | Meas. [W/m²] | N.d.Meas [W/m²] | UoS [-] | Time [hh:mm] |
|---|---|---|---|---|
| A | 547 | 533 | 1.03 | 15:37 |
| B | 554 | 530 | 1.05 | 15:38 |
| B_2 | 557 | 530 | 1.05 | 15:38 |
| C | 540 | 524 | 1.03 | 15:39 |
| D | 557 | 524 | 1.06 | 15:39 |
| E | 719 | 524 | 1.37 | 15:40 |
| F | 1237 | 524 | 2.36 | 15:40 |
| G | 522 | 523 | 1.00 | 15:41 |
| H | 653 | 523 | 1.25 | 15:41 |
| I | 571 | 522 | 1.09 | 15:42 |
| L | 534 | 522 | 1.02 | 15:42 |
| M | 531 | 519 | 1.02 | 15:43 |
| N | 531 | 519 | 1.02 | 15:43 |
| O | 533 | 518 | 1.03 | 15:44 |
| P | 527 | 517 | 1.02 | 15:45 |

Experimental measurements showed a significant increase in solar irradiance values due to solar reflection, both for specular and scattered materials. For all the geometries, the irradiance values were strictly connected with the façade shape and measurement position. With the white scattering surface, it is possible to notice an overall increase in solar irradiance depending on the distance between the measurement point and the scale models. No significant variations were observed with a change in the geometries of the building models, meaning that for the scattering material, the only significant variable is the distance from the façade.

The analysis of the specular reflective surface showed a completely different pattern. Indeed, outside the reflection area, all the geometries show values equal to the undisturbed one, while inside the reflected area the values recorded are up to five times greater than the solar radiation on the horizontal plane (façade with a concave geometry presented in Table 4).

Considering the concave surface, this unique building geometry concentrates the reflected sunlight in a small focal point characterized by a strong increase of perceived light (Table 4) and of surface temperature intensity, which can only be estimated due to the nature of the urban environment that is also influenced by the surface's thermal mass, solar absorbance, and emissivity, such as transient local parameters (such as wind velocity and water presence, as in [28,34]).

Indeed, for the flat and 10% tilted façade, the values doubled the undisturbed solar irradiance for the concave shape. This was due to the Fresnel effect on light reflections, which is greater than five times in the focal point (no precise value has been measured, as it was higher than the full scale of the Table 4. For this reason, Tables 3 and 4 show the results obtained for the flat reflective and the concave reflective façades in three parts of the day.

Compared to scattering material, where the solar irradiance curves at ground level and presents a flat upward shift with respect to horizontal irradiance, for high specular material, it is possible to notice peaks depending on the measurement points and solar position. For the flat reflective façade and in every analyzed position (Figure 2a), a peak value between 1200–1350 W/m² was reached during different hours of the day, and a result was obtained that is two times more than the measured horizontal irradiance (Table 3).

Curved façades behave like a solar concentrator, generating reflection tracks and high irradiance values on the ground. Inside the reflection path, the irradiance values exceed 40–60% of the undisturbed radiation values, while on the shape edges it is possible to reach 100–130% higher irradiance values. The critical area for collector shapes is the focus; near the focus the irradiance value can reach 1800–2200 W/m² compared to the global horizontal irradiance of 650–700 W/m². Inside the focus, values higher than 3000 W/m² have been reached, and at noon, the instrument limit (4000 W/m$^2$) was overtaken, implying that it reached values far above this threshold. Table 4 shows different light-track geometries and the respective recorded values for each measurement position.

Figure 10 shows a comparison, assuming the same type of façade (flat) and same measurement position, between the scattered and the specular façade materials. In Figure 10a, it is possible to see that during the day, the increase of reflectance of the scattering material is affected by the distance from the façade. However, in Figure 10b, it is possible to see that regarding the specular material, when the measurement point is inside the reflected area there is twice the irradiance. When outside, the values are the same as those of the undisturbed one.

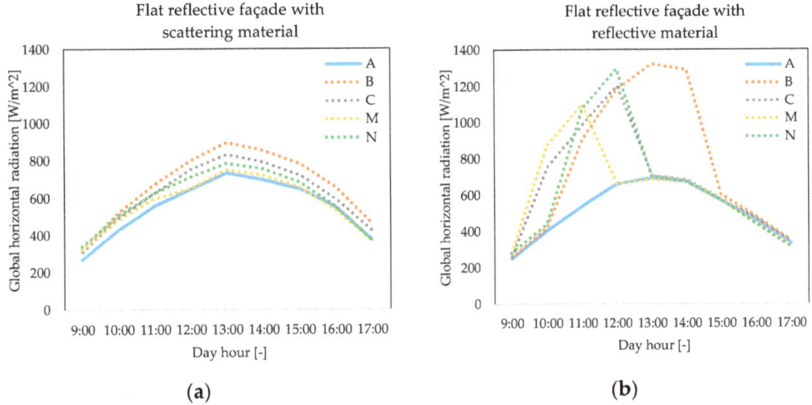


**Figure 10.** Time-dependency value of solar radiation for five measurement points (A, B, C, M, N) for scattering (**a**) and specular (**b**) planar façade geometry (21 September).

The use of the selected light-diffusing material shows radiation values constantly increasing in all the directions around the building model with an intensity that changes only in relation to the distance from the façade (the measured peak is ~133% at 15 cm distance from the façade, equivalent to a H/D ratio of 6.66 in a real building scenario and in which D is the Distance from the façade and H the Height).

Regarding the reflective materials, the behavior of the three geometries is completely different. For the flat geometry, the value of solar irradiance in the reflected area reached a peak of ~200%; while outside of the "reflection zone", the values were almost similar to the irradiance measured on the horizontal plane. The curved façade can have an easy prediction of the caustic shape. The curvature of the façade itself must be adequately large to reduce the intensity of the solar radiation in the focal point, otherwise, the extent of the reflective or specularly reflective façade material should be reduced. The behavior of the 10% tilted flat geometry is similar to that of the flat vertical one, with a similar shape of the reflected area, but with less extension from the building façade (due to its 10% inclination) and radiation values inside that are slightly higher. The convex curved geometry produced a focal point in which the solar irradiance reached values higher than ~300%.

Some authors [57] suggest the use of alternate finishings on the façade, or different materials to avoid reflections problems over the pedestrians. A standard geometry building facing an urban canyon was considered as the reference example. In this case, the use of reflective materials under the fourth floor was discouraged, while the use of retro-reflective or purely diffuse materials was suggested.

A further measurement carried out was thermography. Thanks to these measurements, it was possible to indirectly identify the temperature reached by the surface (with the grey coating shown in Figure 4) due to reflection. With this approach, it is also possible to identify the behavior of the concave façade from a quantitative point of view, as shown in Figure 11. In the focus of the parabola, describing the geometry of the parabolic façade, a severe concentration of solar radiation is present, but on a very limited portion of the surrounding plane. A zone in which the surface temperature considerably exceeds the reference of the temperature scale, set at 100 °C, is clearly identifiable. The distribution of temperatures in the other cases examined is uniform over a larger area of reflection and with values between 60 and 80 °C. The surface temperature results for the curved vertical façade are comparable with the literature findings.

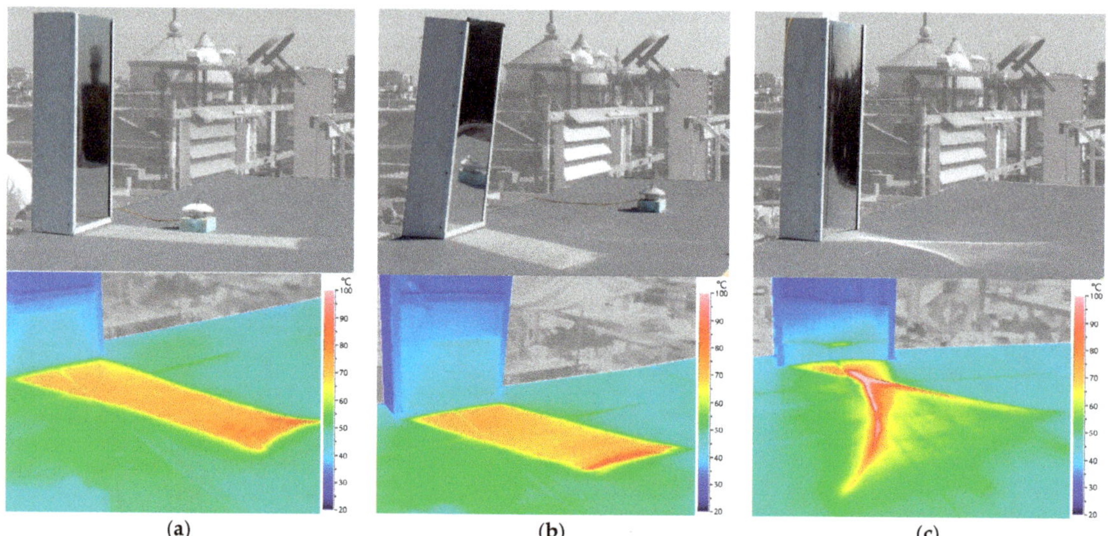

**Figure 11.** (**a**) The reflection shapes (photo on the top and thermography on the bottom) of the skyscraper scale model with a flat vertical façade coated by the reflective film. (**b**) The reflection shapes (photo on the top and thermography on the bottom) of the skyscraper scale model with a flat 10% tilted façade coated by the reflective film. (**c**) The reflection shapes (photo on the top and thermography on the bottom) of the skyscraper scale model with a curved vertical façade coated by the reflective film.

*3.2. Simulated Frequency Distribution of Solar Reflection Occurrences: Hourly Annual Distribution*

Among the results of the preliminary assessment of the building masses, and obtainable through the developed script, it is possible to derive a temporal evaluation of the hours and days during the year in which a generic façade glazed tile and part of the façade meshes are likely to be subject to phenomena of reflection and/or concentration of solar radiation. This can be considered a preliminary risk assessment that depends on geometry, latitude, longitude, and orientation. The following results are exemplificative of the potential risk of a glazed vertical façade module.

The analysis was carried out to determine all the possible angles of incidence of solar radiation that annually can insist on a building with a flat vertical façade facing South (S), East (E), or West (W), in addition to the two intermediate positions, South-East (SE) and South-West (SW). This model is representative of one of the test cases analyzed during the experimental campaign. This hourly analysis was developed considering the latitude and longitude of Milan.

A representative hourly annual distribution of the solar reflection occurrences is presented in Figure 12 for the South-exposed façade.

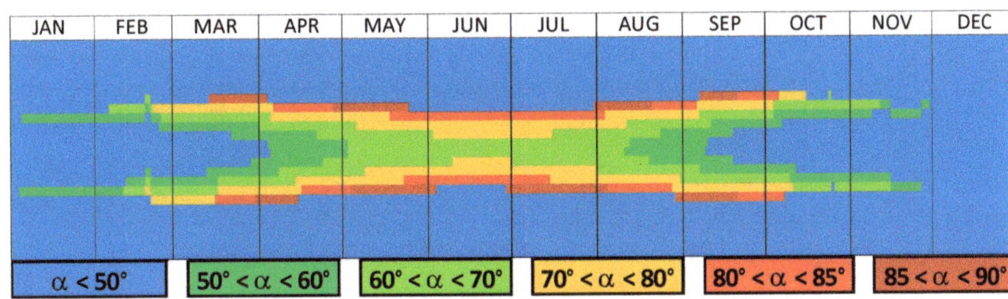



**Figure 12.** Annual hourly distribution of the incidence angles of the solar radiation over the façade of the classical vertical South-oriented skyscraper model.

In the following carpet graph, the different days of the month are reported on the x-Axis and the different hours of the day on the y-Axis. It is then possible to highlight that the first and the last hours of the day during the winter months are critical, and particular attention must be paid to the central hours of the day during the summer months when the incidence angles distributions are greater than 80°.

In general, we considered the 70° angle of radiation incidence as a threshold that could generate negative phenomena of concentrated reflection of solar radiation on the surrounding context. Since the facade of the building model considered is planar and vertical, it is possible to assume that each module of the facade, i.e., each portion of it, has a uniform and homogeneous behavior, respecting the previously identified rules for possible unwanted reflections.

Table 5 shows the incidence angle frequencies compared to the total number of hours of light during the year. The results show that the phenomenon is not negligible, since it afflicts the façade of the building between 22% and 28% of the time, in the same way. We note that the most critical exposure for this type of façade geometry is South, followed by the East (or West), and finally the couple SE/SW. Other façade geometries and alternative locations could lead to different distributions of the angles of incidence, increasing or decreasing the number of critical hours.

**Table 5.** Percentage of solar radiation incidence angles per orientation over the flat façade of the vertical building. The percentages are related only to sunlight hours.

| Incidence Angle $\alpha$ of the Solar Radiation Per Orientation—Flat Vertical Façade | | | |
|---|---|---|---|
| $\alpha$ | E/W | SE/SW | S |
| $\alpha < 50°$ | 51% | 42% | 34% |
| $50° < \alpha < 60°$ | 15% | 22% | 16% |
| $60° < \alpha < 70°$ | 11% | 15% | 21% |
| $70° < \alpha < 80°$ | 8% | 11% | 16% |
| $80° < \alpha < 85°$ | 10% | 5% | 7% |
| $85° < \alpha < 90°$ | 5% | 5% | 6% |
| Total > 70° | 23% | 21% | 29% |

*3.3. Simulated Frequency Distribution of Solar Reflection Occurrences: Spatial Surface Distribution*

In case of any complex façade surfaces, and for every single mesh, which describes the panelized surface of the generic building mass, the number of hours per year, or the possible occurrences of unwanted reflection, they can be represented using a false color scale. The model is simplified, not considering the solar deflection of the glass, which could modify or amplify the occurrence of the phenomenon.

Unlike the analysis presented in the previous section, it is not possible to know which hours and periods of the year unwanted reflections phenomena occur. However, it is possible to identify which portions of the building mass surface are characterized by the highest number of negative occurrences. The script, in this case, was therefore used as a pre-assessment of the proposed geometry, favoring the designer's activity in suggesting possible variations to provide effective proposals for the modification of the geometry during the design phase or to provide local treatments to mitigate the effects.

Therefore, it is possible to make qualitative deductions comparable to the following:

- In the southern-exposed facades with parabolic sections, the possibility of negative effects becomes greater while moving away from the geometric focus of the parabola that describes the surface (Figure 13a);
- In the case of east-facing exposed parabolic surfaces, the area that is close to the focus of the parabola seems more critical (Figure 13b).

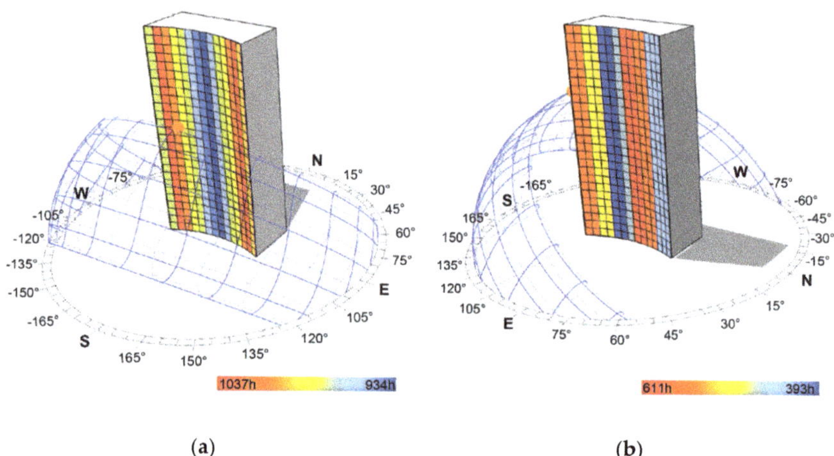

(a)       (b)

**Figure 13.** (**a**) Building mass with a parabolic façade facing South. (**b**) Building mass with a parabolic façade facing East. Both buildings are located in Milan.

Further developments will be the subject of analysis in future publications, in which the script will be expanded in its possibilities of use.

## 4. Discussion

Our results quantify the risk of solar concentration posed by buildings with specularly reflective facades, especially with a concave façade geometry facing the equator. A single high rise building with a flat diffusive façade with an albedo of 0.86 (i.e., an unsoiled white) causes an increase of the incident solar irradiance on the ground by a maximum of 20% (i.e., 1.2 suns at pedestrian level). This is still a significant increase, although the hardware model represents a worst-case scenario, without windows or overhangs.

As in [16], if we consider a building of indefinite length, we can compute the fraction of incident radiation that returns to the sky, which is 36% for a Lambertian wall and pavement with albedo equal to 0.60 and 0.20, respectively. In our case, the fraction of solar radiation returning to the sky was 58% [computed as 0.86 wall albedo × (0.5 sky view factor + 0.5 ground view factor × 0.36 ground albedo)]. Levinson et al. computed that a retro-reflective wall with albedo of 0.60 increases the solar radiation escaping the city to 55% (with a Lambertian street pavement with albedo = 0.20) [16].

Instead, the mirror-like finish, both in the flat and concave configuration, leads to peak irradiances reaching the ground, even three times the incoming irradiance (i.e., 3 suns). This does not approach the degree of solar concentration achieved by Fresnel reflectors designed for high concentration photovoltaics [58], which can exceed 1000 suns. However, local effects and some polished metal cladding might achieve higher values than those we measured, possibly approaching the 25–50 suns of the early developments in solar concentration PV [59]. Indeed, the façades of buildings might perform as modular Fresnel lenses. Therefore, our results do not constitute a worst-case scenario, but rather a realistic scenario of downward reflection of solar radiation with a specularly reflective façade.

Some authors [57] suggest the use of alternate finishings on the façade, or different materials to avoid reflection problems for pedestrians. A standard geometry building facing an urban canyon was considered as the reference example. In this case, the use of reflective materials under the fourth floor was discouraged, while the use of retro-reflective or purely diffuse materials was suggested.

Retro-reflective materials have been proposed as a solution. However, like high-albedo materials [60], they are subject to ageing problems, which decrease their reflectance and retro-reflection over time. Their performance is almost fully recovered after cleaning only by prism-type retro-reflectors [61]. However, only a limited number of studies have been performed on the durability of retro-reflectors. Furthermore, retro-reflectors display the maximum upward to downward reflection ratios for low angles of incidence [62]. This may provide a positive performance at sunrise and sunset, but the retro-reflection ratio is limited to 30–50% during peak hours. Therefore, the application of retro-reflective materials cannot be a panacea for solving careless design. Much attention has been paid in the literature to the potential negative effects on thermal comfort of pedestrians that would be produced by high-albedo diffusely reflecting walls [11,12], while the solar concentration produced by specular reflective materials is of an order of magnitude greater, as demonstrated in this paper and by the empirical evidence from relevant case studies listed in Table 1.

Herein, we argue that outdoor thermal comfort models should be enhanced to represent the directional components of the reflection of solar radiation. Furthermore, development control plans and building codes should include a threshold on solar concentration by buildings in order to avert a radiatively induced urban heat island. In fact, a review of more than 220 projects reports a peak ambient temperature reduction by approximately 2 °C when the albedo of an urban area is increased by 0.3 [9]. This also means that decreasing the amount of solar radiation that escapes the urban canopy layer due to downward reflection increases the ambient temperature. The ambient temperature increase caused by specular reflective (glazed) facades is to be determined, but the canyon albedo with specular reflectors and glazed facades is known [63]. If the façade is fully glazed, with a high window to wall ratio, the canyon albedo is lower than 0.05 with high solar elevations [63], while it is more than 0.15 with wall albedos of 0.50 (as also documented experimentally [64]). Additionally, values lower than 0.10 for the canyon albedo are computed when walls are covered by purely specular reflectors [65]. Therefore, fully glazed facades may decrease the urban albedo by approximately 0.10–0.15, thus leading to ambient temperature increases of the order of magnitude of 0.6–0.7 °C, based on the mitigation reported in the literature with increases in urban albedo [9].

## 5. Conclusions

We analyzed the impact of building geometries and reflection behavior of finishing materials (i.e., diffuse or specular) on the irradiance values measured at street level. The three selected representative geometries (curved convex, vertical flat, and 10% tilted flat) coupled with two different façade materials suggested avoiding caustic curve formation and that the effect of highly reflective surfaces is perceived differently in accordance with the distance from the building.

The measurements taken were an example of the two extreme possible behaviors for approximately Lambertian and specular reflective materials. In a real scenario, the glazing facade material has a behavior that is not perfectly matching both of the boundaries, but rather is in the middle. For a high angle of solar incident radiation, its behavior became almost similar to that of the specular one.

For all these reasons, it is recommended that designers perform a detailed study of the consequences of the materials used for the building envelope during the design phase in accordance with the building context to avoid severe comfort and visual issues in the surroundings of the buildings. In the case of severe impairment, any kind of post-construction mitigation strategy could be impossible, inadequately expensive, or incompatible with the building architecture. All the façades can be critical, and the coupled effect of the wrong geometry and material choices can magnify glare and solar concentration problems. For these reasons, all the surrounding areas particularly sensitive to the reflected light must be carefully identified to understand the impact that reflected sunlight can have on safety and comfort issues.

Among the alternatives, a speditive evaluation (both in the preliminary and construction phase) can be performed by using a parametric script to evaluate any surface under a general orientation only by changing the cone amplitude, as described in Section 2.3.2. This approach can enhance the design of complex envelopes, providing the possibility to evaluate the amount of reflected radiation and, indirectly, to know how much energy is passing through the surface in case of a glazed façade.

The proposed method helps to understand which glazed units can be problematic and for hour many hours, enhancing the design by developing shading strategies.

Future developments of the test and procedure presented can facilitate the analysis of other geometries and materials and be applied to shading strategies such as local overhangs, fins, and external shading systems that can mitigate or exclude critical sunlight reflections, considering sensitive areas as local constraints.

The relevance of this study concerns two aspects. We evaluated the impact of tall buildings and their geometry, with the quantification of solar concentration by archetypical combinations of façade geometry and materials and the identification of a measurement protocol. This research also sheds light on the need for considering solar concentration in research on urban overheating. With an increasing use of specularly reflective materials in the built environment, urban climate models need to embed this capability beyond what has already been reported in studies supporting the performance analysis of retro-reflective materials.

**Author Contributions:** Conceptualization, A.S., A.G.M., A.Z., R.P. and T.P. (Tiziana Poli); methodology, A.S., A.G.M., A.Z., R.P. and T.P. (Tiziana Poli); validation, A.S., A.G.M., A.Z. and R.P.; formal analysis, A.S., A.G.M., A.Z., R.P. and T.P. (Tommaso Pagnacco); investigation, A.S., A.G.M., A.Z., R.P., T.P. (Tommaso Pagnacco) and T.P. (Tiziana Poli); data curation, A.S., A.G.M., A.Z. and R.P.; writing—original draft preparation, A.S., A.G.M., A.Z. and R.P.; writing—review and editing, A.S., A.G.M., A.Z., R.P., T.P. (Tommaso Pagnacco) and T.P. (Tiziana Poli); visualization, A.S., A.G.M., A.Z. and T.P. (Tommaso Pagnacco); supervision, T.P. (Tiziana Poli); project administration, T.P. (Tiziana Poli); funding acquisition, T.P. (Tiziana Poli). All authors have read and agreed to the published version of the manuscript.

**Funding:** This work was funded in part by Politecnico di Milano with the program "Fondi ricerca di base", in part by the Italian Ministry for Economic Development with the projects "Valutazione delle prestazioni di cool materials esposti all'ambiente urbano" and "Sviluppo di materiali e tecnologie per la riduzione degli effetti della radiazione solare", and in part by the Italian Ministry of Education, University and Research with the project "PRINSense".

**Institutional Review Board Statement:** Not applicable.

**Informed Consent Statement:** Not applicable.

**Data Availability Statement:** All the data collected are reported in Appendices A–F.

**Acknowledgments:** This work has been made possible thanks to the measurement instruments given by Elisabetta Rosina and by SEED Lab and the partnership of PA&CO Architecture for the mock-up design and production. The authors thank Osservatorio Meteo Milano Duomo for the validation of weather data and solar radiation data. A special thanks to Michele Alghisi, who helped us during all the measurement phases, and to Vittoria Sykopetrites for reviewing the manuscript.

**Conflicts of Interest:** The authors declare no conflict of interest.

## Appendix A. Flat Diffusive Façade

**Figure A1.** Flat diffusive façade—Experimental setup and measurement points position (Figure 2).

**Table A1.** Irradiance and Unit of sun per each measurement point over the sample plane. All the measurements have been performed on the 19 of September with the exception of the underlined number that refers to the 21 September.

| Start Time [hh:mm] | End Time [hh:mm] | MIN N.d. [W/m$^2$] | MAX N.d. [W/m$^2$] | A | B | C | D | E | F | G | H | I | L | M | N | O | P |
|---|---|---|---|---|---|---|---|---|---|---|---|---|---|---|---|---|---|
| | | | | | | | | | | | | | | | | | |
| 9:00 | 9:15 | 254 | 294 | 267 | 323 | 310 | 292 | 339 | 337 | 333 | 328 | 333 | 316 | 321 | 338 | 346 | 336 |
| | | | | | 1.27 | 1.21 | 1.11 | 1.27 | 1.24 | 1.21 | 1.16 | 1.17 | 1.09 | 1.09 | 1.13 | 1.13 | 1.09 |
| 10:00 | 10:16 | 420 | 458 | 430 | 521 | 495 | 458 | 504 | 517 | 506 | 490 | 495 | 478 | 483 | 505 | 509 | 492 |
| | | | | | 1.24 | 1.16 | 1.06 | 1.16 | 1.18 | 1.15 | 1.10 | 1.10 | 1.06 | 1.06 | 1.10 | 1.10 | 1.05 |
| 11:00 | 11:17 | 549 | 578 | 556 | 673 | 628 | 585 | 627 | 657 | 640 | 614 | 620 | 600 | 597 | 625 | 625 | 602 |
| | | | | | 1.23 | 1.14 | 1.06 | 1.13 | 1.18 | 1.14 | 1.09 | 1.10 | 1.06 | 1.05 | 1.09 | 1.09 | 1.05 |
| _12:03_ | _12:20_ | _654_ | _676_ | _647_ | _799_ | _753_ | _695_ | _717_ | _748_ | _743_ | _723_ | _732_ | _710_ | _650_ | _711_ | _739_ | _703_ |
| | | | | | _1.22_ | _1.15_ | _1.06_ | _1.09_ | _1.13_ | _1.12_ | _1.09_ | _1.10_ | _1.06_ | _0.97_ | _1.06_ | _1.10_ | _1.04_ |
| 13:02 | 13:18 | 546 | 734 | 732 | 895 | 832 | 761 | 773 | 830 | 798 | 754 | 760 | 747 | 749 | 784 | 790 | 755 |
| | | | | | 1.64 | 1.45 | 1.27 | 1.18 | 1.22 | 1.13 | 1.03 | 1.04 | 1.02 | 1.02 | 1.07 | 1.08 | 1.04 |
| 14:00 | 14:16 | 690 | 712 | 695 | 852 | 789 | 727 | 713 | 766 | 732 | 706 | 720 | 713 | 721 | 754 | 761 | 727 |
| | | | | | 1.20 | 1.11 | 1.03 | 1.01 | 1.09 | 1.05 | 1.01 | 1.04 | 1.03 | 1.04 | 1.09 | 1.10 | 1.05 |
| 15:00 | 15:15 | 631 | 656 | 648 | 782 | 720 | 680 | 655 | 696 | 671 | 649 | 667 | 656 | 655 | 684 | 691 | 672 |
| | | | | | 1.19 | 1.10 | 1.04 | 1.01 | 1.07 | 1.04 | 1.01 | 1.04 | 1.03 | 1.03 | 1.08 | 1.09 | 1.06 |
| 16:00 | 16:18 | 510 | 570 | 550 | 656 | 591 | 575 | 527 | 550 | 525 | 512 | 530 | 520 | 535 | 553 | 560 | 540 |
| | | | | | 1.15 | 1.05 | 1.03 | 0.97 | 1.02 | 0.99 | 0.97 | 1.01 | 1.00 | 1.03 | 1.07 | 1.09 | 1.05 |
| 17:00 | 17:17 | 347 | 398 | 374 | 461 | 421 | 420 | 364 | 376 | 356 | 344 | 360 | 361 | 365 | 374 | 378 | 361 |
| | | | | | 1.16 | 1.07 | 1.08 | 0.94 | 0.98 | 0.94 | 0.91 | 0.96 | 0.98 | 1.00 | 1.05 | 1.07 | 1.03 |

The Upper Value Refers to the Measure in Each Point [W/m$^2$] While the Lower Is the Ratio [-] with the Undisturbed Measure (Unit of Sun)

Appendix B. Flat Reflective Façade

**Figure A2.** Flat reflective façade—Experimental setup and measurement points position (Figure 2).

**Table A2.** Irradiance and Unit of sun per each measurement point over the sample plane. All the measurements have been performed on the 25 September.

| Start Time | End Time | MIN N.d. | MAX N.d. | A | B | C | D | E | F | G | H | I | L | M | N | O | P |
|---|---|---|---|---|---|---|---|---|---|---|---|---|---|---|---|---|---|
| [hh:mm] | [hh:mm] | [W/m²] | [W/m²] | The Upper Value Refers to the Measure in Each Point [W/m²] While the Lower Is the Ratio [-] with the Undisturbed Measure (Unit of Sun) | | | | | | | | | | | | | |
| 9:00 | 9:11 | 256 | 279 | 248 | 255 | 263 | 467 | 304 | 282 | 285 | 294 | 297 | 403 | 281 | 281 | 280 | 278 |
| | | | | | 1.00 | 1.02 | 1.80 | 1.15 | 1.06 | 1.06 | 1.08 | 1.09 | 1.47 | 1.01 | 1.00 | 0.99 | 0.98 |
| 10:00 | 10:10 | 416 | 439 | 406 | 422 | 754 | 431 | 467 | 445 | 444 | 450 | 458 | 439 | 870 | 441 | 438 | 437 |
| | | | | | 1.02 | 1.81 | 1.03 | 1.11 | 1.05 | 1.04 | 1.05 | 1.06 | 1.02 | 2.00 | 1.01 | 0.99 | 0.98 |
| 11:00 | 11:11 | 550 | 563 | 540 | 905 | 987 | 557 | 590 | 575 | 575 | 580 | 592 | 572 | 1099 | 1075 | 564 | 557 |
| | | | | | 1.65 | 1.79 | 1.01 | 1.06 | 1.04 | 1.03 | 1.03 | 1.05 | 1.02 | 1.95 | 1.91 | 0.99 | 0.97 |
| 12:05 | 12:16 | 653 | 662 | 661 | 1179 | 1201 | 657 | 680 | 677 | 676 | 673 | 688 | 675 | 663 | 1294 | 1190 | 662 |
| | | | | | 1.81 | 1.84 | 1.01 | 1.04 | 1.03 | 1.02 | 1.02 | 1.04 | 1.02 | 1.00 | 1.96 | 1.80 | 1.00 |
| 13:06 | 13:17 | 676 | 686 | 693 | 1321 | 705 | 698 | 694 | 708 | 711 | 691 | 697 | 686 | 685 | 704 | 1327 | 1374 |
| | | | | | 1.92 | 1.03 | 1.02 | 1.01 | 1.03 | 1.04 | 1.01 | 1.02 | 1.01 | 1.01 | 1.04 | 1.94 | 2.01 |
| 14:01 | 14:1 | 660 | 669 | 672 | 1290 | 680 | 682 | 667 | 1242 | 1259 | 665 | 688 | 677 | 671 | 675 | 684 | 667 |
| | | | | | 1.92 | 1.02 | 1.02 | 1.00 | 1.86 | 1.90 | 1.01 | 1.04 | 1.03 | 1.02 | 1.02 | 1.03 | 1.01 |
| 15:18 | 15:24 | 558 | 565 | 569 | 602 | 576 | 585 | 558 | 1075 | 565 | 1309 | 582 | 578 | 573 | 573 | 571 | 563 |
| | | | | | 1.05 | 1.02 | 1.04 | 1.00 | 1.93 | 1.01 | 2.33 | 1.03 | 1.02 | 1.02 | 1.02 | 1.02 | 1.01 |
| 16:14 | 16:23 | 428 | 449 | 462 | 481 | 468 | 486 | 455 | 782 | 428 | 635 | 453 | 445 | 442 | 440 | 440 | 435 |
| | | | | | 1.06 | 1.04 | 1.08 | 1.02 | 1.75 | 0.97 | 1.45 | 1.03 | 1.02 | 1.02 | 1.03 | 1.03 | 1.02 |
| 16:57 | 17:08 | 308 | 334 | 334 | 347 | 339 | 359 | 525 | 314 | 296 | 297 | 462 | 314 | 313 | 310 | 311 | 308 |
| | | | | | 1.03 | 1.01 | 1.09 | 1.60 | 0.97 | 0.92 | 0.93 | 1.46 | 1.00 | 1.00 | 1.01 | 1.02 | 1.02 |

## Appendix C. Flat 10% Tilted Diffusive Façade

**Figure A3.** Flat 10% tilted diffusive façade—Experimental setup and measurement points position (Figure 2).

**Table A3.** Irradiance and Unit of sun per each measurement point over the sample plane. All the measurements have been performed on the 21 September.

| Start Time | End Time | MIN N.d. | MAX N.d. | A | B | C | D | E | F | G | H | I | L | M | N | O | P |
|---|---|---|---|---|---|---|---|---|---|---|---|---|---|---|---|---|---|
| [hh:mm] | [hh:mm] | [W/m²] | [W/m²] | The Upper Value Refers to the Measure in Each Point [W/m²] While the Lower Is the Ratio [-] with the Undisturbed Measure (Unit of Sun) | | | | | | | | | | | | | |
| 9:03 | 9:19 | 257 | 302 | 268 | 329 | 311 | 289 | 342 | 340 | 343 | 333 | 329 | 316 | 311 | 328 | 329 | 318 |
| | | | | | 1.28 | 1.20 | 1.10 | 1.29 | 1.26 | 1.25 | 1.20 | 1.16 | 1.10 | 1.07 | 1.12 | 1.11 | 1.05 |
| 10:00 | 10:21 | 426 | 474 | 430 | 534 | 500 | 457 | 500 | 516 | 518 | 499 | 498 | 485 | 482 | 509 | 513 | 490 |
| | | | | | 1.28 | 1.17 | 1.07 | 1.15 | 1.18 | 1.18 | 1.13 | 1.12 | 1.08 | 1.06 | 1.11 | 1.11 | 1.05 |
| 11:18 | 11:33 | 590 | 605 | 595 | 744 | 687 | 614 | 652 | 685 | 674 | 639 | 656 | 638 | 642 | 671 | 678 | 636 |
| | | | | | 1.27 | 1.16 | 1.04 | 1.10 | 1.15 | 1.13 | 1.07 | 1.10 | 1.06 | 1.07 | 1.12 | 1.12 | 1.05 |
| 12:23 | 12:39 | 685 | 702 | 692 | 848 | 798 | 715 | 725 | 795 | 775 | 735 | 745 | 727 | 728 | 771 | 774 | 733 |
| | | | | | 1.24 | 1.16 | 1.04 | 1.05 | 1.14 | 1.11 | 1.06 | 1.07 | 1.04 | 1.04 | 1.10 | 1.10 | 1.04 |
| 13:00 | - | - | - | - | - | - | - | - | - | - | - | - | - | - | - | - | - |
| | | | | - | - | - | - | - | - | - | - | - | - | - | - | - | - |
| 14:06 | 14:51 | 554 | 690 | 686 | 850 | 790 | 735 | 713 | 775 | 735 | 702 | 710 | 699 | 694 | 726 | 730 | 690 |
| | | | | | 1.23 | 1.15 | 1.07 | 1.04 | 1.13 | 1.33 | 1.24 | 1.20 | 1.16 | 1.12 | 1.15 | 1.14 | 1.08 |
| 15:01 | 15:16 | 620 | 630 | 632 | 686 | 720 | 673 | 652 | 702 | 673 | 641 | 648 | 642 | 643 | 675 | 683 | 651 |
| | | | | | 1.09 | 1.15 | 1.07 | 1.04 | 1.12 | 1.07 | 1.02 | 1.03 | 1.02 | 1.03 | 1.08 | 1.10 | 1.05 |
| 16:01 | 16:32 | 425 | 498 | 516 | 646 | 587 | 566 | 524 | 557 | 530 | 443 | 456 | 453 | 468 | 487 | 498 | 463 |
| | | | | | 1.30 | 1.19 | 1.16 | 1.09 | 1.17 | 1.14 | 1.04 | 1.06 | 1.05 | 1.07 | 1.10 | 1.14 | 1.07 |
| 17:00 | 17:13 | 333 | 356 | 340 | 426 | 392 | 395 | 343 | 369 | 351 | 339 | 348 | 348 | 349 | 358 | 363 | 349 |
| | | | | | 1.16 | 1.10 | 1.11 | 0.97 | 1.05 | 1.00 | 0.97 | 1.01 | 1.01 | 1.02 | 1.05 | 1.07 | 1.04 |

## Appendix D. Flat 10% Tilted Reflective Façade

**Figure A4.** Flat 10% tilted diffusive façade—Experimental setup and measurement points position (Figure 2).

**Table A4.** Irradiance and Unit of sun per each measurement point over the sample plane. All the measurements have been performed on the 25 September.

| Start Time | End Time | MIN N.d. | MAX N.d. | A | B | C | D | E | F | G | H | I | L | M | N | O | P |
|---|---|---|---|---|---|---|---|---|---|---|---|---|---|---|---|---|---|
| [hh:mm] | [hh:mm] | [W/m²] | [W/m²] | The Upper Value Refers to the Measure in Each Point [W/m²] While the Lower Is the Ratio [-] with the Undisturbed Measure (Unit of Sun) | | | | | | | | | | | | | |
| 9:17 | 9:28 | 302 | 328 | 292 | 312 | 509 | 487 | 363 | 337 | 337 | 344 | 346 | 437 | 331 | 332 | 331 | 329 |
| | | | | | 1.05 | 1.68 | 1.59 | 1.18 | 1.08 | 1.07 | 1.08 | 1.08 | 1.36 | 1.02 | 1.01 | 1.00 | 0.98 |
| 10:20 | 10:31 | 469 | 490 | 455 | 494 | 846 | 488 | 526 | 503 | 504 | 508 | 518 | 500 | 906 | 512 | 498 | 492 |
| | | | | | 1.06 | 1.80 | 1.04 | 1.11 | 1.05 | 1.05 | 1.06 | 1.08 | 1.03 | 1.86 | 1.04 | 1.01 | 0.99 |
| 11:20 | 11:31 | 588 | 603 | 581 | 1155 | 1113 | 596 | 628 | 616 | 614 | 615 | 627 | 613 | 1002 | 1146 | 616 | 605 |
| | | | | | 1.97 | 1.89 | 1.01 | 1.06 | 1.05 | 1.04 | 1.03 | 1.05 | 1.02 | 1.67 | 1.90 | 1.02 | 0.99 |
| 12:20 | 12:31 | 663 | 675 | 654 | 1292 | 1337 | 669 | 680 | 696 | 699 | 687 | 697 | 685 | 677 | 1313 | 1288 | 680 |
| | | | | | 1.95 | 2.02 | 1.00 | 1.02 | 1.04 | 1.05 | 1.03 | 1.05 | 1.03 | 1.00 | 1.95 | 1.91 | 1.01 |
| 13:21 | 13:32 | 642 | 685 | 685 | 1229 | 696 | 682 | 686 | 722 | 1299 | 686 | 702 | 695 | 688 | 708 | 1314 | 694 |
| | | | | | 1.79 | 1.02 | 1.03 | 1.06 | 1.08 | 1.95 | 1.01 | 1.03 | 1.01 | 1.03 | 1.10 | 1.93 | 1.02 |
| 14:20 | 14:31 | 650 | 657 | 656 | 1219 | 675 | 675 | 658 | 1249 | 1232 | 663 | 676 | 667 | 662 | 669 | 679 | 659 |
| | | | | | 1.85 | 1.03 | 1.03 | 1.00 | 1.91 | 1.89 | 1.01 | 1.04 | 1.02 | 1.01 | 1.03 | 1.04 | 1.01 |
| 15:27 | 15:35 | 540 | 549 | 546 | 976 | 563 | 579 | 551 | 1159 | 515 | 1063 | 560 | 554 | 552 | 553 | 558 | 547 |
| | | | | | 1.77 | 1.03 | 1.06 | 1.01 | 2.12 | 0.95 | 1.95 | 1.03 | 1.02 | 1.02 | 1.02 | 1.03 | 1.02 |
| 16:27 | 16:38 | 390 | 411 | 420 | 435 | 421 | 443 | 414 | 709 | 388 | 394 | 411 | 406 | 401 | 399 | 401 | 398 |
| | | | | | 1.05 | 1.02 | 1.08 | 1.02 | 1.75 | 0.96 | 0.99 | 1.04 | 1.04 | 1.03 | 1.02 | 1.04 | 1.03 |
| 17:20 | 17:31 | 248 | 270 | 270 | 288 | 280 | 304 | 456 | 263 | 244 | 244 | 604 | 262 | 263 | 259 | 261 | 257 |
| | | | | | 1.05 | 1.04 | 1.14 | 1.74 | 1.01 | 0.95 | 0.95 | 2.39 | 1.05 | 1.05 | 1.04 | 1.05 | 1.05 |

## Appendix E. Concave Diffusive Façade

**Figure A5.** Concave diffusive façade—Experimental setup and measurement points position (Figure 2).

**Table A5.** Irradiance and Unit of sun per each measurement point over the sample plane. All the measurements have been performed on the 19 of September with the exception of the underlined number that refers to the 21 September.

| Start Time [hh:mm] | End Time [hh:mm] | MIN N.d. [W/m$^2$] | MAX N.d. [W/m$^2$] | A | B | B_2 | C | D | E | F | G | H | I | L | M | N | O | P |
|---|---|---|---|---|---|---|---|---|---|---|---|---|---|---|---|---|---|---|
| | | | | The Upper Value Refers to the Measure in Each Point [W/m$^2$] While the Lower Is the Ratio [-] with the Undisturbed Measure (Unit of Sun) | | | | | | | | | | | | | | |
| 9:28 | 9:47 | 335 | 384 | 351 | 413 | - | 402 | 373 | 417 | 416 | 412 | 405 | 410 | 395 | 397 | 417 | 419 | 407 |
| | | | | | 1.23 | - | 1.18 | 1.09 | 1.19 | 1.18 | 1.15 | 1.12 | 1.12 | 1.06 | 1.06 | 1.10 | 1.10 | 1.06 |
| 10:26 | 10:49 | 494 | 530 | 511 | 604 | 625 | 574 | 530 | 573 | 586 | 575 | 559 | 564 | 543 | 545 | 571 | 573 | 550 |
| | | | | | 1.23 | 1.26 | 1.16 | 1.06 | 1.14 | 1.15 | 1.13 | 1.09 | 1.09 | 1.04 | 1.04 | 1.09 | 1.09 | 1.04 |
| 11:41 | 11:57 | 623 | 645 | 618 | 744 | 798 | 725 | 664 | 693 | 731 | 715 | 686 | 693 | 658 | 665 | 694 | 708 | 677 |
| | | | | | 1.20 | 1.28 | 1.16 | 1.06 | 1.10 | 1.16 | 1.13 | 1.08 | 1.09 | 1.03 | 1.04 | 1.08 | 1.10 | 1.05 |
| 12:45 | 13:01 | 694 | 708 | 713 | 815 | 841 | 792 | 729 | 754 | 801 | 779 | 734 | 751 | 726 | 720 | 762 | 760 | 692 |
| | | | | | 1.17 | 1.21 | 1.14 | 1.05 | 1.09 | 1.15 | 1.12 | 1.05 | 1.07 | 1.03 | 1.02 | 1.08 | 1.07 | 0.98 |
| 13:30 | 13:44 | 716 | 720 | 733 | 865 | 898 | 809 | 741 | 740 | 795 | 773 | 740 | 751 | 739 | 740 | 775 | 780 | 746 |
| | | | | | 1.20 | 1.25 | 1.12 | 1.03 | 1.03 | 1.11 | 1.08 | 1.03 | 1.05 | 1.03 | 1.03 | 1.08 | 1.09 | 1.04 |
| 14:31 | 14:46 | 663 | 668 | 669 | 800 | 831 | 746 | 701 | 682 | 734 | 711 | 682 | 690 | 683 | 690 | 720 | 730 | 701 |
| | | | | | 1.20 | 1.24 | 1.12 | 1.05 | 1.02 | 1.10 | 1.07 | 1.03 | 1.04 | 1.03 | 1.04 | 1.09 | 1.10 | 1.06 |
| 16:28 | 16:44 | 441 | 476 | 480 | 572 | 576 | 506 | 498 | 447 | 472 | 447 | 432 | 447 | 449 | 456 | 470 | 476 | 458 |
| | | | | | 1.18 | 1.20 | 1.06 | 1.05 | 0.95 | 1.01 | 0.97 | 0.95 | 0.99 | 1.00 | 1.02 | 1.06 | 1.08 | 1.04 |
| 17:30 | 17:44 | 262 | 296 | 295 | 341 | 345 | 304 | 314 | 258 | 274 | 256 | 251 | 264 | 273 | 275 | 279 | 282 | 270 |
| | | | | | 1.13 | 1.15 | 1.03 | 1.07 | 0.90 | 0.97 | 0.92 | 0.91 | 0.97 | 1.01 | 1.03 | 1.05 | 1.08 | 1.04 |

## Appendix F. Concave Reflective Façade

**Figure A6.** Concave reflective façade—Experimental setup and measurement points position (Figure 2).

**Table A6.** Irradiance and Unit of sun per each measurement point over the sample plane. All the measurements have been performed on the 25 September.

| Start Time | End Time | MIN N.d. | MAX N.d. | A | B | B_2 | C | D | E | F | G | H | I | L | M | N | O | P |
|---|---|---|---|---|---|---|---|---|---|---|---|---|---|---|---|---|---|---|
| [hh:mm] | [hh:mm] | [W/m²] | [W/m²] | The Upper Value Refers to the Measure in Each Point [W/m²] While the Lower Is the Ratio [-] with the Undisturbed Measure (Unit of Sun) | | | | | | | | | | | | | | |
| 9:35 | 9:53 | 350 | 397 | 341 | 370 | 375 | 394 | 450 | 419 | 396 | 395 | 404 | 409 | 422 | 464 | 389 | 386 | 384 |
| | | | | | 1.01 | 1.0207 | 1.07 | 1.21 | 1.12 | 1.05 | 1.05 | 1.07 | 1.07 | 1.1 | 1.2 | 1 | 0.99 | 0.99 |
| 10:42 | 10:55 | 512 | 539 | 517 | 526 | 533 | 1402 | 865 | 562 | 545 | 544 | 549 | 560 | 608 | 668 | 553 | 536 | 529 |
| | | | | | 1.02 | 1.03 | 2.71 | 1.67 | 1.08 | 1.05 | 1.04 | 1.05 | 1.07 | 1.15 | 1.26 | 1.04 | 1.00 | 0.99 |
| 11:41 | 11:56 | 620 | 634 | 631 | 651 | 667 | 690 | 631 | 657 | 650 | 649 | 648 | 663 | 649 | 853 | 1022 | 647 | 637 |
| | | | | | 1.04 | 1.07 | 1.10 | 1.00 | 1.04 | 1.03 | 1.03 | 1.03 | 1.05 | 1.03 | 1.35 | 1.62 | 1.02 | 1.00 |
| 12:42 | 12:56 | 678 | 689 | 689 | 1286 | 1521 | 696 | 680 | 694 | 701 | 702 | 696 | 713 | 701 | 687 | 718 | 1103 | 812 |
| | | | | | 1.90 | 2.24 | 1.03 | 1.00 | 1.02 | 0.00 | 1.03 | 1.03 | 1.01 | 1.03 | 1.02 | 1.00 | 1.05 | 1.62 |
| 13:40 | 13:48 | 674 | 682 | 692 | 1055 | 1823 | 695 | 689 | 686 | 702 | 1492 | 686 | 704 | 694 | 687 | 693 | 752 | 1089 |
| | | | | | 1.55 | 2.68 | 1.02 | 1.01 | 1.01 | 1.03 | 2.18 | 1.00 | 1.03 | 1.02 | 1.01 | 1.02 | 1.11 | 1.61 |
| 15:07 | 15:17 | 573 | 601 | 611 | 620 | 627 | 606 | 617 | 602 | 1095 | 607 | 681 | 623 | 590 | - | - | 592 | - |
| | | | | | 1.03 | 1.05 | 1.01 | 1.04 | 1.01 | 1.85 | 1.04 | 1.17 | 1.07 | 1.02 | - | - | 1.04 | - |
| 15:37 | 15:44 | 518 | 533 | 547 | 554 | 557 | 540 | 557 | 719 | 1237 | 522 | 653 | 571 | 534 | 531 | 531 | 533 | 527 |
| | | | | | 1.05 | 1.06 | 1.03 | 1.06 | 1.37 | 2.37 | 1.00 | 1.25 | 1.09 | 1.03 | 1.02 | 1.02 | 1.03 | 1.02 |
| 16:45 | 16:55 | 346 | 372 | 379 | 383 | 384 | 376 | 396 | 380 | 363 | 340 | 569 | 433 | 360 | 358 | 354 | 355 | 351 |
| | | | | | 1.04 | 1.05 | 1.03 | 1.09 | 1.04 | 1.00 | 0.95 | 1.60 | 1.23 | 1.02 | 1.03 | 1.02 | 1.03 | 1.02 |
| 17:37 | 17:47 | 197 | 228 | 237 | 237 | 231 | 251 | 238 | 203 | 192 | 196 | 225 | 214 | 213 | 208 | 210 | 208 | |
| | | | | | 1.04 | 1.04 | 1.03 | 1.14 | 1.10 | 0.94 | 0.91 | 0.94 | 1.09 | 1.05 | 1.05 | 1.04 | 1.07 | 1.06 |

## References


1. Santamouris, M.; Vasilakopoulou, K. Present and Future Energy Consumption of Buildings: Challenges and Opportunities towards Decarbonisation. *e-Prime* **2021**, *1*, 100002. [CrossRef]
2. Zhang, J.; Mohegh, A.; Li, Y.; Levinson, R.; Ban-Weiss, G. Systematic Comparison of the Influence of Cool Wall versus Cool Roof Adoption on Urban Climate in the Los Angeles Basin. *Environ. Sci. Technol.* **2018**, *52*, 11188–11197. [CrossRef] [PubMed]
3. Zhang, Y.; Long, E.; Li, Y.; Li, P. Solar radiation reflective coating material on building envelopes: Heat transfer analysis and cooling energy saving. *Energy Explor. Exploit.* **2017**, *35*, 748–766. [CrossRef]
4. Akbari, H.; Pomerantz, M.; Taha, H. Cool surfaces and shade trees to reduce energy use and improve air quality in urban areas. *Sol. Energy* **2001**, *70*, 295–310. [CrossRef]
5. Akbari, H. Measured energy savings from the application of reflective roofs in two small non-residential buildings. *Energy* **2003**, *28*, 953–967. [CrossRef]
6. Akbari, H.; Konopacki, S. Calculating energy-saving potentials of heat-island reduction strategies. *Energy Policy* **2005**, *33*, 721–756. [CrossRef]
7. Synnefa, A.; Santamouris, M.; Akbari, H. Estimating the effect of using cool coatings on energy loads and thermal comfort in residential buildings in various climatic conditions. *Energy Build.* **2007**, *39*, 1167–1174. [CrossRef]
8. Santamouris, M. Cooling the cities—A review of reflective and green roof mitigation technologies to fight heat island and improve comfort in urban environments. *Sol. Energy* **2014**, *103*, 682–703. [CrossRef]
9. Santamouris, M.; Ding, L.; Fiorito, F.; Oldfield, P.; Osmond, P.; Paolini, R.; Prasad, D.; Synnefa, A. Passive and active cooling for the outdoor built environment—Analysis and assessment of the cooling potential of mitigation technologies using performance data from 220 large scale projects. *Sol. Energy* **2017**, *154*, 14–33. [CrossRef]
10. Zinzi, M. Exploring the potentialities of cool facades to improve the thermal response of Mediterranean residential buildings. *Sol. Energy* **2016**, *135*, 386–397. [CrossRef]
11. Lee, H.; Mayer, H. Thermal comfort of pedestrians in an urban street canyon is affected by increasing albedo of building walls. *Int. J. Biometeorol.* **2018**, *62*, 1199–1209. [CrossRef] [PubMed]
12. Nazarian, N.; Dumas, N.; Kleissl, J.; Norford, L. Effectiveness of cool walls on cooling load and urban temperature in a tropical climate. *Energy Build.* **2019**, *187*, 144–162. [CrossRef]
13. Rossi, F.; Pisello, A.L.; Nicolini, A.; Filipponi, M.; Palombo, M. Analysis of retro-reflective surfaces for urban heat island mitigation: A new analytical model. *Appl. Energy* **2014**, *114*, 621–631. [CrossRef]
14. Rossi, F.; Morini, E.; Castellani, B.; Nicolini, A.; Bonamente, E.; Anderini, E.; Cotana, F. Beneficial effects of retroreflective materials in urban canyons: Results from seasonal monitoring campaign. *J. Phys. Conf. Ser.* **2015**, *655*, 12012. [CrossRef]
15. Yuan, J.; Farnham, C.; Emura, K. Development of a retro-reflective material as building coating and evaluation on albedo of urban canyons and building heat loads. *Energy Build.* **2015**, *103*, 107–117. [CrossRef]
16. Levinson, R.; Chen, S.; Slack, J.; Goudey, H.; Harima, T.; Berdahl, P. Design, characterization, and fabrication of solar-retroreflective cool-wall materials. *Sol. Energy Mater. Sol. Cells* **2020**, *206*, 110117. [CrossRef]
17. Yang, S.-H.; Matzarakis, A. Study on human thermal comfort for architecture—The Example of Shanghai, China. In Proceedings of the ICUC9—9th International Conference on Urban Climate, Toulouse, France, 20–24 July 2015.
18. Matzarakis, A.; Fröhlich, D.; Gangwisch, M.; Ketterer, C.; Peer, A.; Freiburg, A.; Freiburg, D. Developments and applications of thermal indices in urban structures by RayMan and SkyHelios model. In Proceedings of the ICUC9—9th International Conference on Urban Climate, Toulouse, France, 20–24 July 2015.
19. Schiler, M.; Valmont, E. Microclimatic impact: Glare around the Walt Disney Concert Hall. In Proceedings of the Solar World Congress 2005 Joint American Solar Energy Society/International Solar Energy Conference, Orlando, 6–12 August 2005; p. 511.
20. Furler, R.A. Angular Dependence of Optical Properties of Homogeneous Glasses. *ASHRAE Trans.* **1991**, *98*, 1.
21. CityA.M. Dazzling Shard Has Inadvertent Effect of Blinding Train Drivers. 2012. Available online: https://www.cityam.com/dazzling-shard-has-inadvertent-effect-blinding-train-drivers/ (accessed on 20 January 2022).
22. Rose, T.; Wollert, A. The dark side of photovoltaic—3D simulation of glare assessing risk and discomfort. *Environ. Impact Assess. Rev.* **2015**, *52*, 24–30. [CrossRef]
23. Ho, C.K.; Ghanbari, C.M.; Diver, R.B. Methodology To Assess Potential Glint and Glare Hazards From Concentrating Solar Power Plants: Analytical Models and Experimental Validation. *J. Sol. Energy Eng.* **2011**, *133*, 3102. [CrossRef]
24. Las Vegas Review Journal. Vdara Visitor: 'Death Ray' Scorched Hair. 2010. Available online: https://www.reviewjournal.com/news/vdara-visitor-death-ray-scorched-hair/ (accessed on 20 January 2022).
25. Pexels. Available online: https://www.pexels.com/ (accessed on 16 March 2022).
26. Campbell-Dollaghan, K. A Brief History of Buildings That Melt Things. Available online: https://gizmodo.com/a-brief-history-of-buildings-that-melt-things-1247657178 (accessed on 20 January 2022).
27. Danks, R.; Good, J. Urban Scale Simulations of Solar Reflections in the Built Environment: Methodology & Validation. In Proceedings of the Symposium on Simulation for Architecture & Urban Design 2016, London, UK, 16–18 May 2016; pp. 19–26.
28. Danks, R.; Good, J.; Sinclair, R. Assessing reflected sunlight from building facades: A literature review and proposed criteria. *Build. Environ.* **2016**, *103*, 193–202. [CrossRef]

29. Development Control Plan. Section 3 General Provisions, Sydney, AU. 2012. Available online: https://www.cityofsydney.nsw.gov.au/-/media/corporate/files/2020-07-migrated/files_s-1/section3-dcp2012-170619.pdf?download=true (accessed on 16 March 2022).

30. Buildings Department. *Guidelines on Design and Construction Requirements for Energy Efficiency of Residential Buildings*; Buildings Department: Hong Kong, 2014.

31. Dwyer, C. *Planning Advice Note Solar Glare: Guidelines and Best Practice for Assessing Solar Glare in the City of London*; The City of London Corporation: London, UK, 2017.

32. Suk, J.Y.; Schiler, M.; Kensek, K. Is Exterior Glare Problematic? Investigation on Visual Discomfort Caused by Reflected Sunlight on Specular Building Facades. In Proceedings of the PLEA 2016 Conference on Passive and Low energy Architecture.Cities, buildings, People: Towards Regenerative Environments, Los Angeles, CA, USA, 11–13 July 2016.

33. Shih, N.J.; Huang, Y.S. A study of reflection glare in Taipei. *Build. Res. Inf.* **2001**, *29*, 30–39. [CrossRef]

34. Danks, R.; Good, J. Urban Solar Reflection Identification, Simulation, Analysis and Mitigation: Learning From Case Studies. In Proceedings of the eSim 2016 Building Performance Simulation Conference eSim 2016, Hamilton, ON, Canada, 3–6 May 2016; pp. 309–318.

35. Brzezicki, M. The Influence of Reflected Solar Glare Caused by the Glass Cladding of a Building: Application of Caustic Curve Analysis. *Comput. Civ. Infrastruct. Eng.* **2012**, *27*, 347–357. [CrossRef]

36. Yang, X.; Grobe, L.; Stephen, W. Simulation of Reflected Daylight From Building Envelopes. In Proceedings of the 13th Conference of International Building Performance Simulation Association, Chambéry, France, 26–28 August 2013; pp. 3673–3680.

37. Ou, Y. Quantitative study of reflection of sunlight by a glass curtain wall resulting in a visual masking effect. *Appl. Opt.* **2014**, *53*, 6893. [CrossRef] [PubMed]

38. Shih, N.J.; Huang, Y.-S. An analysis and simulation of curtain wall reflection glare. *Build. Environ.* **1999**, *36*, 619–626. [CrossRef]

39. Vollmer, M.; Möllmann, K.P. Caustic effects due to sunlight reflections from skyscrapers: Simulations and experiments. *Eur. J. Phys.* **2012**, *33*, 1429–1455. [CrossRef]

40. POV-Ray. Available online: http://www.povray.org/ (accessed on 16 March 2022).

41. OpticStudio. Available online: https://www.zemax.com/pages/opticstudio (accessed on 16 March 2022).

42. Ishak, N.M.; Hien, W.N.; Jenatabadi, H.S.; Mustafa, N.A.; Zawawi, E.M.A. Effect of reflective building façade on pedestrian visual comfort. *IOP Conf. Ser. Earth Environ. Sci.* **2019**, *385*, 12059. [CrossRef]

43. Achsani, R.A.; Wonorahardjo, S. Studies on Visual Environment Phenomena of Urban Areas: A Systematic Review. *IOP Conf. Ser. Earth Environ. Sci.* **2020**, *532*, 12016. [CrossRef]

44. Wong, J.S.J. A comprehensive ray tracing study on the impact of solar reflections from glass curtain walls. *Environ. Monit. Assess.* **2016**, *188*, 16. [CrossRef]

45. Danks, R.; Good, J.; Sinclair, R. Avoiding The Dreaded Death Ray: Controlling Facade Reflections through Purposeful Design. In Proceedings of the Façade Tectonics World Congress, Los Angeles, CA, USA, 2016; Volume 2, ISBN 978-1-882352-43-2. Available online: https://www.researchgate.net/publication/309136628_Avoiding_The_Dreaded_Death_Ray_Controlling_Facade_Reflections_Through_Purposeful_Design (accessed on 16 March 2022).

46. McNeel Rhinoceros. Available online: https://www.rhino3d.com/ (accessed on 16 March 2022).

47. Grasshopper for Rhinoceros. Available online: https://www.grasshopper3d.com/ (accessed on 16 March 2022).

48. Barr, J. Skyscraper Height. *J. Real Estate Financ. Econ.* **2012**, *45*, 723–753. [CrossRef]

49. UNI—Ente Italiano di Normazione. *UNI EN 14500—Benessere Termico E Visivo, Metodi Di Prova E Di Calcolo*; UNI—Ente Italiano di Normazione: Milano, Italy, 2021.

50. ASTM International (ASTM). *ASTM E903—Standard Test Method for Solar Absorptance, Reflectance, and Transmittance of Materials Using Integrating Spheres*; ASTM International: West Conshohocken, PA, USA, 2020; p. 17.

51. Masson, V.; Grimmond, C.S.B.; Oke, T.R. Evaluation of the Town Energy Balance (TEB) Scheme with Direct Measurements from Dry Districts in Two Cities. *J. Appl. Meteorol.* **2002**, *41*, 1011–1026. [CrossRef]

52. Lemonsu, A.; Grimmond, C.S.B.; Masson, V. Modeling the Surface Energy Balance of the Core of an Old Mediterranean City: Marseille. *J. Appl. Meteorol.* **2004**, *43*, 312–327. [CrossRef]

53. Ban-Weiss, G.A.; Woods, J.; Levinson, R. Using remote sensing to quantify albedo of roofs in seven California cities, Part 1: Methods. *Sol. Energy* **2015**, *115*, 777–790. [CrossRef]

54. LBNL WINDOW Version 7.7.16. Available online: https://windows.lbl.gov/software/window (accessed on 16 March 2022).

55. González, J.; Fiorito, F. Daylight Design of Office Buildings: Optimisation of External Solar Shadings by Using Combined Simulation Methods. *Buildings* **2015**, *5*, 560–580. [CrossRef]

56. Duffie, J.; Beckman, W. *Solar Engineering of Thermal Processes*, 4th ed.; John Wiley & Sons: Hoboken, NJ, USA, 2013; ISBN 9780470873663.

57. Takebayashi, H. High-reflectance technology on building façades: Installation guidelines for pedestrian comfort. *Sustainability* **2016**, *8*, 785. [CrossRef]

58. Shanks, K.; Ferrer-Rodriguez, J.P.; Fernàndez, E.F.; Almonacid, F.; Pérez-Higueras, P.; Senthilarasu, S.; Mallick, T. A >3000 suns high concentrator photovoltaic design based on multiple Fresnel lens primaries focusing to one central solar cell. *Sol. Energy* **2018**, *169*, 457–467. [CrossRef]

59. Gordon, J.M. A 100-sun linear photovoltaic solar concentrator design from inexpensive commercial components. *Sol. Energy* **1996**, *57*, 301–305. [CrossRef]
60. Paolini, R.; Terraneo, G.; Ferrari, C.; Sleiman, M.; Muscio, A.; Metrangolo, P.; Poli, T.; Destaillats, H.; Zinzi, M.; Levinson, R. Effects of soiling and weathering on the albedo of building envelope materials: Lessons learned from natural exposure in two European cities and tuning of a laboratory simulation practice. *Sol. Energy Mater. Sol. Cells* **2020**, *205*, 110264. [CrossRef]
61. Wang, J.; Liu, S.; Meng, X.; Gao, W.; Yuan, J. Application of retro-reflective materials in urban buildings: A comprehensive review. *Energy Build.* **2021**, *247*, 111137. [CrossRef]
62. Yuan, J.; Emura, K.; Farnham, C. Evaluation of retro-reflective properties and upward to downward reflection ratio of glass bead retro-reflective material using a numerical model. *Urban Clim.* **2021**, *36*, 100774. [CrossRef]
63. Tsangrassoulis, A.; Santamouris, M. Numerical estimation of street canyon albedo consisting of vertical coated glazed facades. *Energy Build.* **2003**, *35*, 527–531. [CrossRef]
64. Kotopouleas, A.; Giridharan, R.; Nikolopoulou, M.; Watkins, R.; Yeninarcilar, M. Experimental investigation of the impact of urban fabric on canyon albedo using a 1:10 scaled physical model. *Sol. Energy* **2021**, *230*, 449–461. [CrossRef]
65. Aida, M.; Gotoh, K. Urban albedo as a function of the urban structure—A two-dimensional numerical simulation—Part II. *Bound. Layer Meteorol.* **1982**, *23*, 415–424. [CrossRef]

*Review*

# Advancement on Thermal Comfort in Educational Buildings: Current Issues and Way Forward

**Giulia Lamberti** [1,*], **Giacomo Salvadori** [1], **Francesco Leccese** [1], **Fabio Fantozzi** [1] **and Philomena M. Bluyssen** [2]

[1]  School of Engineering, University of Pisa, 56122 Pisa, Italy; giacomo.salvadori@unipi.it (G.S.); francesco.leccese@unipi.it (F.L.); fabio.fantozzi@unipi.it (F.F.)

[2]  Faculty of Architecture and the Built Environment, Delft University of Technology, 2628 Delft, The Netherlands; P.M.Bluyssen@tudelft.nl

*  Correspondence: giulia.lamberti@phd.unipi.it

**Abstract:** The thermal environment in educational buildings is crucial to improve students' health and productivity, as they spend a considerable amount of time in classrooms. Due to the complexity of educational buildings, research performed has been heterogeneous and standards for thermal comfort are based on office studies with adults. Moreover, they rely on single dose-response models that do not account for interactions with other environmental factors, or students' individual preferences and needs. A literature study was performed on thermal comfort in educational buildings comprising of 143 field studies, to identify all possible confounding parameters involved in thermal perception. Educational stage, climate zone, model adopted to investigate comfort, and operation mode were then selected as confounding parameters and discussed to delineate the priorities for future research. Results showed that children often present with different thermal sensations than adults, which should be considered in the design of energy-efficient and comfortable educational environments. Furthermore, the use of different models to analyse comfort can influence field studies' outcomes and should be carefully investigated. It is concluded that future studies should focus on a more rational evaluation of thermal comfort, also considering the effect that local discomfort can have on the perception of an environment. Moreover, it is important to carefully assess possible relationships between HVAC systems, building envelope, and thermal comfort, including their effect on energy consumption. Since several studies showed that the perception of the environment does not concern thermal comfort only, but it involves the aspects of indoor air, acoustic, and visual quality, their effect on the health and performance of the students should be assessed. This paper provides a way forward for researchers, which should aim to have an integrated approach through considering the positive effects of indoor exposure while considering possible individual differences.

**Keywords:** thermal comfort; indoor environmental quality; educational buildings; energy consumptions; local discomfort

**Citation:** Lamberti, G.; Salvadori, G.; Leccese, F.; Fantozzi, F.; Bluyssen, P.M. Advancement on Thermal Comfort in Educational Buildings: Current Issues and Way Forward. *Sustainability* **2021**, *13*, 10315. https://doi.org/10.3390/su131810315

Academic Editors: Mitja Košir and Manoj Kumar Singh

Received: 27 July 2021
Accepted: 9 September 2021
Published: 15 September 2021



## 1. Introduction

Students spend a good part of the day in schools, and, especially when considering children, they are particularly exposed to an unfavourable indoor environmental quality (IEQ) [1]. Therefore, the relations between classroom characteristics and comfort should be carefully investigated [2]. As the aim of educational buildings is to provide the best learning conditions for students and teachers [3], classrooms should be designed to improve concentration and to stimulate the learning process [4–6], but also be climate-responsive [7]. Since the thermal environment can largely affect students' wellbeing, it is also fundamental to ensure thermal comfort in classrooms to improve students' health and productivity [8].

For the assessment of thermal comfort, several indices have been developed [9], but Fanger's rational (or heat-balance) [10] and the adaptive models [11–13] are the most commonly used. Indeed, it is necessary to raise questions regarding students' possibility to adapt, as at different educational levels adaptation may differ, and, especially at low educational levels, teachers are the only ones who can actively modify the thermal environment [14]. Nevertheless, it should be noted that children and adults do not always have the same thermal perception, therefore pupils' preference on the thermal environment should be considered, as it could help to co-design classrooms [15]. Furthermore, in educational buildings, different activities are carried out, which can influence the thermal comfort evaluation especially at a high metabolic rate [16].

In educational buildings, the duration of field studies varied largely from less than a week to a whole year [17], and they were performed according to three classes [18]: (i) Class III, based on measurements of indoor temperature and humidity at a certain height; (ii) Class II, including field measurements of the six basic parameters in one location at a certain height; (iii) Class I, comprising the measurement of all the environmental parameters at three different heights (0.1, 0.6, 1.2 m) to evaluate local discomfort. Most studies in schools were performed according to Class II [17], while Class III was used for investigations of the adaptive model [19,20]. Only a few papers have been based on Class I [21–24], and in these cases draught, radiant asymmetry, vertical air temperature difference, and floor temperature were measured [22,25,26]. Alternatively, other IEQ aspects were included in the investigation, such as $CO_2$ concentration [4,27,28] or other factors of IEQ, such as noise level [29–31] or illumination level on the work plane [32–34]. Furthermore, as the goal for the buildings of tomorrow is to combine the aspects of energy efficiency and comfort [35], studies were also focused on the impact of thermal comfort on energy efficiency [22,36]. Indeed, due to the complexity of the parameters that influence buildings' performance and indoor environment, it is crucial to focus on the aspects that contribute to determining the health and wellbeing of the occupants, also in relation to architectural and HVAC system design, towards a multi-objective approach to building performance.

The measurement of environmental parameters has been often combined with subjective measurement, which consisted of various types of questionnaires [17]. The first ones included questions regarding thermal sensation and preference, while recent studies also include the evaluation of local thermal comfort, humidity sensation and preference, air velocity sensation and preference, personal regulation, preferred adaptive strategies, information on the clothing worn, and the activity performed prior to the survey [25,37,38]. Simplified questionnaires for children were also provided, to ensure the correctness of the collected data [39,40]. Recently, questionnaires have also included aspects of health and performance of students [6,41,42]. Both longitudinal and transversal surveys were used by researchers, but it was never defined how long the survey should be and how many respondents are necessary for the evaluation of thermal comfort in educational buildings [17,43].

Given the complexity of these environments, there is a lack of standards dealing with thermal comfort in educational buildings, as current regulations such as ISO 7730 [44], ASHRAE 55 [45], and EN 16798-1 [46] seem to be not sufficient to provide comfortable conditions for students and teachers. Indeed, these standards refer to data recorded in laboratories [44] or field studies using comfort data recorded on healthy adults in buildings across the world [11,13], which do not take into account student and teacher individual preferences. Indeed, standards were often developed for environments such as offices, thus, they do not include the peculiarities of educational buildings and they are often based on dose-response models that are not able to explain people's individual preferences and needs.

In conclusion, thermal comfort in educational buildings has been largely investigated and there are many models and indices that have been used with this purpose. However, there are still problems that should be solved, which do not emerge clearly due to the rapid growth of scientific literature. Studies have been often carried out based on the experience of single researchers, rather than adopting a coordinated effort of predetermined directions to develop consistent solutions and guidelines. There is then the need for a collection, a rationalised classification, and analysis of these studies to inspect the present state, aimed at identifying the current issues and to guide future research towards solutions to such problems. This paper aims at filling this gap, highlighting the current issues in thermal comfort studies, and proposing new directions for research with the purpose of integrating the interactions between humans and the environment.

## 2. Methods

### 2.1. Search Strategy

The literature search was performed on the electronic databases Scopus, ScienceDirect, Google Scholar, and Researchgate in the period from March to November 2020. The search keywords used in the databases were {"thermal comfort"} AND {"classroom" OR "class" OR "educational buildings"}, using an integrated search in the title, keywords, and abstract of the papers. Moreover, the selected references were analysed individually to extract relevant information. The following inclusion criteria were selected: (i) original peer-reviewed articles; (ii) field studies investigating thermal comfort and related aspects of IEQ in educational buildings; (iii) full text published in English. Exclusion criteria were: (i) studies not focused on building engineering (e.g., articles focused on physiological or psychological aspects only); (ii) articles investigating energy consumptions only; (iii) simulation studies which did not include field measurements; (iv) books, book chapters, and conference reviews. Review articles were inspected, but not included in the classification of the studies in educational buildings.

Thanks to the possibility to analyse the number of resulting articles, Scopus was used as the primary database to continue this review. From the initial search using the selected keywords, 958 documents were extracted, and following removal of the ones without an English full-text, 916 articles remained. Of these, 445 were research articles, 409 conference papers, 27 were reviews, 18 book chapters, 2 books, and 15 conference reviews. In total, 854 papers (research and conference articles) were then analysed, as well as the 27 review articles. From the title and abstract inspection, the final number of articles considered was 143. Figure 1 shows the geographical distribution of thermal comfort studies in educational buildings, while Figure 2 shows the increase in thermal comfort studies in educational buildings over time.

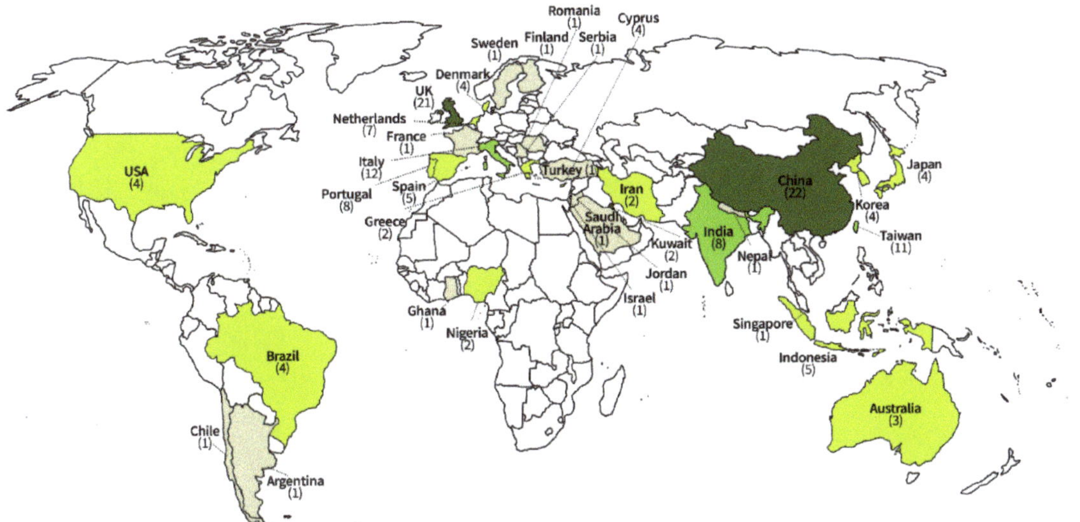



**Figure 1.** Geographical distribution of the studies on thermal comfort in educational buildings over time.

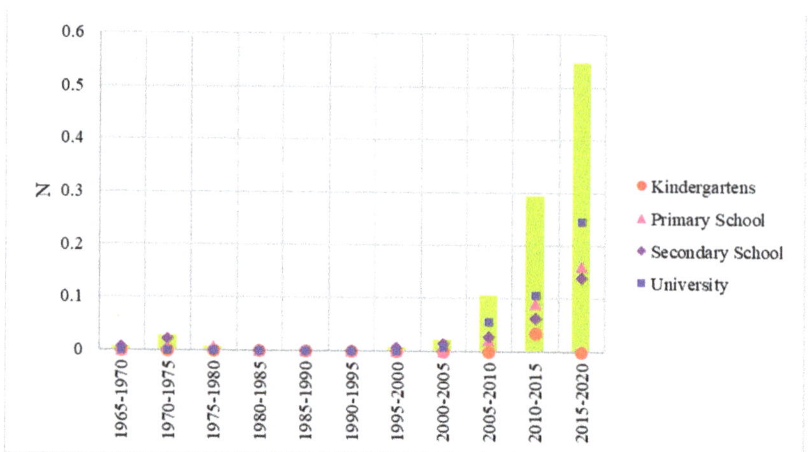

**Figure 2.** Normalized number of studies (N) on thermal comfort in educational buildings. The normalization is done with respect to the total number (143) of considered publications.

*2.2. Data Extraction and Analysis*

Data were extracted from the selected articles, analysing the full text. For the discussion, the year, location of the study, educational stage, climate zone, model used to determine thermal comfort, building's operation mode, and period of the survey were derived, when available. These characteristics were selected because they can possibly explain individual differences in thermal perception. From the analysis of the existing literature, the current issues were identified, and new directions of research were proposed (Figure 3).

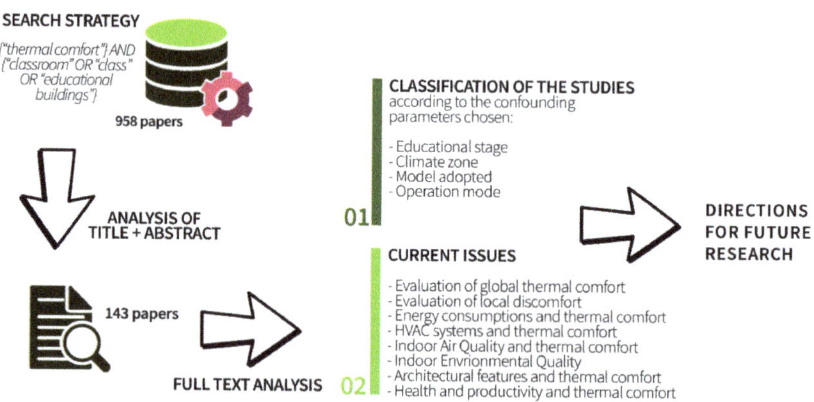


**Figure 3.** Methodology and supporting material used to define the directions for future research.

## 3. Classification of the Studies

By conducting a detailed analysis of the obtained results, it was possible to identify the main categories in which the studies, currently present in the literature, can be grouped. The investigation was based on manual grouping, which was conducted after an accurate inspection of the full text of the selected papers. It resulted that there were four main confounding parameters often identified by researchers as the main causes of differences in thermal sensations among students: the educational stage, climate zone, the model adopted, and operation mode. Table 1 shows the variability of comfort temperatures between the different categories found in the analysed studies. The influence of these confounding parameters on the thermal perception and the results of the analysis are discussed below.

**Table 1.** Variability of comfort temperatures between the different categories found in the analysed studies. The number of papers considered are reported in the parenthesis. Data extracted from [47].

| Characteristics | | Comfort Temperature | |
| --- | --- | --- | --- |
| | | Min (°C) | Max (°C) |
| Educational stage | Kindergarten (4) | 20.7 | 26.0 |
| | Primary school (40) | 14.7 | 30 |
| | Secondary school (39) | 14.7 | 35 |
| | University (60) | 15.5 | 31.5 |
| Climate zone | Group A (21) | 20.0 | 31.0 |
| | Group B (13) | 14.7 | 25.0 |
| | Group C (97) | 14.7 | 35.0 |
| | Group D (12) | 16.0 | 26.0 |
| Model adopted | Rational (38) | 15.0 | 30.7 |
| | Adaptive (23) | 14.7 | 29.2 |
| | Both (34) | 16.0 | 31.0 |
| | Others (48) | 14.7 | 35.0 |
| Operation mode | Air conditioned (38) | 14.7 | 26.9 |
| | Naturally ventilated (51) | 14.7 | 31.5 |
| | Mixed mode (45) | 15.7 | 30.0 |

*3.1. Educational Stage*

The educational stage is the most important aspect that should be evaluated when considering thermal comfort in educational buildings. At different educational stages, students present various ages, diverse possibilities to adapt, and carry out different activities, which can influence their metabolic rate and their capability to respond correctly to questionnaires regarding thermal sensation and preference. Indeed, the age can be a crucial factor for the perception of the thermal environment, also due to the different physiological and psychological characteristics of the pupils. Furthermore, the presence of outdoor activities and stationary or transient conditions, which are determined by the duration of lectures, is also a function of the educational stage. Finally, the density of the classroom can also affect the perception of the thermal environment. Most studies were carried out in universities, followed by secondary and primary schools, and kindergartens (Figure 4).

The evaluation of thermal comfort in kindergartens is a recent topic, developed for the first time in 2012 [48], and only a few works can be found in the literature. The focus of previous studies was mostly on the development of new comfort models for the children, both rational [49] and adaptive [50]. Since children at that age do not present with reading or writing skills yet, researchers also aimed at creating a specific questionnaire for thermal comfort assessment [51]. Studies indicated that pre-school children present comfort temperatures 0.5 °C lower in summer and 3.3 °C in winter [52].

In primary schools, the first work on thermal comfort dates back to 1975 [19], which estimated thermal neutrality from over 6000 assessments obtained. At this educational stage, in most cases children prefer cooler environments, showing that the PMV model underestimates mean thermal sensation up to 1.5 scale point [53], and children present comfort temperatures about 4 °C to 2 °C lower than the predictions from rational and adaptive models, respectively [54]. It is also important to highlight the differences between children's and teachers' thermal sensation [55], as it is more difficult to reach thermal comfort for pupils. Indeed, children present with a lower comfort temperature [56], at least 3 °C lower than adults during cooling seasons [57], and they are also less sensitive to temperature changes than adults [11]. This different perception can be attributed to children's higher metabolic rate, as their activities involve several games including physical exercise, as well as their limited possibility to adapt [52], the influence of the characteristics of their home environments, and outdoor playing, which may alter their thermal perception [47]. It was also demonstrated that social background and behaviour can influence children's thermal preference [58]. Furthermore, thermal comfort models seem to predict inaccurately the thermal sensation of pupils, as the PMV index usually overestimates the perception of scholars, while the adaptive model predicts higher comfort temperatures than the actual ones [59–61].

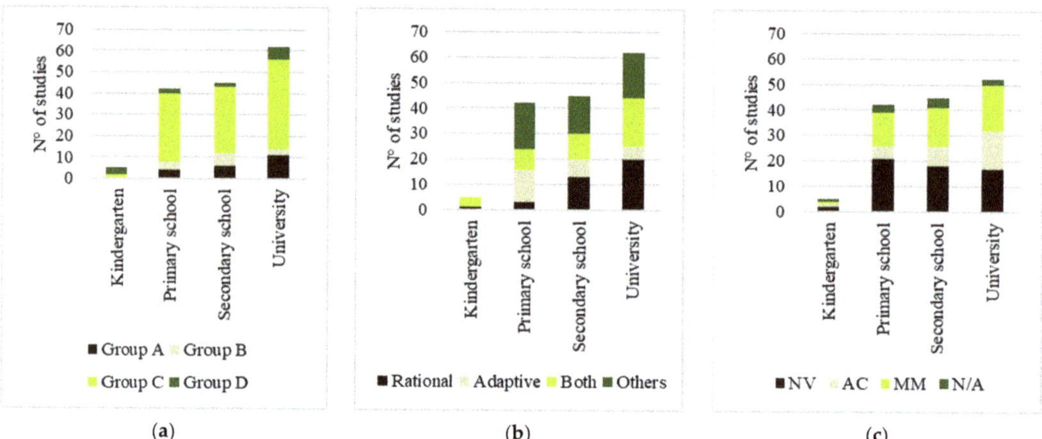

**Figure 4.** Number of studies divided according to the educational stage in different climate zones (**a**), the model adopted (**b**), and operation mode (**c**). Legend: NV = naturally ventilated, AC = air-conditioned, MM = mixed-mode, N/A = not available.

In secondary schools, students can give more reliable information on their thermal state and preference, therefore it is easier for researchers to compare their subjective response to the objective analysis. At this educational stage, several studies showed differences in thermal sensations and neutral temperatures despite the climate, season, and operation modes being the same [17], which can be a specific problem when trying to set the comfort temperatures for the achievement of students' wellbeing. For example, children preferred cooler environments than adults [62–64] even in the tropics [65], and thermal preference changed during summer months [66]. Moreover, students generally accepted cool thermal sensations faster than warm thermal sensations [67]. In the past, more emphasis has been given to the importance of energy savings rather than learning conditions [68,69], therefore there are gaps in the literature considering the improvement of students' performance in relation to the indoor environmental conditions. For air-conditioned buildings, there is evidence that HVAC systems do not always enhance comfort, but they may also be a cause of global and local thermal discomfort [70].

In universities, students have a greater possibility to adapt, and they may be in transient conditions since the duration of the lectures is shorter than at other educational stages. However, even though Fanger's theory was based on experiments carried out on university students, researchers found some divergencies between the predicted thermal sensation obtained with PMV and the real thermal sensation from questionnaires. This can be attributed to several problems related to students' possibility to adapt, adjusting their clothing [24,38,71], or controlling the environment [38,71,72]. Even psychological adaptation can play a fundamental role in adapting to the thermal environment [24].

Furthermore, students may be in transient conditions, as the time they remain in the classroom is limited and they often move outdoors. Differences in neutral temperatures were found in laboratories and classrooms [73,74], in different seasons [23,75,76], and due to gender differences [77].

### 3.2. Climate Zone

The classification per climate zone is relevant, since different climatic conditions can influence the thermal perception and preference of students due to adaptive processes. Considering that thermal history can affect students' thermal comfort [78], the classification per climate zone is relevant in the perspective of improving environmental conditions and reducing energy consumptions in the global warming era. Indeed, the current challenge for building designers is to provide low-energy buildings while enhancing thermal comfort, especially under warmer conditions caused by climate change [79]. According to the Köppen–Geiger climate classification [80], there are five different climate groups (from A to E), divided according to the seasonal precipitation and temperature patterns. The climate zone can influence students' possibility to acclimatise, influences the indoor environment from the outdoor conditions, but also affects the thermal insulation of the clothing worn by students. Most studies have been carried out in the climate Group C, as can be noticed in Figure 5.

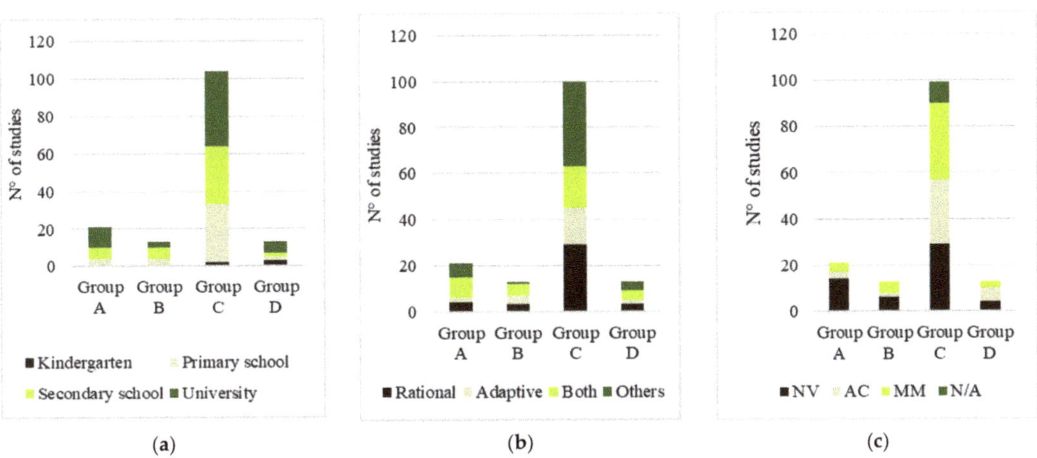


**Figure 5.** Number of studies divided according to the climate zone at different educational stages (**a**), model adopted (**b**) and operation mode (**c**). Legend: NV = naturally ventilated, AC = air-conditioned, MM = mixed-mode, N/A = not available.

Group A includes the tropical climates and about 15% of the studies (22 studies) were carried out in this climate, mostly in naturally ventilated classrooms, applying both rational and adaptive models (Figure 5). Studies were carried out in India, Brazil, Indonesia, Malaysia, Singapore, Nigeria, USA, and Ghana, as reported in Appendix A. In Group A, the range of comfort temperature varied between 20.0 °C and 31.0 °C, indicating large differences in the same climate zones [17], which makes the comparison of neutral temperatures difficult. In Group A, students had a higher heat tolerance and they adapted to the thermal environment, while the temperatures largely exceeded the comfort range given by the standards [17]. These facts are particularly relevant, as they can have a consistent impact on energy conservation strategies, although some studies show that in the past twenty years the comfort temperature has decreased due to the increasing use of air conditioners [81].

Group B includes dry (arid and semi-arid), hot, and cold climates. In this climate zone, only 9% (13 studies) of the studies were carried out. Thermal comfort was investigated during the whole year and used both rational and adaptive models (Figure 5). Studies were carried out in Jordan, Cyprus, Chile, Iran, China, Kuwait, and Saudi Arabia (Appendix A). In Group B, comfort temperature varied between 14.7 °C and 25.0 °C, indicating less of a difference than in Group A [17]. Nevertheless, these values also exceed the comfort range given by standards, probably due to adaptation.

Group C includes temperate climates and comprises various types of climates, therefore temperature variations and adaptability within it can be large. Most studies were carried out in this group (69%, 99 studies) (Figure 5), and included investigations in several countries (Appendix A). Studies were carried out in all the seasons in naturally ventilated and air-conditioned buildings, using both adaptive and rational models (Figure 5). The first studies were carried out in the UK [66,82], which also presents the highest number of investigations along with China and Italy. The comfort temperatures varied between 14.7 °C and 35.0 °C [17]. This variation is probably because in Group C there is a wide range of climates, therefore students are exposed to various weather conditions. In this climate zone, students showed a great capability to adapt, especially the ones exposed to wider climate variations [83].

Group D includes continental climates, and limited research has been performed in this climate zone (8%, 12 studies). Studies were conducted in naturally ventilated and air-conditioned buildings during all the seasons using rational and adaptive models (Figure 5). They were carried out in China, Korea, Romania, Sweden, Finland, Turkey, and Nepal (Appendix A). The comfort temperature ranged between 16.0 °C and 26.0 °C [17], showing a large variability in students' preferences.

Group E includes polar and arctic climates, with an average temperature below 10 °C. No study on thermal comfort in educational buildings was found for this climate zone.

### 3.3. Model Adopted

The model adopted to analyse thermal comfort in educational buildings is relevant, as it can influence the predicted thermal sensation. Studies were carried out using the rational model (27%, 38 studies), adaptive model (16%, 23 studies) separately or together (24%, 34 studies), or other indicators (33%, 48 studies) (Figure 6). Indeed, different models can lead to diverse conclusions on the thermal state of students; therefore, it is fundamental to choose the model that is the closest to their real thermal sensation, according to their diverse characteristics and needs. Studies showed that none of the models accurately predict the thermal sensation of students. Rational and adaptive models should be combined to improve the prediction of the thermal environment, as they are complementary and not contradictory [18].

The rational model is generally applicable to air-conditioned spaces where occupants are in steady-state conditions with limited possibility to adapt. However, these conditions do not always occur in educational buildings, which leads to an overestimation or underestimation of the thermal sensation. Furthermore, the rational model is often incompatible in temperate and tropical climates [17]. To overcome these problems, corrections of PMV index, such as ePMV [63,67] or aPMV [84] have been provided and used by researchers to assess the thermal environment in educational buildings (Appendix A). Even though this model seems to be, in most cases, too inaccurate to predict thermal comfort in educational buildings, it is still widely used by researchers. However, it should be noted that the reliability of the PMV index largely depends on the precision of the assessment input parameters [85], which must be carefully evaluated to avoid misleading results.

In educational buildings, students often have adaptive opportunities. Most studies reported higher comfort temperatures than the ones predicted by the adaptive model, while lower comfort temperatures were found in secondary schools and compatible results in universities [17]. These results are consistent with the assumption of the adaptive model that considers occupants as able to modify their environment to achieve thermal comfort. This type of adaptation is typical of university students, which show comfort temperatures in line with the ones predicted by the adaptive model [17].

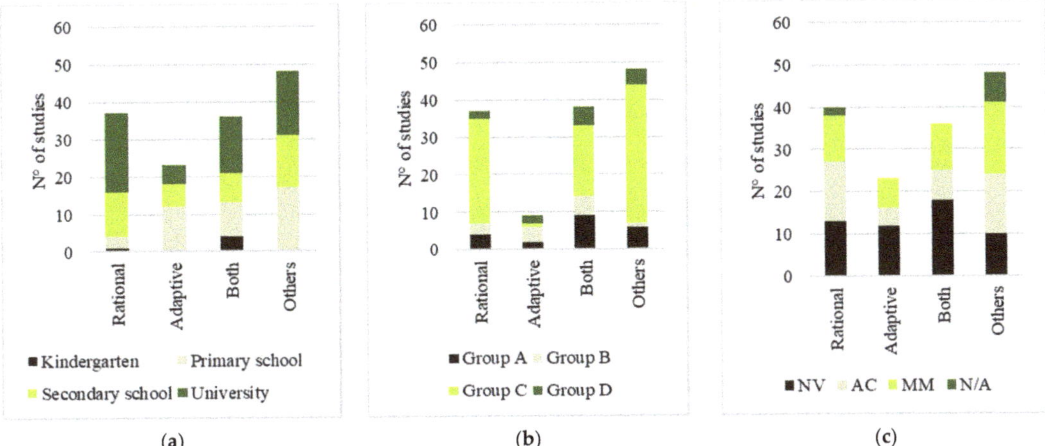

**Figure 6.** Number of studies divided according to the model adopted, at different educational stages (**a**), climate zone (**b**) and operation mode (**c**). Legend: NV = naturally ventilated, AC = air-conditioned, MM = mixed-mode, N/A = not available.

### 3.4. Operation Mode

The operation mode of the classroom can also influence the thermal perception of the students, as they experience diverse thermal environments in them (i.e., steady-state in air-conditioned, transient in free-running buildings). The operation mode influences the indoor environmental conditions also as a function of the outdoor climate, and different comfort temperatures can be found in relation to the operation mode. Furthermore, it defines the adaptation of the students and their control of the thermal environment, which can regulate their thermal perception. Most studies were carried out in naturally ventilated (NV) classrooms, followed by mixed-mode (MM) and air-conditioned (AC) schools (Figure 7).

In naturally ventilated buildings, the possibility to adapt is limited to the opening/closing of doors and windows. Several studies on primary schools were performed in naturally ventilated classrooms, and less in secondary schools, universities, and kindergartens (Figure 7). In climates A and B, natural ventilation is often used as the operation mode of classrooms, and less in climates C and D, as the colder climates need the presence of HVAC systems to achieve comfort in buildings (Figure 7). In most studies, the adaptive and rational models were both used, while when considered separately, the adaptive model was used in more cases than the rational (Figure 7). In particular, in hot-humid climates, observed comfort temperatures in naturally ventilated classrooms were found to be about 1.7 °C lower than the ASHRAE-recommended value [86], showing the importance of carefully evaluating comfort temperature also for energy savings. Furthermore, in naturally ventilated classrooms, students expressed comfort even when the environmental parameters were out of the standard's comfort zone [67,87]. These results are consistent with the adaptive hypothesis [13], which implies that humans can adapt behaviourally, physiologically, and psychologically to the environmental conditions to which they are subjected.

About 30% of the studies (38 papers) were performed in air-conditioned classrooms (Figure 7), where HVAC systems were switched on during the investigation period. The highest number of studies in air-conditioned classrooms was in universities, followed by secondary, primary schools, and kindergartens (Figure 7). The heating system was switched on during the winter in temperate climates and longer in colder climates (Group D), while the increasing use of cooling systems during summer, especially in warmer climates, could be detected. Climate Groups C and D were often investigating air-conditioned classrooms, and researchers usually used the rational model to assess thermal comfort. Only a few studies used the adaptive model as well as both models (Figure 7), even if adaptive behaviours were observed in air-conditioned schools [87]. Field studies in air-conditioned schools were also carried out to design energy efficient classrooms [88], with evident implications on students' wellbeing and energy consumptions.

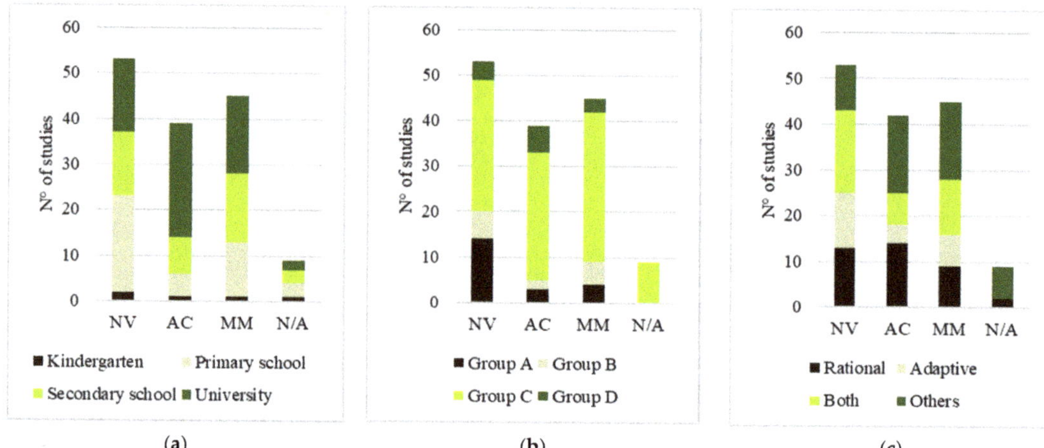

**Figure 7.** Number of studies divided according to the operation mode, at different educational stages (**a**), climate zone (**b**) and model adopted (**c**). Legend: NV = naturally ventilated, AC = air-conditioned, MM = mixed-mode, N/A = not available.

Studies in mixed-mode buildings were often investigating the thermal environment during several seasons (Appendix A). Most studies in mixed-mode classrooms were equally distributed in secondary, primary schools, and universities (Figure 7). Climate Group C was the most investigated and both rational and adaptive models were used, both separately or together (Figure 7). This operation mode should be carefully evaluated, as the comfort temperatures and needs of the occupants vary according to the season of investigation and the operation mode of the classroom. Differences in students' performance between conditioned and naturally ventilated classrooms were also evaluated [89], but no significant difference was found. In this direction, studies comparing air-conditioned and naturally ventilated spaces did not show different neutral temperatures, preferred temperatures, and thermal acceptability [75].

## 4. Current Issues

In this section, the complete selection of the studies was analysed to identify the most common issues that are present in the evaluation of thermal comfort in classrooms, which are reported in the following paragraphs.

### 4.1. Evaluation of Global Thermal Comfort

Most of the investigations on global thermal comfort have been focused on field measurements only, both objective and subjective, while few studies compared on-site measurements with simulations. There is no agreement on how global thermal comfort should be assessed, as different models and indices to assess thermal comfort were applied (Figure 8). Among them, the most common were the rational and the adaptive models (Appendix A). Some researchers were focused on the assessment of air temperature and relative humidity only in Portuguese school buildings [68,69], evaluating the impact of refurbishments on the indoor environment [69] and also in relation to other aspects of IEQ [2,34]. Indices such as effective temperature ET [19], new effective temperature ET* [73,77], corrected effective temperature CET [19], tropical summer index TSI [90], or the PMV correction for naturally ventilated buildings [91] were also calculated. The corrections of PMV index, ePMV [63,67] and aPMV [84], were also applied. In several studies, the operative temperature was assessed, and the neutral temperature was derived from the subjective responses of the students [75,92,93]. It was deduced that pupils' thermal sensation is higher than adults [59], and that their thermal sensation is not related to indoor temperature only, but also to their home environment [58]. Furthermore, thermal sensation changed during the class hour and adaptation occurred after about 20 min after entering the classroom [94]. In some cases, the outdoor environmental conditions were considered [49,58,95], as they can also influence the perception of the indoor environment. The main purpose of these studies was to assess the environmental conditions in educational buildings [51]. For this reason, new algorithms and models were provided for scholars [49,96,97], and adaptive comfort equations obtained from regression analysis were developed for children [50,61] to assess the comfort temperatures in different educational stages and all classrooms [47]. Ranges of comfort temperatures were also provided for classrooms, considering different educational stages and climate zones [17,47]. To consider the applicability and the differences between predictive models of thermal comfort, some researchers performed a comparison between them [63]. However, it lacks agreement among researchers in regards to the model that should be used to evaluate global thermal comfort, which is one of the current issues that should be faced.

### 4.2. Evaluation of Local Discomfort

Recently, the assessment of local discomfort has become increasingly relevant, as it can have a great influence on people's wellbeing, but also on energy consumption. However, it lacks sufficient scientific evidence regarding the importance of local discomfort on students' wellbeing. The main causes of local discomfort found in classrooms were draught risk [21,31,98], vertical air temperature difference [21,99,100], warm and cool floors [21,99,101], and radiant asymmetry [21,25,98,99,101]. Studies were performed using the UCB Berkeley model for thermal comfort [26], questionnaires including specific questions on local discomfort [21] and its causes [25], and in some cases, occupants were asked to report the part of the body that was subjected to local discomfort [22]. The purpose of these studies varies from evaluating the percentage of dissatisfied students due to local discomfort [98] and the difference between global and local thermal comfort [22], to investigating the relation between local thermal comfort and productivity [25]. Even if a growing interest in this topic can be detected, only a few studies were carried out, and broader knowledge on the issue of local discomfort is needed.

### 4.3. Energy Consumptions and Thermal Comfort

The impact of thermal comfort on energy consumption is a debated topic that was faced by researchers in different ways. The evaluation of energy consumption has been carried out through on-site measurements, simulations, or in climate chambers approximating the conditions of the typical classrooms. In some cases, researchers investigated the direct relationship between energy consumption and thermal comfort [22,36,102]. Energy consumption was also compared to thermal comfort in association with indoor air quality [103] and visual comfort [104,105]. The concerns about energy savings were compared to ventilation strategies [95,106], HVAC systems operation, and architectural features [107,108]. Researchers also compared refurbished and non-refurbished educational buildings, which resulted in a reduction of consumption for renovated schools [93,102,109]. An investigation into the influence of shading devices on indoor environmental quality and energy consumption was also performed [110]. Finally, an algorithm to improve thermal comfort and indoor air quality and reduce consumption, using the least amount of energy from air conditioning and ventilation fans, was developed [111]. This analysis showed the importance of combining the issue of consumption with thermal comfort studies to enhance students' wellbeing and reduce energy requirements. However, it is still difficult to propose solutions that reduce energy consumption without compromising thermal comfort for children and adults.

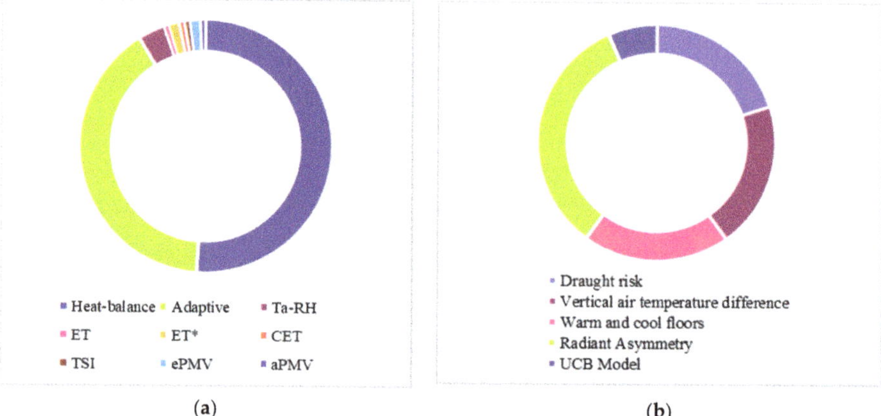



(**a**)                    (**b**)

**Figure 8.** Indices used by researchers to investigate global (**a**) and local (**b**) thermal comfort in classrooms.

### 4.4. HVAC Systems and Thermal Comfort

The evaluation of thermal comfort is fundamental for heating, ventilation, and air conditioning system installation and settings, and has become increasingly important in recent years [17]. Regarding heating and cooling systems, studies have been carried out to compare traditional and innovative systems in terms of thermal comfort improvement [112]. Other investigations have been carried out to design better conditioning systems in educational buildings [88]. Through assessing comfort temperatures, it is possible to reduce the need for heating and cooling systems and maximise the energy savings without impairing thermal comfort [113]. Concerning ventilation strategies, their impact on thermal comfort and IAQ was investigated [27]; as well as the effectiveness of different types of ventilation, such as natural, hybrid ventilation, and air conditioning [114]; the acceptance of thermal conditions; and energy use considering different ventilation strategies and exhaust configurations [106]. Thermal comfort was also analysed considering stratum ventilation using a pulsating air supply [115], and in relation to gender differences [116], showing an improved thermal comfort in comparison to a conventional constant air supply. Then, field studies were carried out to compare the conditions found in educational buildings with the criteria in the standards [117]. Since ventilation systems often create a non-uniformity into the environment due to the air distribution, researchers analysed the possibility to improve comfort through managing these non-uniformities, considering thermal preferences [118]. The investigations of the HVAC systems' impact on thermal comfort are very heterogeneous because different types of conditioning systems exist. To provide guidelines for their design and installation, it is necessary to understand their impact on students' wellbeing, while being aware of their individual needs, and not only focusing on energy consumption.

### 4.5. Indoor Air Quality and Thermal Comfort

Indoor air quality (IAQ) has been often investigated together with thermal comfort, as they are important aspects to ensure health and wellbeing in classrooms. $CO_2$ concentration was the parameter measured most frequently [68,119] and was compared with the threshold limit values given by standards. In some cases, the $CO_2$ concentration was found to be below the threshold values given by standards [117,120,121], and, in other cases, no correlation was found between $CO_2$ concentration and number of people [94]. Instead, in naturally ventilated buildings, $CO_2$ concentration was found to be very high when windows were closed [122]. No correlation was found in classrooms between $CO_2$ values and students' feelings of tiredness [59]. From $CO_2$ concentration decay in classrooms, the air change per hour and the ventilation effectiveness in these spaces were determined [123].

From a comparison between refurbished and non-refurbished buildings, it was seen that the renovated constructions improved thermal comfort, but increased the $CO_2$ concentration, and therefore reduced indoor air quality [109]. The impact of different ventilation modes on thermal comfort and $CO_2$ levels was also evaluated [27,69,124], as well as the relation between IAQ and thermal comfort [2]. In some cases, the concentration of other pollutants, such as VOCs, $NO_2$, and CO was measured and correlated to the outdoor conditions [125–127]. The subjective perception of air quality was correlated to the environmental conditions, which showed that the perceived environmental quality was highly correlated to parameters, such as air temperature and ventilation rates [6]. There is a need to understand the relationship between indoor air quality and thermal comfort perception, and to understand how to improve them without increasing energy consumption.

### 4.6. Indoor Environmental Quality

Indoor environmental quality (IEQ), which includes thermal comfort, air quality, visual, and acoustic quality, can affect students' health, comfort, and productivity. Indeed, research is increasingly focusing on multi-domain approaches to indoor environmental perception and behaviour [128,129], inspecting the various aspects of IEQ on people's comfort and satisfaction [130]. The evaluation of IEQ through objective and subjective measurements has been often carried out, while also considering the psychological and physiological impact on occupants' comfort [4,33], and pupils' performance [131] or symptomatology [132]. Indeed, the perception of the environment includes the four aspects of IEQ and can have an impact on students' health and learning abilities [41,70,131], but also on their wellbeing [6,32]. In some cases, the subjective assessment was performed through questionnaires to evaluate the perception, preferences, and needs regarding IEQ in classrooms [133], but also statistical surveys have been carried out [134]. Thermal comfort was frequently associated with lighting quality, considering different configurations of architectural characteristics [32,110] and the impact on energy consumption [105,108]. Since thermal perception is strongly related to acoustic, visual, and air quality, researchers should not focus on the evaluation of thermal comfort only, but on all the aspects of IEQ.

### 4.7. Architectural Features and Thermal Comfort

Thermal comfort in buildings is closely related to their architectural features, including dimensions, window-wall ratio, presence of shading systems, building orientation, articulation of classrooms, and the properties of the building envelope [17]. Most researchers considered classrooms as uniform spaces, however, due to solar radiation or to the position of ventilation systems, classrooms are usually non-uniform environments, and therefore thermal discomfort may occur locally. The influence of buildings' envelope on thermal comfort was evaluated [110], considering the effect of different types of insulated roofs (e.g., PCM, composite) on the possibility of overheating of classrooms [135]. In addition, dynamic characteristics of the envelope were evaluated and through the application of films that allow the control of solar radiation [136,137]. In some studies, researchers assessed the improved conditions of a renovation [138] or compared different types of school building constructions (light weight and medium weight) [60]. The relation between classroom characteristics and thermal comfort was also investigated [2,139], including the influence of a façade design to prevent overheating and improve daylight requirements [140], and the use of natural ventilation and ceiling fans to improve comfort [141]. The influence of shading systems and window configuration, including the glass ratio and glass properties, on occupants' comfort and energy demands were also assessed [97,142]. Additionally, studies were performed to allow building designers to choose the best configurations to improve comfort and reduce energy demands [108], in addition to considering the influence of the local climate on the architectural project [107]. Finally, the effect of building design on learning rate and perception was investigated in some works [143,144]. Despite its importance, studies on thermal comfort involving these aspects are not frequent and should be increased to provide guidance to building designers, aimed at improving indoor conditions and reducing consumptions.

*4.8. Health and Productivity and Thermal Comfort*

In the past, students' productivity has been inspected in regard to air quality [27] through the analysis of their performance in relation to different aspects of schoolwork (numerical or language-based) under different ventilation rates [145], or through providing them diverse tests measuring arithmetic concentration, performance speed, task performance accuracy, and visual memory [124]. Students' learning efficiency was also evaluated in relation to the characteristics of the shading systems [142] and increased classroom temperature [146]. Scholars' performance has been related to all the aspects of indoor environmental quality [41,131,147] and to the thermal environment only [148,149]. The influence of IEQ on health and productivity has been demonstrated as a function of Fanger's PMV and personal factors [25]. Researchers showed that students' health and productivity depend on buildings' features [2,143]. Thermal sensation as well as IEQ have been correlated to health-related issues that can occur as students pass a considerable amount of time in educational buildings [6,42,150]. It must be noted that the assessment of the influence of the thermal environment on health and productivity is not easy, as several variables can influence students' performance and wellbeing. Researchers have tried to determine students' performance through questionnaires, in which they were self-reporting their productivity [41], or through the measurement of speed errors [149]. Furthermore, the effect of the microclimate in classrooms was assessed through the measurement of cardiac autonomic control (ECG) and cognitive performances of the students [151]. This issue remains a very interesting and debated topic, as it still lacks a generally accepted scientific method to assess the influence of the thermal environment on students' health and productivity.

## 5. Conclusions and Directions for Future Research

The investigation of the current literature showed that researchers focused on different issues, adopting diverse models and indices to investigate thermal comfort in classrooms. However, to provide healthy and human-centred buildings, it is important to focus on students' individual needs and preferences, and not only on single dose-response relationships that have been established for the average adult. It is also clear that the focus should be on preventing negative effects as well as creating positive effects for human health. Indeed, even if the environmental conditions comply with guidelines, in several cases the prolonged stay indoors is not healthy [152]. An integrated approach that considers the positive and negative effects of indoor exposure is therefore needed, including the individual preferences and needs of the occupants [152]. To achieve this integrated approach, several aspects must be accounted for (Figure 9), while considering the individual differences that may be present in relation to the diverse educational stages, climate zone, model adopted, and operation mode.

From this analysis, it was possible to outline the current issues and delineate the directions for future research. However, it is important to note the limitations that the present study may have. The manual grouping of the confounding parameters, which was useful for the direct control of the information contained in the scientific literature, could be combined with statistic methods to test other possible classifications (e.g., country in which the study was carried out, period of investigation, year). Furthermore, the grouping per climate zone may present some limitations, as they include different countries that are characterised by diverse sociocultural backgrounds, and therefore the variability of comfort temperature can be very high. Even the different educational systems might affect the opportunities to adapt and therefore the comfort temperature. These considerations highlight that it is difficult to generalise indications regarding thermal comfort in educational buildings. On the contrary, from this review, it emerged the necessity of studies aiming at meeting the needs of the students, in order to provide human-centred buildings.

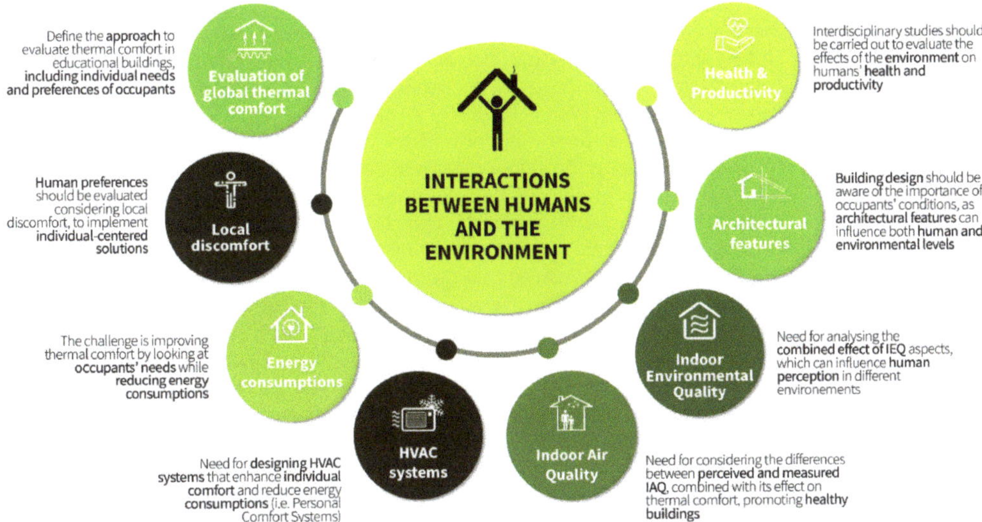


**Figure 9.** Directions for future research on thermal comfort in educational buildings.

The evaluation of global thermal comfort was carried out using different models and indices, but, in several cases, individual needs were not analysed. Given the variety of indices that have been used for comfort assessment, it is important to define a model for the evaluation of global thermal comfort in classrooms, which should take into account the individual needs and preferences of occupants.

The same concerns are related to local thermal discomfort, even if few studies were carried out on this topic. From this review it can be concluded that local discomfort is an important issue that should be assessed not only to inspect students' satisfaction but also to investigate the reasons for the higher productivity of students located in certain positions in the classrooms, to provide solutions that are more individual-centred.

Regarding energy demands, the literature showed that high consumptions are often connected to the characteristics of the building's envelope and to the improper control of HVAC systems. Students' dissatisfaction was often related to warm sensations during winter and cold sensations in summer, due to the extensive use of HVAC systems [47]. Since thermal comfort is also a function of the ventilation strategy, building designers need to consider the configurations that minimise consumptions and improve occupants' wellbeing. The aim is then to provide comfort in classrooms and reduce energy demands by looking at occupants' needs and preferences [153].

In this direction, HVAC systems can have a great impact on people's comfort and energy consumptions. Studies including different ventilation regimes are necessary. Moreover, in order to inspect people's needs, studies on personal comfort systems (PCS), which operate at the individual level are needed.

However, thermal comfort evaluation should not overlook the interaction with other environmental aspects since they all contribute to the human perception of the environment. The integration of IAQ in thermal comfort studies is necessary and should include objective and subjective measurements, working towards human-centred buildings. There is a need to consider the differences between perceived and measured IAQ, combined with its effect on thermal comfort.

Moreover, future studies should focus on the healing power of indoor environments that involve all the IEQ aspects, including thermal comfort, air, visual, and acoustic qualities. Indeed, people are subjected to a combination of them, and only through the analysis of their combined effect is it possible to understand humans' perception.

This review showed the need for research in order to understand the relation between thermal comfort and architectural features, to improve indoor environmental conditions and wellbeing in classrooms. This is necessary to guide building designers, which should be aware of the importance of occupants' conditions, as architectural features can influence the perception both at human (i.e., influence of personal control on perception) and environmental (i.e., providing uniform/non-uniform environments) levels.

Furthermore, as the indoor environment can affect students' health and productivity, it is fundamental to investigate it not as a single dose-response system only, but include interactions at both human and environmental levels to define a methodology for understanding the impact of the indoor environment on them.

All these issues showed that thermal comfort in educational buildings is still a very debated topic. In this way, there is the need to analyse them, considering their effect on individuals and their interactions with the environment. After the COVID-19 pandemic, the importance of ensuring healthy environments became even more evident, also due to the increasing amount of time that people spend indoors. Indeed, the pandemic period revealed the difficulties in providing sufficient indoor air quality, which should be enhanced without compromising thermal comfort and energy consumptions. There is the necessity to adapt educational buildings to the pandemic and post-pandemic periods, which should be considered together with climate change issues and needs identified before the pandemic.

This review, which critically analyses the studies according to different confounding parameters, highlighted the current issues and defined a way forward in research, represented a contribution in this direction, and will guide researchers and building designers towards a human-centred approach, which is currently lacking.

**Author Contributions:** Conceptualization, G.L., G.S., F.L., F.F. and P.M.B.; methodology, G.L., G.S., F.L., F.F. and P.M.B.; formal analysis, G.L., G.S., F.L., F.F. and P.M.B.; writing—original draft preparation, G.L.; writing—review and editing, G.S., F.L., F.F. and P.M.B.; funding acquisition, G.S., F.L., F.F. All authors have read and agreed to the published version of the manuscript.

**Funding:** The research was partially funded by the University of Pisa with the "Ateneo 2020 Funds", assigned to individual researchers following an annual evaluation of the research products of the last 4 years by a Selected Scientific Commission.

**Institutional Review Board Statement:** Not applicable.

**Informed Consent Statement:** Not applicable.

**Conflicts of Interest:** The authors declare no conflict of interest.

## Appendix A. Studies on Thermal Comfort in Educational Buildings

In this Appendix, the analysis of the 143 selected studies is reported. Table A1 shows the studies on thermal comfort in educational buildings extracted from the literature, including the relevant information necessary for the analysis:

- Author(s).
- Year of publication.
- Location of the study (country).
- Educational stage, which comprises kindergartens, primary, secondary schools, and universities.
- Climate zone, which is analysed according to the Köppen–Geiger classification (where A: tropical climates; B: dry (arid and semi-arid) hot and cold climates; C: temperate climates; D: continental climates; E: polar and arctic climates).
- Model adopted, which includes rational, adaptive, both (rational + adaptive), and others (where other indices or models were adopted, as described in Section 4).
- Operation mode, which consists of naturally ventilated (NV), air-conditioned (AC), and mixed-mode (MM) buildings.
- Period of the survey, which expresses the season of measurements.
- Reference.

**Table A1.** Studies on thermal comfort in educational buildings extracted from the literature.

| Author(s) | Year | Location | Educational Stage | Climate Zone | Model Adopted | Operation Mode | Time of Survey | References |
|---|---|---|---|---|---|---|---|---|
| Auliciems A. | 1969 | UK | Secondary school | C | Others | AC | Autumn–Spring–Winter | [82] |
| Auliciems A. | 1972 | UK | Secondary school | C | Others | AC | Autumn–Spring–Winter | [148] |
| Auliciems | 1973 | UK | Secondary school | C | Others | NV | Spring–Summer | [66] |
| Humphreys | 1973 | UK | Secondary school | C | Others | NV | Spring–Summer | [62] |
| Auliciems | 1975 | Australia | Primary school | C | Others | AC | Autumn–Winter | [19] |
| Humphreys | 1977 | UK | Primary school | C | Others | NV | Summer | [154] |
| Kwok et al. | 1998 | USA | Secondary school | A | Both | MM | Autumn–Winter | [155] |
| Kwok et al. | 2003 | Japan | Secondary school | C | Adaptive | MM | Summer | [87] |
| Wong et al. | 2003 | Singapore | Secondary school | A | Rational | NV | Summer | [67] |
| Krüger et al. | 2004 | Brazil | University | C | Others | NV | Summer–Winter | [34] |
| Hu et al. | 2006 | China | University | C | Others | MM | Summer–Winter | [73] |
| Hwang et al. | 2006 | Taiwan | University | C | Rational | MM | Summer | [77] |
| Corgnati et al. | 2007 | Italy | University | C | Rational | AC | All seasons | [29] |
| Wargocki et al. | 2007 | Denmark | Secondary school | C | Others | NV | Summer | [145] |
| Wargocki et al. | 2007 | Denmark | Secondary school | C | Rational | NV | Summer | [146] |
| Zhang et al. | 2007 | China | University | C | Rational | NV | Spring | [92] |
| Cheng et al. | 2008 | Taiwan | University | C | Rational | MM | Autumn–Spring–Summer | [75] |
| Theodosiou et al. | 2008 | Greece | Primary school | C | Others | MM | Autumn–Spring–Winter | [103] |
| Al-Rashidi et al. | 2009 | Kuwait | Secondary school | B | Both | AC | - | [63] |
| Buratti et al. | 2009 | Italy | University | C | Both | AC | Spring–Winter | [99] |
| Corgnati et al. | 2009 | Italy | University | C | Both | NV | Autumn–Spring | [98] |
| Hwang et al. | 2009 | Taiwan | Primary + Secondary school | C | Both | NV | Autumn–Winter | [86] |
| Mumovic et al. | 2009 | UK | Secondary school | C | Rational | MM | Winter | [31] |
| Zeiler et al. | 2009 | Netherlands | Primary school | C | Rational | AC | Spring–Winter | [112] |
| Yao et al., | 2010 | China | University | C | Both | NV | Spring | [24] |
| Cao et al. | 2011 | China | University | D | Rational | AC | Summer–Winter | [76] |

**Table A1.** *Cont.*

| Author(s) | Year | Location | Educational Stage | Climate Zone | Model Adopted | Operation Mode | Time of Survey | References |
|---|---|---|---|---|---|---|---|---|
| Jung et al. | 2011 | South Korea | University | C | Both | MM | Autumn–Spring | [156] |
| Liu et al. | 2011 | Japan | University | C | Others | MM | Summer–Winter | [123] |
| Al-Rashidi et al. | 2012 | Kuwait | Primary school | B | Others | MM | Spring | [27] |
| Conceicao et al. | 2012 | Portugal | Kindergarten | C | Both | MM | Summer–Winter | [49] |
| De Giuli et al. | 2012 | Italy | Primary school | C | Adaptive | NV | Spring | [4] |
| Lee et al. | 2012 | China | University | C | Rational | AC | - | [70] |
| Liang et al. | 2012 | Taiwan | Primary +Secondary school | C | Adaptive | NV | Autumn–Winter | [97] |
| Maki et al. | 2012 | Japan | University | C | Others | AC | Spring–Summer | [104] |
| Puteh et al. | 2012 | Malaysia | Secondary school | A | Others | NV | - | [150] |
| Teli et al. | 2012 | UK | Primary school | C | Both | NV | Spring–Summer | [53] |
| Barbhuiya et al. | 2013 | UK | University | C | Others | AC | Spring–Winter | [95] |
| Barrett et al. | 2013 | UK | Primary school | C | Others | MM | All seasons | [151] |
| D'Ambrosio et al. | 2013 | Italy | Primary + Secondary | C | Rational | NV | Summer–Winter | [21] |
| Fabbri et al. | 2013 | Italy | Kindergarten | C | Rational | - | Autumn | [50] |
| Pereira et al. | 2013 | Portugal | Secondary school | C | Rational | MM | Spring–Winter | [157] |
| Teli et al. | 2013 | UK | Primary school | C | Others | NV | Spring–Summer | [59] |
| Wargocki et al. | 2013 | Denmark | Secondary school | C | Rational | AC | Summer | [149] |
| Yang et al. | 2013 | USA | University | C | Others | AC | Winter | [144] |
| Baruah et al. | 2014 | India | University | C | Rational | NV | Spring–Winter | [91] |
| Choi et al. | 2014 | USA | University | C | Others | AC | - | [41] |
| De Giuli et al. | 2014 | Italy | Primary school | C | Both | MM | Spring–Summer–Winter | [33] |
| Gao et al. | 2014 | Denmark | Primary school | C | Adaptive | AC | All seasons | [42] |
| Katafygiotou et al. | 2014 | Cyprus | Secondary school | C | Others | - | - | [36] |
| Mishra et al. | 2014 | India | University | A | Both | NV | Spring–Winter | [72] |
| Mishra et al. | 2014 | India | University | A | Both | NV | Spring–Winter | [5] |
| Pereira et al. | 2014 | Portugal | Secondary school | C | Rational | NV | Spring | [158] |
| Serghides et al. | 2014 | Cyprus | University | C | Rational | AC | Summer–Winter | [105] |
| Teli et al. | 2014 | UK | Primary school | C | Both | NV | Summer | [60] |
| Turunen et al. | 2014 | Finland | Primary school | D | Others | AC | Spring–Summer | [6] |
| Wang et al. | 2014 | China | University | D | Adaptive | AC | Spring–Winter | [23] |
| Yun et al. | 2014 | Korea | Kindergarten | D | Both | NV | Spring–Summer | [48] |
| Almeida et al. | 2015 | Portugal | Secondary school | C | Others | MM | Spring–Summer–Winter | [68] |

**Table A1.** *Cont.*

| Author(s) | Year | Location | Educational Stage | Climate Zone | Model Adopted | Operation Mode | Time of Survey | References |
|---|---|---|---|---|---|---|---|---|
| Almeida et al. | 2015 | Portugal | Secondary school | C | Adaptive | MM | Spring | [69] |
| Barrett et al. | 2015 | UK | Secondary school | C | Others | MM | All seasons | [147] |
| De Dear et al. | 2015 | Australia | Secondary school | C | Both | MM | Summer | [83] |
| Fong et al. | 2015 | China | University | C | Rational | AC | - | [106] |
| Huang et al. | 2015 | Taiwan | Primary school | C | Adaptive | NV | Autumn–Spring–Summer | [159] |
| Huang et al. | 2015 | Taiwan | Primary school | C | Adaptive | MM | Autumn–Spring | [114] |
| Mishra et al. | 2015 | India | University | A | Rational | MM | Autumn–Summer | [89] |
| Nam et al. | 2015 | Korea | Kindergarten | D | Both | AC | Spring–Summer | [51] |
| Nico et al. | 2015 | Italy | University | C | Both | NV | - | [160] |
| Almeida et al. | 2016 | Portugal | Primary + Secondary + University | C | Both | NV | Spring | [101] |
| Liu et al. | 2016 | China | Secondary school | B | Rational | NV | Winter | [161] |
| Vittal et al. | 2016 | India | University | A | Both | NV | Winter | [90] |
| Castilla et al. | 2017 | Spain | University | C | Others | - | - | [139] |
| Hadad et al. | 2017 | Iran | Primary school | B | Both | MM | Autumn–Summer–Winter | [55] |
| Martinez-Molina et al. | 2017 | Spain | Primary school | C | Rational | MM | Autumn–Winter | [54] |
| Mishra et al. | 2017 | Netherlands | University | C | Adaptive | AC | Spring | [94] |
| Montazami et al. | 2017 | UK | Primary school | C | Adaptive | NV | Summer | [56] |
| Montazami et al. | 2017 | UK | Primary school | C | Adaptive | NV | Summer | [58] |
| Stazi et al. | 2017 | Italy | Primary school | C | Adaptive | AC | Spring–Winter | [162] |
| Teli et al. | 2017 | UK | Primary school | C | Adaptive | NV | All seasons | [61] |
| Trebilcock et al. | 2017 | Chile | Primary school | B | Adaptive | NV | Summer–Winter | [163] |
| Wang et al. | 2017 | China | University | D | Both | AC | Autumn–Spring–Winter | [164] |
| Zaki et al. | 2017 | Malaysia, Japan | University | A, C | Both | MM | Winter–Spring | [74] |
| Bajc et al. | 2018 | Serbia | University | C | Rational | MM | Winter | [25] |
| Bluyssen et al. | 2018 | Netherlands | Primary school | C | Others | MM | Spring | [2] |
| Fang et al. | 2018 | China | University | C | Both | AC | Autumn–Summer | [26] |
| Hamzah et al. | 2018 | Indonesia | Secondary school | A | Rational | NV | Summer | [65] |
| Kim et al. | 2018 | Australia | Secondary school | C | Both | MM | Autumn | [37] |
| Kumar et al. | 2018 | India | University | A | Both | NV | Spring–Summer | [71] |

**Table A1.** *Cont.*

| Author(s) | Year | Location | Educational Stage | Climate Zone | Model Adopted | Operation Mode | Time of Survey | References |
|---|---|---|---|---|---|---|---|---|
| Singh et al. | 2018 | India | University | A | Both | NV | Spring–Summer | [38] |
| Aghniaey et al. | 2019 | USA | University | C | Both | AC | Summer | [113] |
| Ali et al. | 2019 | Jordan | Secondary school | B | Both | NV | - | [102] |
| Barbic et al. | 2019 | Italy | University | C | Others | AC | - | [151] |
| Bluyssen et al. | 2019 | Netherlands | Primary school | C | Others | MM | Spring | [165] |
| Branco et al. | 2019 | Portugal | Primary school | C | Others | MM | All seasons | [125] |
| Calama-González et al. | 2019 | Spain | Secondary school | C | Adaptive | NV | All seasons | [110] |
| Campano et al. | 2019 | Spain | Secondary school | C | Both | MM | Autumn–Summer–Winter | [166] |
| Chen et al. | 2019 | Taiwan | Primary school | C | Others | NV | Summer | [142] |
| Chen et al. | 2019 | Taiwan | University | C | Rational | MM | Spring | [167] |
| Chitaru et al. | 2019 | Romania | Secondary school | D | Rational | NV | Summer–Winter | [122] |
| Colinart et al. | 2019 | France | Secondary school | C | Others | - | All seasons | [138] |
| Costa et al. | 2019 | Brazil | University | A | Others | MM | Summer | [107] |
| Fabozzi et al. | 2019 | Italy | University | C | Both | MM | Summer | [168] |
| Haddad et al. | 2019 | Iran | Primary school | B | Adaptive | NV | Autumn–Spring–Winter | [169] |
| Hamzah et al. | 2019 | Indonesia | University | A | Others | AC | Spring | [88] |
| Heracleous et al. | 2019 | Cyprus | University | B | Adaptive | MM | Winter | [170] |
| Huang et al. | 2019 | China | University | D | Others | AC | Spring | [171] |
| Jindal | 2019 | India | Secondary school | A | Adaptive | NV | All seasons | [96] |
| Jing et al. | 2019 | China | University | C | Rational | AC | Winter | [22] |
| Karyono et al. | 2019 | Indonesia | University | A | Others | AC | - | [81] |
| Korateng et al. | 2019 | Ghana | University | A | Adaptive | NV | Spring–Summer | [172] |
| Lawrence et al. | 2019 | UK | University | C | Rational | MM | Summer–Winter | [93] |
| Li et al. | 2019 | China | University | C | Others | AC | Summer | [116] |
| Liu et al. | 2019 | China | University | B | Rational | NV | Winter | [84] |
| Liu et al. | 2019 | China | University | C | Rational | NV | Winter | [173] |
| Monna et al. | 2019 | Palestina | Secondary school | C | Others | MM | All seasons | [174] |
| Ranjbar | 2019 | Turkey | University | D | Others | MM | Summer–Winter | [124] |
| Shen et al. | 2019 | China | University | C | Rational | NV | Summer–Winter | [175] |
| Shrestha et al. | 2019 | Nepal | Secondary school | D | Adaptive | NV | Autumn | [120] |
| Simanic et al. | 2019 | Sweden | Primary school | D | Others | MM | All seasons | [117] |
| Tian et al. | 2019 | China | University | C | Rational | AC | Summer | [115] |
| Toyinbo et al. | 2019 | Nigeria | Primary school | A | Others | NV | - | [127] |

**Table A1.** *Cont.*

| Author(s) | Year | Location | Educational Stage | Climate Zone | Model Adopted | Operation Mode | Time of Survey | References |
|---|---|---|---|---|---|---|---|---|
| Vallarades et al. | 2019 | Taiwan | University | C | Rational | AC | Summer | [111] |
| Zhang et al. | 2019 | Netherlands | Primary school | C | Others | - | Spring | [14] |
| Zhang et al. | 2019 | Netherlands | Primary school | C | Others | - | Spring | [133] |
| Al-Khatri et al. | 2020 | Arabia | Secondary school | B | Both | AC | Summer | [176] |
| Barbosa et al. | 2020 | Portugal | Secondary school | C | Others | MM | Spring–Winter | [109] |
| Boutet et al. | 2020 | Argentina | Primary + Secondary school | C | Others | - | All seasons | [32] |
| Campano-Laborda et al. | 2020 | Spain | Secondary school | C | Others | MM | Spring–Winter | [132] |
| da Silva Júnior et al. | 2020 | Brazil | Secondary school | A | Rational | AC | Summer | [177] |
| Hamzah et al. | 2020 | Indonesia | Primary school | A | Both | NV | Spring | [178] |
| Heracleous et al. | 2020 | Cyprus | Secondary school | B | Adaptive | MM | Summer–Winter | [179] |
| Jiang et al. | 2020 | China | Secondary school | B | Both | MM | Winter | [119] |
| Jowkar et al. | 2020 | UK | University | C | Others | MM | - | [78] |
| Jowkar et al. | 2020 | UK | University | C | Adaptive | MM | Autumn–Winter | [180] |
| Jowkar et al. | 2020 | UK | University | C | Others | AC | Autumn–Spring–Winter | [181] |
| Korsavi et al. | 2020 | UK | Primary school | C | Adaptive | MM | All seasons | [182] |
| Liu et al. | 2020 | China | University | B | Rational | NV | Autumn–Spring | [183] |
| Liu et al. | 2020 | China | University | D | Rational | MM | Autumn–Winter | [100] |
| Munonye et al. | 2020 | Nigeria | Primary school | A | Both | NV | Autumn–Spring | [184] |
| Papadopoulos et. | 2020 | Greece | University | C | Rational | AC | Winter | [126] |
| Pistore et al. | 2020 | Italy | Secondary school | C | Others | - | - | [134] |
| Talarosha et al. | 2020 | Indonesia | Primary school | A | Others | NV | Spring | [121] |
| Wang et al. | 2020 | China | Secondary school | C | Rational | AC | Summer | [131] |
| Zhang et al. | 2020 | China | University | C | Rational | - | - | [118] |

## References

1. Bluyssen, P.M. Health, Comfort and Performance of Children in Classrooms–New Directions for Research. *Indoor Built Environ.* **2016**, *26*, 1040–1050. [CrossRef]
2. Bluyssen, P.M.; Zhang, D.; Kurvers, S.; Overtoom, M.; Ortiz-Sanchez, M. Self-Reported Health and Comfort of School Children in 54 Classrooms of 21 Dutch School Buildings. *Build. Environ.* **2018**, *138*, 106–123. [CrossRef]
3. Mendell, M.; Heath, G. Do Indoor Pollutants and Thermal Conditions in Schools Influence Student Performance? A Critical Review of Literature. *Indoor Air* **2005**, *15*, 27–52. [CrossRef]
4. De Giuli, V.; Da Pos, O.; De Carli, M. Indoor Environmental Quality and Pupil Perception in Italian Primary Schools. *Build. Environ.* **2012**, *56*, 335–345. [CrossRef]
5. Mishra, A.K.; Ramgopal, M. Thermal Comfort Field Study in Undergraduate Laboratories–An Analysis of Occupant Perceptions. *Build. Environ.* **2014**, *76*, 62–72. [CrossRef]
6. Turunen, M.; Toyinbo, O.; Putus, T.; Nevalainen, A.; Shaughnessy, R.; Haverinen-Shaughnessy, U. Indoor Environmental Quality in School Buildings, and the Health and Wellbeing of Students. *Int. J. Hyg. Environ. Health* **2014**, *217*, 733–739. [CrossRef] [PubMed]
7. Fantozzi, F.; Hamdi, H.; Rocca, M.; Vegnuti, S. Use of Automated Control Systems and Advanced Energy Simulations in the Design of Climate Responsive Educational Building for Mediterranean Area. *Sustainability* **2019**, *11*, 1660. [CrossRef]
8. Lamberti, G.; Fantozzi, F.; Salvadori, G. Thermal Comfort in Educational Buildings: Future Directions Regarding the Impact of Environmental Conditions on Students' Health and Performance. In Proceedings of the 2020 IEEE International Conference on Environment and Electrical Engineering and 2020 IEEE Industrial and Commercial Power Systems Europe (EEEIC/I&CPS Europe), Madrid, Spain, 9–12 June 2020; pp. 1–6.
9. Fantozzi, F.; Rocca, M. An Extensive Collection of Evaluation Indicators to Assess Occupants' Health and Comfort in Indoor Environment. *Atmosphere* **2020**, *11*, 90. [CrossRef]
10. Fanger, P.O. *Thermal Comfort. Analysis and Applications in Environmental Engineering*; Danish Technical Press: Copenhagen, Denmark; McGraw-Hill: New York, NY, USA, 1970.
11. Humphreys, M.; Nicol, F.; Roaf, S. *Adaptive Thermal Comfort: Foundations and Analysis*; Routledge: London, UK, 2016; ISBN 978-0-415-6916-1.
12. Nicol, F.; Humphreys, M.; Roaf, S. *Adaptive Thermal Comfort: Principles and Practice*; Routledge: London, UK, 2012; p. 175, ISBN 978-0-203-12301-0.
13. de Dear, R.; Brager, G.S.; Cooper, D. Developing an Adaptive Model of Thermal Comfort and Preference-Final Report on RP-884. *ASHRAE Trans.* **1997**, *104*, 1–13.
14. Zhang, D.; Bluyssen, P.M. Actions of Primary School Teachers to Improve the Indoor Environmental Quality of Classrooms in the Netherlands. *Intell. Build. Int.* **2019**, *13*, 1–13. [CrossRef]
15. Bluyssen, P.M.; Kim, D.H.; Eijkelenboom, A.; Ortiz-Sanchez, M. Workshop with 335 Primary School Children in The Netherlands: What Is Needed to Improve the IEQ in Their Classrooms? *Build. Environ.* **2020**, *168*, 106486. [CrossRef]
16. Fantozzi, F.; Lamberti, G. Determination of Thermal Comfort in Indoor Sport Facilities Located in Moderate Environments: An Overview. *Atmosphere* **2019**, *10*, 769. [CrossRef]
17. Zomorodian, Z.S.; Tahsildoost, M.; Hafezi, M. Thermal Comfort in Educational Buildings: A Review Article. *Renew. Sustain. Energy Rev.* **2016**, *59*, 895–906. [CrossRef]
18. Brager, G.S.; de Dear, R.J. Thermal Adaptation in the Built Environment: A Literature Review. *Energy Build.* **1998**, *27*, 83–96. [CrossRef]
19. Auliciems, A. Warmth and Comfort in the Subtropical Winter: A Study in Brisbane Schools. *J. Hyg.* **1975**, *74*, 339–343. [CrossRef]
20. James, A.-D.; Koranteng, C. An Assessment of Thermal Comfort in a Warm and Humid School Building at Accra, Ghana. *Pelagia Res. Libr. Adv. Appl. Sci. Res.* **2012**, *2012*, 3.
21. D'Ambrosio Alfano, F.R.; Ianniello, E.; Palella, B.I. PMV–PPD and Acceptability in Naturally Ventilated Schools. *Build. Environ.* **2013**, *67*, 129–137. [CrossRef]
22. Jing, S.; Lei, Y.; Wang, H.; Song, C.; Yan, X. Thermal Comfort and Energy-Saving Potential in University Classrooms during the Heating Season. *Energy Build.* **2019**, *202*, 109390. [CrossRef]
23. Wang, Z.; Li, A.; Ren, J.; He, Y. Thermal Adaptation and Thermal Environment in University Classrooms and Offices in Harbin. *Energy Build.* **2014**, *77*, 192–196. [CrossRef]
24. Yao, R.; Liu, J.; Li, B. Occupants' Adaptive Responses and Perception of Thermal Environment in Naturally Conditioned University Classrooms. *Appl. Energy* **2010**, *87*, 1015–1022. [CrossRef]
25. Bajc, T.; Banjac, M.; Todorović, M.; Stevanović, Ž. Experimental and Statistical Survey on Local Thermal Comfort Impact on Working Productivity Loss in University Classrooms. *Therm. Sci.* **2018**, *2018*, 379–392. [CrossRef]
26. Fang, Z.; Zhang, S.; Cheng, Y.; Fong, A.M.L.; Oladokun, M.O.; Lin, Z.; Wu, H. Field Study on Adaptive Thermal Comfort in Typical Air Conditioned Classrooms. *Build. Environ.* **2018**, *133*, 73–82. [CrossRef]
27. Al-Rashidi, K.; Loveday, D.; Al-Mutawa, N. Impact of Ventilation Modes on Carbon Dioxide Concentration Levels in Kuwait Classrooms. *Energy Build.* **2012**, *47*, 540–549. [CrossRef]

28. Franco, A.; Leccese, F. Measurement of $CO_2$ Concentration for Occupancy Estimation in Educational Buildings with Energy Efficiency Purposes. *J. Build. Eng.* **2020**, *32*, 101714. [CrossRef]
29. Corgnati, S.P.; Filippi, M.; Viazzo, S. Perception of the Thermal Environment in High School and University Classrooms: Subjective Preferences and Thermal Comfort. *Build. Environ.* **2007**, *42*, 951–959. [CrossRef]
30. Leccese, F.; Rocca, M.; Salvadori, G. Fast Estimation of Speech Transmission Index Using the Reverberation Time: Comparison between Predictive Equations for Educational Rooms of Different Sizes. *Appl. Acoust.* **2018**, *140*, 143–149. [CrossRef]
31. Mumovic, D.; Palmer, J.; Davies, M.; Orme, M.; Ridley, I.; Oreszczyn, T.; Judd, C.; Critchlow, R.; Medina, H.A.; Pilmoor, G.; et al. Winter Indoor Air Quality, Thermal Comfort and Acoustic Performance of Newly Built Secondary Schools in England. *Build. Environ.* **2009**, *44*, 1466–1477. [CrossRef]
32. Boutet, M.L.; Hernández, A.L.; Jacobo, G.J. Methodology of Quantitative Analysis and Diagnosis of Higro-Thermal and Lighting Monitoring for School Buildings in a Hot-Humid Mid-Latitude Climate. *Renew. Energy* **2020**, *145*, 2463–2476. [CrossRef]
33. De Giuli, V.; Zecchin, R.; Corain, L.; Salmaso, L. Measurements of Indoor Environmental Conditions in Italian Classrooms and Their Impact on Children's Comfort. *Indoor Built Environ.* **2014**, *24*, 689–712. [CrossRef]
34. Krüger, E.L.; Zannin, P.H.T. Acoustic, Thermal and Luminous Comfort in Classrooms. *Build. Environ.* **2004**, *39*, 1055–1063. [CrossRef]
35. Li, Q.; Zhang, L.; Zhang, L.; Wu, X. Optimizing Energy Efficiency and Thermal Comfort in Building Green Retrofit. *Energy* **2021**, *237*, 121509. [CrossRef]
36. Katafygiotou, M.C.; Serghides, D.K. Thermal Comfort of a Typical Secondary School Building in Cyprus. *Sustain. Cities Soc.* **2014**, *13*, 303–312. [CrossRef]
37. Kim, J.; de Dear, R. Thermal Comfort Expectations and Adaptive Behavioural Characteristics of Primary and Secondary School Students. *Build. Environ.* **2018**, *127*, 13–22. [CrossRef]
38. Singh, M.K.; Kumar, S.; Ooka, R.; Rijal, H.B.; Gupta, G.; Kumar, A. Status of Thermal Comfort in Naturally Ventilated Classrooms during the Summer Season in the Composite Climate of India. *Build. Environ.* **2018**, *128*, 287–304. [CrossRef]
39. Fabbri, K. *Indoor Thermal Comfort Perception: A Questionnaire Approach Focusing on Children*; Springer: Cham, Germany, 2015; ISBN 978-3-319-18650-4.
40. Haddad, S.; King, S.; Osmond, P.; Heidari, S. Questionnaire Design to Determine Children's Thermal Sensation, Preference and Acceptability in the Classroom. In Proceedings of the 28th PLEA International Conference, Lima, Peru, 7 November 2012.
41. Choi, S.; Guerin, D.; Kim, H.; Brigham, J.K.; Bauer, T.A. Indoor Environmental Quality of Classrooms and Student Outcomes: A Path Analysis Approach. *J. Learn. Spaces* **2014**, *2*, 2013–2014.
42. Gao, J.; Wargocki, P.; Wang, Y. Ventilation System Type, Classroom Environmental Quality and Pupils' Perceptions and Symptoms. *Build. Environ.* **2014**, *75*, 46–57. [CrossRef]
43. Mishra, A.K.; Ramgopal, M. Field Studies on Human Thermal Comfort—An Overview. *Build. Environ.* **2013**, *64*, 94–106. [CrossRef]
44. ISO 7730. *Ergonomics of the Thermal Environment-Analytical Determination and Interpretation of Thermal Comfort Using Calculation of the PMV and PPD Indices and Local Thermal Comfort Criteria*; ISO: Geneve, Switzerland, 2006.
45. ASHRAE. *ANSI/ASHRAE Standard 55: Thermal Environmental Conditions for Human Occupancy*; ASHRAE: Peachtree Corners, GA, USA, 2004.
46. EN 16798-1. *Energy Performance of Buildings-Ventilation for Buildings-Part 1: Indoor Environmental Input Parameters for Design and Assessment of Energy Performance of Buildings Addressing Indoor Air Quality, Thermal Environment, Lighting and Acoustics*; CEN: Brusselles, Belgium, 2019.
47. Singh, M.K.; Ooka, R.; Rijal, H.B.; Kumar, S.; Kumar, A.; Mahapatra, S. Progress in Thermal Comfort Studies in Classrooms over Last 50 Years and Way Forward. *Energy Build.* **2019**, *188*, 149–174. [CrossRef]
48. Rupp, R.F.; Vásquez, N.G.; Lamberts, R. A Review of Human Thermal Comfort in the Built Environment. *Energy Build.* **2015**, *105*, 178–205. [CrossRef]
49. Yun, H.; Nam, I.; Kim, J.; Yang, J.; Lee, K.; Sohn, J. A Field Study of Thermal Comfort for Kindergarten Children in Korea: An Assessment of Existing Models and Preferences of Children. *Build. Environ.* **2014**, *75*, 182–189. [CrossRef]
50. Conceição, E.Z.E.; Gomes, J.M.M.; Antão, N.H.; Lúcio, M.M.J.R. Application of a Developed Adaptive Model in the Evaluation of Thermal Comfort in Ventilated Kindergarten Occupied Spaces. *Build. Environ.* **2012**, *50*, 190–201. [CrossRef]
51. Fabbri, K. Thermal Comfort Evaluation in Kindergarten: PMV and PPD Measurement through Datalogger and Questionnaire. *Build. Environ.* **2013**, *68*, 202–214. [CrossRef]
52. Nam, I.; Yang, J.; Lee, D.; Park, E.; Sohn, J.-R. A Study on the Thermal Comfort and Clothing Insulation Characteristics of Preschool Children in Korea. *Build. Environ.* **2015**, *92*, 724–733. [CrossRef]
53. Mors, S.; Hensen, J.L.M.; Loomans, M.G.L.C.; Boerstra, A.C. Adaptive Thermal Comfort in Primary School Classrooms: Creating and Validating PMV-Based Comfort Charts. *Build. Environ.* **2011**, *46*, 2454–2461. [CrossRef]
54. Teli, D.; Jentsch, M.F.; James, P.A.B. Naturally Ventilated Classrooms: An Assessment of Existing Comfort Models for Predicting the Thermal Sensation and Preference of Primary School Children. *Energy Build.* **2012**, *53*, 166–182. [CrossRef]
55. Martinez-Molina, A.; Boarin, P.; Tort-Ausina, I.; Vivancos, J.-L. Post-Occupancy Evaluation of a Historic Primary School in Spain: Comparing PMV, TSV and PD for Teachers' and Pupils' Thermal Comfort. *Build. Environ.* **2017**, *117*, 248–259. [CrossRef]

56. Haddad, S.; Osmond, P.; King, S. Revisiting Thermal Comfort Models in Iranian Classrooms during the Warm Season. *Build. Res. Inf.* **2017**, *45*, 457–473. [CrossRef]

57. Montazami, A.; Gaterell, M.; Nicol, F.; Lumley, M.; Thoua, C. Developing an Algorithm to Illustrate the Likelihood of the Dissatisfaction Rate with Relation to the Indoor Temperature in Naturally Ventilated Classrooms. *Build. Environ.* **2017**, *111*, 61–71. [CrossRef]

58. Montazami, A.; Gaterell, M.; Nicol, F.; Lumley, M.; Thoua, C. Impact of Social Background and Behaviour on Children's Thermal Comfort. *Build. Environ.* **2017**, *122*, 422–434. [CrossRef]

59. Teli, D.; James, P.A.B.; Jentsch, M.F. Thermal Comfort in Naturally Ventilated Primary School Classrooms. *Build. Res. Inf.* **2013**, *41*, 301–316. [CrossRef]

60. Teli, D.; Jentsch, M.F.; James, P.A.B. The Role of a Building's Thermal Properties on Pupils' Thermal Comfort in Junior School Classrooms as Determined in Field Studies. *Build. Environ.* **2014**, *82*, 640–654. [CrossRef]

61. Teli, D.; Bourikas, L.; James, P.A.B.; Bahaj, A.S. Thermal Performance Evaluation of School Buildings Using a Children-Based Adaptive Comfort Model. *Sustain. Synerg. Build. Urban Scale* **2017**, *38*, 844–851. [CrossRef]

62. Humphreys, M.A. Clothing and thermal comfort of secondary school children in summertime. In *Thermal Comfort and Moderate Heat Stress*; Langdon: London, UK, 1973.

63. Al-Rashidi, K.E.; Loveday, D.L.; Al-Mutawa, N.K. Investigating the Applicability of Different Thermal Comfort Models in Kuwait Classrooms Operated in Hybrid Air-Conditioning Mode. In *Sustainability in Energy and Buildings*; Howlett, R.J., Jain, L.C., Lee, S.H., Eds.; Springer: Berlin/Heidelberg, Germany, 2013; pp. 347–355, ISBN 978-3-642-36645-1.

64. Kim, J.; Zhou, Y.; Schiavon, S.; Raftery, P.; Brager, G. Personal Comfort Models: Predicting Individuals' Thermal Preference Using Occupant Heating and Cooling Behavior and Machine Learning. *Build. Environ.* **2018**, *129*, 96–106. [CrossRef]

65. Hamzah, B.; Gou, Z.; Mulyadi, R.; Amin, S. Thermal Comfort Analyses of Secondary School Students in the Tropics. *Buildings* **2018**, *8*, 56. [CrossRef]

66. Auliciems, A. Thermal Sensations of Secondary Schoolchildren in Summer. *J. Hyg.* **1973**, *71*, 453–458. [CrossRef]

67. Wong, N.H.; Khoo, S.S. Thermal Comfort in Classrooms in the Tropics. *Energy Build.* **2003**, *35*, 337–351. [CrossRef]

68. Almeida, R.M.S.F.; de Freitas, V.P.; Delgado, J.M.P.Q. Indoor Environmental Quality in Classrooms: Case Studies. In *School Buildings Rehabilitation: Indoor Environmental Quality and Enclosure Optimization*; Almeida, R.M.S.F., de Freitas, V.P., Delgado, J.M.P.Q., Eds.; Springer International Publishing: Cham, Germany, 2015; pp. 31–57, ISBN 978-3-319-15359-9.

69. Almeida, R.M.S.F.; de Freitas, V.P. IEQ Assessment of Classrooms with an Optimized Demand Controlled Ventilation System. *Energy Procedia* **2015**, *78*, 3132–3137. [CrossRef]

70. Lee, M.C.; Mui, K.W.; Wong, L.T.; Chan, W.Y.; Lee, E.W.M.; Cheung, C.T. Student Learning Performance and Indoor Environmental Quality (IEQ) in Air-Conditioned University Teaching Rooms. *Build. Environ.* **2012**, *49*, 238–244. [CrossRef]

71. Kumar, S.; Singh, M.K.; Mathur, A.; Mathur, J.; Mathur, S. Evaluation of Comfort Preferences and Insights into Behavioural Adaptation of Students in Naturally Ventilated Classrooms in a Tropical Country, India. *Build. Environ.* **2018**, *143*, 532–547. [CrossRef]

72. Mishra, A.K.; Ramgopal, M. Thermal Comfort in Undergraduate Laboratories—A Field Study in Kharagpur, India. *Build. Environ.* **2014**, *71*, 223–232. [CrossRef]

73. Hu, P.F.; Liu, W.; Jiang, Z. Study on Indoor Thermal Sensation of Young College Students in the Area Which Is Hot in Summer and Cold in Winter. *Age* **2006**, *18*, 18–23.

74. Zaki, S.A.; Damiati, S.A.; Rijal, H.B.; Hagishima, A.; Abd Razak, A. Adaptive Thermal Comfort in University Classrooms in Malaysia and Japan. *Build. Environ.* **2017**, *122*, 294–306. [CrossRef]

75. Cheng, M.-J.; Hwang, R.-L.; Lin, T.-P. Field Experiments on Thermal Comfort Requirements for Campus Dormitories in Taiwan. *Indoor Built Environ.* **2008**, *17*, 191–202. [CrossRef]

76. Cao, B.; Zhu, Y.; Ouyang, Q.; Zhou, X.; Huang, L. Field Study of Human Thermal Comfort and Thermal Adaptability during the Summer and Winter in Beijing. *Energy Build.* **2011**, *43*, 1051–1056. [CrossRef]

77. Hwang, R.-L.; Lin, T.-P.; Kuo, N.-J. Field Experiments on Thermal Comfort in Campus Classrooms in Taiwan. *Energy Build.* **2006**, *38*, 53–62. [CrossRef]

78. Jowkar, M.; de Dear, R.; Brusey, J. Influence of Long-Term Thermal History on Thermal Comfort and Preference. *Energy Build.* **2020**, *210*, 109685. [CrossRef]

79. Holmes, M.J.; Hacker, J.N. Climate Change, Thermal Comfort and Energy: Meeting the Design Challenges of the 21st Century. *Energy Build.* **2007**, *39*, 802–814. [CrossRef]

80. Beck, H.E.; Zimmermann, N.E.; McVicar, T.R.; Vergopolan, N.; Berg, A.; Wood, E.F. Present and Future Köppen-Geiger Climate Classification Maps at 1-Km Resolution. *Sci. Data* **2018**, *5*, 180214. [CrossRef] [PubMed]

81. Karyono, T.; Heryanto, S.; Faridah, I. Thermal Comfort Study of University Students in Jakarta, Indonesia. In Proceedings of the 8th Windsor Conference, Windsor, UK, 10–13 April 2014; pp. 1276–1285.

82. Auliciems, A. Thermal Requirements of Secondary Schoolchildren in Winter. *J. Hyg.* **1969**, *67*, 59–65. [CrossRef]

83. De Dear, R.; Kim, J.; Candido, C.; Deuble, M. Adaptive Thermal Comfort in Australian School Classrooms. *Build. Res. Inf.* **2015**, *43*, 383–398. [CrossRef]

84. Liu, J.; Yang, X.; Jiang, Q.; Qiu, J.; Liu, Y. Occupants' Thermal Comfort and Perceived Air Quality in Natural Ventilated Classrooms during Cold Days. *Build. Environ.* **2019**, *158*, 73–82. [CrossRef]

85. D'Ambrosio Alfano, F.; Olesen, B.W.; Palella, B.; Riccio, G.; Pepe, D. Fifty Years of PMV Model: Reliability, Implementation and Design of Software for Its Calculation. *Atmosphere* **2019**, *11*, 49. [CrossRef]
86. Hwang, R.-L.; Lin, T.-P.; Chen, C.-P.; Kuo, N.-J. Investigating the Adaptive Model of Thermal Comfort for Naturally Ventilated School Buildings in Taiwan. *Int. J. Biometeorol.* **2009**, *53*, 189–200. [CrossRef]
87. Kwok, A.G.; Chun, C. Thermal Comfort in Japanese Schools. *Sol. Energy* **2003**, *74*, 245–252. [CrossRef]
88. Hamzah, B.; Kusno, A.; Mulyadi, R. Design of Energy Efficient and Thermally Comfortable Air-Conditioned University Classrooms in the Tropics. *Int. J. Sustain. Energy* **2019**, *38*, 382–397. [CrossRef]
89. Mishra, A.K.; Ramgopal, M. A Comparison of Student Performance between Conditioned and Naturally Ventilated Classrooms. *Build. Environ.* **2015**, *84*, 181–188. [CrossRef]
90. Vittal, R. Perceived Thermal Environment of Naturally- Ventilated Classrooms in India. *Creat. Space* **2016**, *3*, 149–166. [CrossRef]
91. Baruah, P.; Singh, M.; Mahapatra, S. Thermal Comfort in Naturally Ventilated Classrooms. In Proceedings of the 30th International Plea Conference, Ahmedabad, India, 16–18 December 2014.
92. Zhang, G.; Zheng, C.; Yang, W.; Zhang, Q.; Moschandreas, D.J. Thermal Comfort Investigation of Naturally Ventilated Classrooms in a Subtropical Region. *Indoor Built Environ.* **2007**, *16*, 148–158. [CrossRef]
93. Lawrence, R.; Elsayed, M.; Keime, C. Evaluation of Environmental Design Strategies for University Buildings. *Build. Res. Inf.* **2019**, *47*, 883–900. [CrossRef]
94. Mishra, A.K.; Derks, M.T.H.; Kooi, L.; Loomans, M.G.L.C.; Kort, H.S.M. Analysing Thermal Comfort Perception of Students through the Class Hour, during Heating Season, in a University Classroom. *Build. Environ.* **2017**, *125*, 464–474. [CrossRef]
95. Barbhuiya, S.; Barbhuiya, S. Thermal Comfort and Energy Consumption in a UK Educational Building. *Build. Environ.* **2013**, *68*, 1–11. [CrossRef]
96. Jindal, A. Investigation and Analysis of Thermal Comfort in Naturally Ventilated Secondary School Classrooms in the Composite Climate of India. *Archit. Sci. Rev.* **2019**, *62*, 466–484. [CrossRef]
97. Liang, H.-H.; Lin, T.-P.; Hwang, R.-L. Linking Occupants' Thermal Perception and Building Thermal Performance in Naturally Ventilated School Buildings. *Appl. Energy* **2012**, *94*, 355–363. [CrossRef]
98. Corgnati, S.P.; Ansaldi, R.; Filippi, M. Thermal Comfort in Italian Classrooms under Free Running Conditions during Mid Seasons: Assessment through Objective and Subjective Approaches. *Build. Environ.* **2009**, *44*, 785–792. [CrossRef]
99. Buratti, C.; Ricciardi, P. Adaptive Analysis of Thermal Comfort in University Classrooms: Correlation between Experimental Data and Mathematical Models. *Build. Environ.* **2009**, *44*, 674–687. [CrossRef]
100. Liu, G.; Jia, Y.; Cen, C.; Ma, B.; Liu, K. Comparative Thermal Comfort Study in Educational Buildings in Autumn and Winter Seasons. *Sci. Technol. Built Environ.* **2020**, *26*, 185–194. [CrossRef]
101. Almeida, R.M.S.F.; Ramos, N.M.M.; de Freitas, V.P. Thermal Comfort Models and Pupils' Perception in Free-Running School Buildings of a Mild Climate Country. *Energy Build.* **2016**, *111*, 64–75. [CrossRef]
102. Ali, H.H.; Al-Hashlamun, R. Assessment of Indoor Thermal Environment in Different Prototypical School Buildings in Jordan. *Alex. Eng. J.* **2019**, *58*, 699–711. [CrossRef]
103. Theodosiou, T.G.; Ordoumpozanis, K.T. Energy, Comfort and Indoor Air Quality in Nursery and Elementary School Buildings in the Cold Climatic Zone of Greece. *Energy Build.* **2008**, *40*, 2207–2214. [CrossRef]
104. Maki, Y.; Shukuya, M. Visual and Thermal Comfort and Its Relations to Exergy Consumption in a Classroom with Daylighting. *Int. J. Energy* **2012**, *11*, 481–492. [CrossRef]
105. Serghides, D.; Chatzinikola, C.K.; Katafygiotou, M. Comparative Studies of the Occupants' Behaviour in a University Building during Winter and Summer Time. *Int. J. Sustain. Energy* **2014**, *34*, 528–551. [CrossRef]
106. Fong, M.L.; Hanby, V.; Greenough, R.; Lin, Z.; Cheng, Y. Acceptance of Thermal Conditions and Energy Use of Three Ventilation Strategies with Six Exhaust Configurations for the Classroom. *Build. Environ.* **2015**, *94*, 606–619. [CrossRef]
107. Costa, M.L.; Freire, M.R.; Kiperstok, A. Strategies for Thermal Comfort in University Buildings-The Case of the Faculty of Architecture at the Federal University of Bahia, Brazil. *J. Environ. Manag.* **2019**, *239*, 114–123. [CrossRef] [PubMed]
108. Bakmohammadi, P.; Noorzai, E. Optimization of the Design of the Primary School Classrooms in Terms of Energy and Daylight Performance Considering Occupants' Thermal and Visual Comfort. *Energy Rep.* **2020**, *6*, 1590–1607. [CrossRef]
109. Barbosa, F.C.; de Freitas, V.P.; Almeida, M. School Building Experimental Characterization in Mediterranean Climate Regarding Comfort, Indoor Air Quality and Energy Consumption. *Energy Build.* **2020**, *212*, 109782. [CrossRef]
110. Calama-González, C.M.; Suárez, R.; Leon-Rodriguez, A.; Ferrari, S. Assessment of Indoor Environmental Quality for Retrofitting Classrooms with An Egg-Crate Shading Device in A Hot Climate. *Sustainability* **2019**, *11*, 1078. [CrossRef]
111. Valladares, W.; Galindo, M.; Gutiérrez, J.; Wu, W.-C.; Liao, K.-K.; Liao, J.-C.; Lu, K.-C.; Wang, C.-C. Energy Optimization Associated with Thermal Comfort and Indoor Air Control via a Deep Reinforcement Learning Algorithm. *Build. Environ.* **2019**, *155*, 105–117. [CrossRef]
112. Zeiler, W.; Boxem, G. Effects of Thermal Activated Building Systems in Schools on Thermal Comfort in Winter. *Build. Environ.* **2009**, *44*, 2308–2317. [CrossRef]
113. Aghniaey, S.; Lawrence, T.M.; Sharpton, T.N.; Douglass, S.P.; Oliver, T.; Sutter, M. Thermal Comfort Evaluation in Campus Classrooms during Room Temperature Adjustment Corresponding to Demand Response. *Build. Environ.* **2019**, *148*, 488–497. [CrossRef]

114. Huang, K.-T.; Hwang, R.-L. Parametric Study on Energy and Thermal Performance of School Buildings with Natural Ventilation, Hybrid Ventilation and Air Conditioning. *Indoor Built Environ.* **2015**, *25*, 1148–1162. [CrossRef]

115. Tian, X.; Zhang, S.; Lin, Z.; Li, Y.; Cheng, Y.; Liao, C. Experimental Investigation of Thermal Comfort with Stratum Ventilation Using a Pulsating Air Supply. *Build. Environ.* **2019**, *165*, 106416. [CrossRef]

116. Li, Y.; Tian, X.; Liao, C.; Cheng, Y. Effects of Gender on Thermal Comfort of Stratum Ventilation with Pulsating Air Supply. In *IOP Conference Series: Materials Science and Engineering*; IOP Publishing: Bristol, UK, 2019.

117. Simanic, B.; Nordquist, B.; Bagge, H.; Johansson, D. Indoor Air Temperatures, $CO_2$ Concentrations and Ventilation Rates: Long-Term Measurements in Newly Built Low-Energy Schools in Sweden. *J. Build. Eng.* **2019**, *25*, 100827. [CrossRef]

118. Zhang, S.; Lu, Y.; Lin, Z. Coupled Thermal Comfort Control of Thermal Condition Profile of Air Distribution and Thermal Preferences. *Build. Environ.* **2020**, *177*, 106867. [CrossRef]

119. Jiang, J.; Wang, D.; Liu, Y.; Di, Y.; Liu, J. A Field Study of Adaptive Thermal Comfort in Primary and Secondary School Classrooms during Winter Season in Northwest China. *Build. Environ.* **2020**, *175*, 106802. [CrossRef]

120. Shrestha, M.; Rijal, H. Study on Adaptive Thermal Comfort in Naturally Ventilated Secondary School Buildings in Nepal. In *IOP Conference Series: Materials Science and Engineering*; IOP Publishing: Bristol, UK, 2019; Volume 294, p. 012062. [CrossRef]

121. Talarosha, B.; Satwiko, P.; Aulia, D.N. Air Temperature and $CO_2$ Concentration in Naturally Ventilated Classrooms in Hot and Humid Tropical Climate. In *IOP Conference Series: Earth and Environmental Science*; IOP Publishing: Bristol, UK, 2020; Volume 402, p. 012008. [CrossRef]

122. Chitaru, G.M.; Istrate, A.; Catalina, T. Numerical Analysis of the Impact of Natural Ventilation on the Indoor Air Quality and Thermal Comfort in a Classroom. *E3S Web Conf.* **2019**, *111*, 01023. [CrossRef]

123. Liu, S.; Yoshino, H.; Mochida, A. A Measurement Study on the Indoor Climate of a College Classroom. *Int. J. Vent.* **2011**, *10*, 251–262. [CrossRef]

124. Ranjbar, A. Analysing the Effects of Thermal Comfort and Indoor Air Quality in Design Studios and Classrooms on Student Performance. In *IOP Conference Series: Earth and Environmental Science*; IOP Publishing: Bristol, UK, 2019; Volume 609, p. 042086. [CrossRef]

125. Branco, P.; Ferraz, M.; Martins, F.; Sousa, S. Quantifying Indoor Air Quality Determinants in Urban and Rural Nursery and Primary Schools. *Environ. Res.* **2019**, *176*, 108534. [CrossRef]

126. Papadopoulos, G.; Panaras, G.; Tolis, E.I. Thermal Comfort and Indoor Air Quality Assessment in University Classrooms. In *IOP Conference Series: Earth and Environmental Science*; IOP Publishing: Bristol, UK, 2020; Volume 410, p. 012095. [CrossRef]

127. Toyinbo, O.; Phipatanakul, W.; Shaughnessy, R.; Haverinen-Shaughnessy, U. Building and Indoor Environmental Quality Assessment of Nigerian Primary Schools: A Pilot Study. *Indoor Air* **2019**, *29*, 510–520. [CrossRef]

128. Schweiker, M.; Ampatzi, E.; Andargie, M.S.; Andersen, R.K.; Azar, E.; Barthelmes, V.M.; Berger, C.; Bourikas, L.; Carlucci, S.; Chinazzo, G.; et al. Review of Multi-domain Approaches to Indoor Environmental Perception and Behaviour. *Build. Environ.* **2020**, *176*, 106804. [CrossRef]

129. Leccese, F.; Rocca, M.; Salvadori, G.; Belloni, E.; Buratti, C. Towards a Holistic Approach to Indoor Environmental Quality Assessment: Weighting Schemes to Combine Effects of Multiple Environmental Factors. *Energy Build.* **2021**, *245*, 111056. [CrossRef]

130. Bourikas, L.; Gauthier, S.; Khor Song En, N.; Xiong, P. Effect of Thermal, Acoustic and Air Quality Perception Interactions on the Comfort and Satisfaction of People in Office Buildings. *Energies* **2021**, *14*, 333. [CrossRef]

131. Wang, D.; Song, C.; Wang, Y.; Xu, Y.; Liu, Y.; Liu, J. Experimental Investigation of the Potential Influence of Indoor Air Velocity on Students' Learning Performance in Summer Conditions. *Energy Build.* **2020**, *219*, 110015. [CrossRef]

132. Campano Laborda, M.; Domínguez-Amarillo, S.; Fernandez-Aguera, J.; Acosta, I. Indoor Comfort and Symptomatology in Non-University Educational Buildings: Occupants' Perception. *Atmosphere* **2020**, *11*, 357. [CrossRef]

133. Zhang, D.; Ortiz, M.A.; Bluyssen, P.M. Clustering of Dutch School Children Based on Their Preferences and Needs of the IEQ in Classrooms. *Build. Environ.* **2019**, *147*, 258–266. [CrossRef]

134. Pistore, L.; Pittana, I.; Cappelletti, F.; Romagnoni, P.; Gasparella, A. Analysis of Subjective Responses for the Evaluation of the Indoor Environmental Quality of an Educational Building. *Sci. Technol. Built Environ.* **2020**, *26*, 195–209. [CrossRef]

135. Chang, S.F.; Hwang, R.L.; Huang, K.T. Improvement of Thermal Comfort in Naturally Ventilated Classrooms by Phase Change Material Roofs in Taiwan. In *IOP Conference Series: Materials Science and Engineering*; IOP Publishing: Bristol, UK, 2019; Volume 609, p. 042034. [CrossRef]

136. Orosa, J.A.; Oliveira, A.C. A Field Study on Building Inertia and Its Effects on Indoor Thermal Environment. *Renew. Energy* **2012**, *37*, 89–96. [CrossRef]

137. Salandin, A.; Vettori, M.; Vettori, S. Thin Solar Film Application for Improving Thermal Comfort in Classrooms. In *Construction and Building Research*; Springer: Dordrecht, The Netherlands, 2014; pp. 531–538, ISBN 978-94-007-7789-7.

138. Colinart, T.; Bendouma, M.; Glouannec, P. Building Renovation with Prefabricated Ventilated Façade Element: A Case Study. *Energy Build.* **2019**, *186*, 221–229. [CrossRef]

139. Castilla, N.; Llinares, C.; Bravo, J.M.; Blanca, V. Subjective Assessment of University Classroom Environment. *Build. Environ.* **2017**, *122*, 72–81. [CrossRef]

140. Kiil, M.; Simson, R.; de Luca, F.; Thalfeldt, M.; Kurnitski, J. Overheating and Daylighting Evaluation for Free-Running Classroom Designs. In *IOP Conference Series: Materials Science and Engineering*; IOP Publishing: Bristol, UK, 2019; Volume 352, p. 012059. [CrossRef]
141. Camacho-Montano, S.C.; Cook, M.; Wagner, A. Avoiding Overheating in Existing School Buildings through Optimized Passive Measures. *Build. Res. Inf.* **2020**, *48*, 349–363. [CrossRef]
142. Chen, Y.H.; Hwang, R.L.; Huang, K.T. Sensitivity Analysis of Envelope Design on the Summer Thermal Comfort of Naturally Ventilated Classrooms in Taiwan. In *IOP Conference Series: Materials Science and Engineering*; IOP Publishing: Bristol, UK, 2019; Volume 609, p. 042035. [CrossRef]
143. Barrett, P.; Zhang, Y.; Moffat, J.; Kobbacy, K. A Holistic, Multi-Level Analysis Identifying the Impact of Classroom Design on Pupils' Learning. *Build. Environ.* **2013**, *59*, 678–689. [CrossRef]
144. Yang, Z.; Becerik-Gerber, B.; Mino, L. A Study on Student Perceptions of Higher Education Classrooms: Impact of Classroom Attributes on Student Satisfaction and Performance. *Build. Environ.* **2013**, *70*, 171–188. [CrossRef]
145. Wargocki, P.; Wyon, D.P. The Effects of Outdoor Air Supply Rate and Supply Air Filter Condition in Classrooms on the Performance of Schoolwork by Children (RP-1257). *HVACR Res.* **2007**, *13*, 165–191. [CrossRef]
146. Wargocki, P.; Wyon, D.P. The Effects of Moderately Raised Classroom Temperatures and Classroom Ventilation Rate on the Performance of Schoolwork by Children (RP-1257). *HVACR Res.* **2007**, *13*, 193–220. [CrossRef]
147. Barrett, P.; Davies, F.; Zhang, Y.; Barrett, L. The Impact of Classroom Design on Pupils' Learning: Final Results of a Holistic, Multi-Level Analysis. *Build. Environ.* **2015**, *89*, 118–133. [CrossRef]
148. Auliciems, A. Classroom Performance as a Function of Thermal Comfort. *Int. J. Biometeorol.* **1972**, *16*, 233–246. [CrossRef] [PubMed]
149. Wargocki, P.; Wyon, D.P. Providing Better Thermal and Air Quality Conditions in School Classrooms Would Be Cost-Effective. *Build. Environ.* **2013**, *59*, 581–589. [CrossRef]
150. Puteh, M.; Ibrahim, M.H.; Adnan, M.; Che'Ahmad, C.N.; Noh, N.M. Thermal Comfort in Classroom: Constraints and Issues. *Procedia-Soc. Behav. Sci.* **2012**, *46*, 1834–1838. [CrossRef]
151. Barbic, F.; Minonzio, M.; Cairo, B.; Shiffer, D.; Dipasquale, A.; Cerina, L.; Vatteroni, A.; Urechie, V.; Verzelletti, P.; Badilini, F.; et al. Effects of Different Classroom Temperatures on Cardiac Autonomic Control and Cognitive Performances in Undergraduate Students. *Physiol. Meas.* **2019**, *40*, 054005. [CrossRef] [PubMed]
152. Bluyssen, P. The Need for Understanding the Indoor Environmental Factors and Its Effects on Occupants through an Integrated Analysis. In *IOP Conference Series: Materials Science and Engineering*; IOP Publishing: Bristol, UK, 2019; Volume 609, p. 022001. [CrossRef]
153. Ortiz, M.A.; Bluyssen, P.M. Developing Home Occupant Archetypes: First Results of Mixed-Methods Study to Understand Occupant Comfort Behaviours and Energy Use in Homes. *Build. Environ.* **2019**, *163*, 106331. [CrossRef]
154. Humphreys, M.A. A Study of the Thermal Comfort of Primary School Children in Summer. *Build. Environ.* **1977**, *12*, 231–239. [CrossRef]
155. Kwok, A.G. Thermal Comfort in Tropical Classrooms. *ASHRAE Trans.* **1998**, *104*, 1031–1047.
156. Jung, G.J.; Song, S.K.; Ahn, Y.C.; Oh, G.S.; Im, Y.B. Experimental Research on Thermal Comfort in the University Classroom of Regular Semesters in Korea. *J. Mech. Sci. Technol.* **2011**, *25*, 503–512. [CrossRef]
157. Dias Pereira, L.; Cardoso, E.; Gameiro da Silva, M. Indoor Air Quality Audit and Evaluation on Thermal Comfort in a School in Portugal. *Indoor Built Environ.* **2013**, *24*, 256–268. [CrossRef]
158. Dias Pereira, L.; Raimondo, D.; Corgnati, S.P.; Gameiro da Silva, M. Assessment of Indoor Air Quality and Thermal Comfort in Portuguese Secondary Classrooms: Methodology and Results. *Build. Environ.* **2014**, *81*, 69–80. [CrossRef]
159. Huang, K.T.; Huang, W.P.; Lin, T.P.; Hwang, R.L. Implementation of Green Building Specification Credits for Better Thermal Conditions in Naturally Ventilated School Buildings. *Build. Environ.* **2015**, *86*, 141–150. [CrossRef]
160. Nico, M.A.; Liuzzi, S.; Stefanizzi, P. Evaluation of Thermal Comfort in University Classrooms through Objective Approach and Subjective Preference Analysis. *Appl. Ergon.* **2015**, *48*, 111–120. [CrossRef]
161. Liu, Y.; Jiang, J.; Wang, D.; Liu, J. The Indoor Thermal Environment of Rural School Classrooms in Northwestern China. *Indoor Built Environ.* **2016**, *26*, 662–679. [CrossRef]
162. Stazi, F.; Naspi, F.; Ulpiani, G.; Di Perna, C. Indoor Air Quality and Thermal Comfort Optimization in Classrooms Developing an Automatic System for Windows Opening and Closing. *Energy Build.* **2017**, *139*, 732–746. [CrossRef]
163. Trebilcock, M.; Soto-Muñoz, J.; Yañez, M.; Figueroa-San Martin, R. The Right to Comfort: A Field Study on Adaptive Thermal Comfort in Free-Running Primary Schools in Chile. *Build. Environ.* **2017**, *114*, 455–469. [CrossRef]
164. Wang, Z.; Ning, H.; Zhang, X.; Ji, Y. Human Thermal Adaptation Based on University Students in China's Severe Cold Area. *Sci. Technol. Built Environ.* **2017**, *23*, 413–420. [CrossRef]
165. Bluyssen, P.M.; Zhang, D.; Krooneman, A.-J.; Freeke, A. The Effect of Wall and Floor Colouring on Temperature and Draught Feeling of Primary School Children. *E3S Web Conf.* **2019**, *111*, 02032. [CrossRef]
166. Campano Laborda, M.; Domínguez-Amarillo, S.; Fernandez-Aguera, J.; Sendra, J. Thermal Perception in Mild Climate: Adaptive Thermal Models for Schools. *Sustainability* **2019**, *11*, 3948. [CrossRef]
167. Chen, P.-Y.; Chan, Y.-C. Developing the Methodology to Investigate the Thermal Comfort of Hot-Humid Climate under Different Ventilation Modes. *J. Phys. Conf. Ser.* **2019**, *1343*, 012149. [CrossRef]

168. Fabozzi, M.; Dama, A. Field Study on Thermal Comfort in Naturally Ventilated and Air-Conditioned University Classrooms. *Indoor Built Environ.* **2019**, *29*, 851–859. [CrossRef]
169. Haddad, S.; Osmond, P.; King, S. Application of Adaptive Thermal Comfort Methods for Iranian School children. *Build. Res. Inf.* **2019**, *47*, 173–189. [CrossRef]
170. Heracleous, C.; Michael, A. Experimental Assessment of the Impact of Natural Ventilation on Indoor Air Quality and Thermal Comfort Conditions of Educational Buildings in the Eastern Mediterranean Region during the Heating Period. *J. Build. Eng.* **2019**, *26*, 100917. [CrossRef]
171. Huang, M.; Zhang, G. Research on Thermal Environment of University Classrooms in Severe Cold Areas Based on Different Seating Rate. In *IOP Conference Series: Earth and Environmental Science*; IOP Publishing: Bristol, UK, 2019; Volume 330, p. 032064.
172. Koranteng, C.; Simons, B.; Essel, C. Climate Responsive Buildings: A Comfort Assessment of Buildings on KNUST Campus, Kumasi. *J. Eng. Des. Technol.* **2019**, *17*, 862–877. [CrossRef]
173. Liu, J.; Luo, Q.; Cai, T. Students Responses to Thermal Environments in University Classrooms in Zunyi, China. In *IOP Conference Series: Earth and Environmental Science*; IOP Publishing: Bristol, UK, 2019; Volume 592, p. 012168. [CrossRef]
174. Monna, S.; Baba, M.; Juaidi, A.; Barlet, A.; Bruneau, D. Improving Thermal Environment for School Buildings in Palestine, the Role of Passive Design. *J. Phys. Conf. Ser.* **2019**, *1343*, 012190. [CrossRef]
175. Shen, J.Y.; Kojima, S.; Ying, X.Y.; Hu, X.J. Influence of Thermal Experience on Thermal Comfort in Naturally Conditioned University Classrooms. Lowland Technology International. *Lowl. Technol. Int.* **2019**, *21*, 107–122.
176. Al-Khatri, H.; Alwetaishi, M.; Gadi, M. Exploring Thermal Comfort Experience and Adaptive Opportunities of Female and Male High School Students. *J. Build. Eng.* **2020**, *31*, 101365. [CrossRef]
177. Da Silva Júnior, A.; Mendonça, K.C.; Vilain, R.; Pereira, M.L.; Mendes, N. On the Development of a Simplified Model for Thermal Comfort Control of Split Systems. *Build. Environ.* **2020**, *179*, 106931. [CrossRef]
178. Hamzah, B.; Mulyadi, R.; Amin, S.; Kusno, A. Adaptive Thermal Comfort of Naturally Ventilated Classrooms of Elementary Schools in the Tropics. In *IOP Conference Series: Earth and Environmental Science*; IOP Publishing: Bristol, UK, 2020; Volume 402, p. 012021. [CrossRef]
179. Heracleous, C.; Michael, A. Thermal Comfort Models and Perception of Users in Free-Running School Buildings of East-Mediterranean Region. *Energy Build.* **2020**, *215*, 109912. [CrossRef]
180. Jowkar, M.; Rijal, H.B.; Brusey, J.; Montazami, A.; Carlucci, S.; Lansdown, T.C. Comfort Temperature and Preferred Adaptive Behaviour in Various Classroom Types in the UK Higher Learning Environments. *Energy Build.* **2020**, *211*, 109814. [CrossRef]
181. Jowkar, M.; Rijal, H.B.; Montazami, A.; Brusey, J.; Temeljotov-Salaj, A. The Influence of Acclimatization, Age and Gender-Related Differences on Thermal Perception in University Buildings: Case Studies in Scotland and England. *Build. Environ.* **2020**, *179*, 106933. [CrossRef]
182. Korsavi, S.S.; Montazami, A. Children's Thermal Comfort and Adaptive Behaviours; UK Primary Schools during Non-Heating and Heating Seasons. *Energy Build.* **2020**, *214*, 109857. [CrossRef]
183. Liu, J.; Yang, X.; Liu, Y. A Thermal Comfort Field Study of Naturally Ventilated Classrooms during Spring and Autumn in Xi'an, China. In *Proceedings of the 11th International Symposium on Heating, Ventilation and Air Conditioning (ISHVAC 2019)*; Wang, Z., Zhu, Y., Wang, F., Wang, P., Shen, C., Liu, J., Eds.; Springer: Singapore, 2020; pp. 1073–1081.
184. Munonye, C.; Ji, Y. Evaluating the Perception of Thermal Environment in Naturally Ventilated Schools in a Warm and Humid Climate in Nigeria. *Build. Serv. Eng. Res. Technol.* **2020**, *42*, 5–25. [CrossRef]

*Systematic Review*

# Variables That Affect Thermal Comfort and Its Measuring Instruments: A Systematic Review

Tamara Mamani [1], Rodrigo F. Herrera [1,*], Felipe Muñoz-La Rivera [1,2,3] and Edison Atencio [1]

1   School of Civil Engineering, Pontificia Universidad Católica de Valparaíso, Valparaíso 2340000, Chile; tamara.mamani.q@mail.pucv.cl (T.M.); felipe.munoz@pucv.cl (F.M.-L.R.); edison.atencio@pucv.cl (E.A.)
2   International Centre for Numerical Methods in Engineering (CIMNE), C/Gran Capitán S/N UPC Campus Nord, Edifici C1, 08034 Barcelona, Spain
3   School of Civil Engineering, Universitat Politècnica de Catalunya, Carrer de Jordi Girona, 1, 08034 Barcelona, Spain
*   Correspondence: rodrigo.herrera@pucv.cl

**Abstract:** Thermal comfort can impact the general behavior of the occupants, and considering that humans currently perform 90% of their daily work indoors, it is necessary to improve the accuracy of thermal comfort assessments, and a correct selection of variables could make this possible. However, no review integrates all the variables that could influence thermal comfort evaluation, which relates them to their respective capture devices. For this reason, this research identifies all the variables that influence the thermal comfort of a building, together with the measurement tools for these variables, evaluating the relevance of each one in the research carried out to date. For this purpose, a systematic literature review was carried out by analyzing a set of articles selected under certain defined inclusion/exclusion criteria. In this way, it became evident that the most used variables to measure thermal comfort are the same as those used by the predicted mean vote (PMV) model; however, research focused on the behavior of the occupants has focused on new variables that seek to respond to individual differences in human thermal perception.

**Keywords:** thermal comfort; capture devices; building; variables

**Citation:** Mamani, T.; Herrera, R.F.; Muñoz-La Rivera, F.; Atencio, E. Variables That Affect Thermal Comfort and Its Measuring Instruments: A Systematic Review. *Sustainability* **2022**, *14*, 1773. https://doi.org/10.3390/su14031773

Academic Editors: Mitja Košir and Manoj Kumar Singh

Received: 31 December 2021
Accepted: 1 February 2022
Published: 4 February 2022

**Publisher's Note:** MDPI stays neutral with regard to jurisdictional claims in published maps and institutional affiliations.

## 1. Introduction

The quality of the environment has a relevant influence on people's physical and mental health. Therefore, and considering that 90% of human beings perform 90% of their daily work in indoor environments, it is important to control indoor air quality and comfort for the occupants [1]. Thus, it has been sought to achieve greater comfort and satisfaction with the indoor environment to provide an environment that does not affect the performance of the activities performed by occupants, considering alternatives that have a lower energy cost [2]. Thermal comfort is defined according to the American Society of Heating, Refrigerating, and Air-Conditioning Engineers (ASHRAE) as "the mental condition that expresses satisfaction with the thermal environment and is evaluated subjectively" [3]. This can impact occupants' behavior, for example, in a classroom, where dissatisfaction with thermal comfort can cause a reduction in concentration and productivity, essential elements to improve the performance of students and teachers [4,5].

There are two main models used to study and analyze thermal comfort: the stationary and adaptive models. The stationary model, based on Fanger's theoretical basis, is based on the thermal balance that the human body undergoes with the environment, under the same air conditioning conditions throughout the study, where thermal comfort is evaluated as the combination of environmental factors and individual factors employing an equation called predicted mean vote (PMV). This is an index that predicts and represents the mean vote of thermal sensation on a standard scale for many people under certain combinations of variables given by the indoor thermal environment [3,6]. On the other hand, in the adaptive model, the user is an active actor who interacts with their environment, adapting it according to their preferences and comfort, considering that environmental conditions can vary and do not remain static, giving way to naturally ventilated spaces. In this way, human thermal comfort combines the subjective sensation of a group of people and the objective interaction with the surrounding environment [2,7].

Now, in terms of using each model, different approaches to indoor climate research are adopted depending on the researcher's specialty. First, engineers are inclined toward deterministic research results with a stationary approach since it fits their view of the role of a stable climate in an indoor environment, where occupants are not expected to attenuate it or interact significantly with their environment. On the other hand, architects are inclined to consider occupants interacting with their environment, adapting it better to their needs and expectations, through strategies such as opening windows or building orientation rather than considering static and unchanging environments [8,9].

However, both the adaptive and stationary models do not consider the individuality of occupants in thermal comfort, as both approaches have been developed based on the responses of large groups of people. These two models do not use several static and dynamic factors that influence occupants' satisfaction with their thermal environment [10]. Static factors, e.g., gender, are independent of time, whereas dynamic factors, e.g., acclimatization or age, contribute to the change in thermal comfort over time [11].

Thus, more and more research has focused on studying these new variables, where the understanding of the causes of individual differences in human thermal perception has gained popularity in recent studies on comfort [12]. For example, in Wang L. et al. [13], a model is developed based on the neutral clothing variable, looking at the influence of contextual factors. On the other hand, Wang X. et al. [14] address individual occupant differences in thermal comfort, the variables that produce it, and possible solutions to address these problems.

Thus, understanding the importance and effectiveness of variables will facilitate the development of solutions to minimize negative impacts. A correct selection of variables could improve the accuracy of predictions when using the results of traditional analytical models. Added to this, the acquisition of data from the environment is the first step on the way to obtain information from the environment—i.e., through different capture devices for the variables encountered, it is possible to obtain information with useful and relevant meaning, an implication or input for decisions or actions [15].

Currently, the ASHRAE Global Thermal Comfort Database II [16] is an online database covering many variables that influence thermal comfort, containing objective climatic observations from field studies in indoor environments with the corresponding subjective evaluations measured at the time of building occupants [17]. However, no review integrates and reviews all the variables that influence thermal comfort, both the stationary model and the adaptive and dynamic ones.

In this context, this research aimed to identify all the variables that influence the thermal comfort of a building through a literature review that can represent the relevance of each factor in various investigations of thermal comfort. In addition, it sought to identify the devices and tools for capturing these variables, to discuss the conditions that should be considered at the time of making the measurements. In this way, this article aims to contribute to future research that seeks to include new variables in studies. By analyzing the most relevant variables in the literature, both the traditional ones used in existing models and the variables attributable to occupants, we seek to contribute to improving the accuracy of thermal comfort measurements.

## 2. Materials and Methods

To meet the aim of this research, a bibliometric study was carried out and, subsequently, a systematic literature review. Through a qualitative and quantitative analysis, a set of selected articles were analyzed under certain parameters that are in accordance with the objectives of this study research. Complementarily, through the review carried out, the metrics of the associated publications were analyzed, in addition to an analysis and discussion through a manual review of the variables that affect thermal comfort.

The work is divided into three main stages: (1) identification of variables affecting thermal comfort, (2) frequency study of variables in the literature, and (3) identification of devices to capture the variables. Figure 1 shows the tools, activities, and deliverables for each of the stages this research addresses, where the acronym TCV refers to "thermal comfort variables".

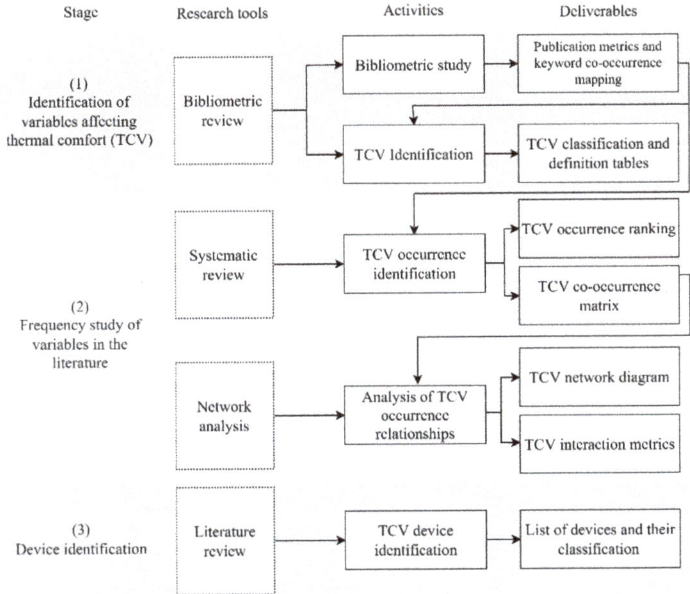


**Figure 1.** Research methodology.

*2.1. Identification of Variables Affecting Thermal Comfort*

First, a systematic review was carried out to identify, select, and include the articles to be evaluated in this study, following the guidelines of the Preferred Reporting Items for Systematic Reviews and Meta-Analyses (PRISMA) methodology. Figure 2 shows the phases followed to obtain the publications analyzed in this study.

For the first phase, the keywords and the database for the search were chosen. The Web of Science database was used to obtain the articles in this study since it offers a wide variety of disciplines related to the topic to be addressed in this work. The keywords selected were (a) thermal comfort, (b) building, (c) variables, and d) management. Thus, combinations I ("thermal comfort" AND "building" AND "variable") and II ("thermal comfort" AND "building" AND "management") were used to search for articles. Then, a series of filters was applied in the database to work with the study areas of interest to evaluate articles in this work, corresponding to (1) construction building technologies and green sustainable science technology and (2) civil and environmental engineering. Additionally, a filter was applied with the time of publications from 2001 to the present, giving 761 articles.

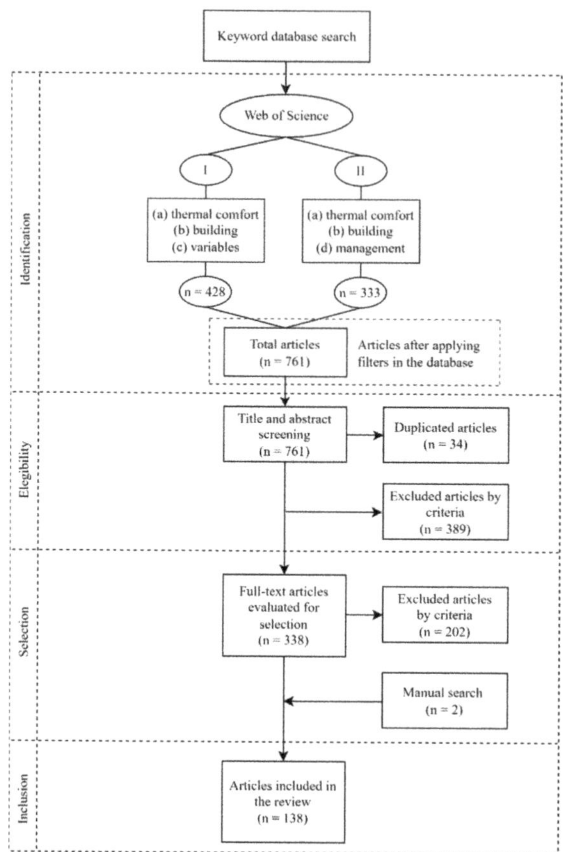

**Figure 2.** PRISMA flow diagram for the systematic review.

For the next eligibility phase, 34 duplicate publications were eliminated from reviewing titles after applying combinations I and II. Then, the articles considered relevant for evaluation in this study were filtered, where a series of inclusion and exclusion criteria were defined. Inclusion criteria included (1) articles that could study variables that affect thermal comfort, (2) case studies of buildings where thermal comfort is evaluated, (3) methodologies to evaluate thermal comfort, (4) studies that address capture technologies for thermal comfort variables, and (5) publications in journals, conferences, and reviews. On the other hand, the exclusion criteria were (1) studies in structures other than offices, residences, and educational buildings; (2) studies that address the evaluation for phases other than the operative phase of the structure; (3) studies in languages other than English or Spanish; and 4) studies without full text available.

In all, 389 articles were excluded through a review of title and abstract, leaving a total of 338 articles, which were filtered again through a full-text reading, thus excluding 202 articles that did not meet the inclusion/exclusion criteria to be considered in this study. In addition to this, a manual search was included within the same references of the selected studies since there were articles that could contribute to the evaluation of variables that affect thermal comfort. Thus, 2 articles were added in the inclusion phase of the publications, leaving a total of 138, which were considered for this research.

Once the set of articles was selected, a bibliometric study of the associated publications was carried out, reviewing the sources, countries with the highest incidence in research, and annual distribution of the publication. In addition, using the VOSviewer visualization tool, a map of co-occurrence of keywords of the authors was represented to analyze and rectify the central theme covered in this research and deepen the area of study. For the elaboration of the map, first, 82 terms were reduced to 55 by replacing those terms that alluded to the same concept. For example, adaptive model, adaptive thermal comfort, and adaptive comfort were unified into a single term called adaptive model, or those that alluded to the same term in plural or singular, such as educational buildings, which was unified to educational building.

Then, a systematic literature review of the articles was carried out. In this way, based on certain studies, the TCVs were defined and classified, analyzing possible repetitions or variables with the same name, to have a defined list and subsequently perform a manual search of the TCVs in the literature.

### 2.2. Frequency Study of LCTs in the Literature

After identifying the TCVs, a systematic literature review was carried out by searching for each variable in the articles. An Excel spreadsheet was used to create a matrix of occurrence of TCVs in the literature, where the TCVs were arranged in a row and then the articles in a column, considering the relationship between each one in case the variable was used for research purposes and not only mentioned. This allowed a count to be carried out of the TCVs in the total number of articles, with the help of which, finally, in order to obtain the ranking, the frequency of the TCVs was plotted and ordered in a histogram.

On the other hand, with the matrix of occurrence of variables in the literature, a co-occurrence matrix of TCVs was developed, corresponding to a triangular matrix where the variables are ordered in the first row. A column was used to count how many times the variable appears together in the same document, considering the matrix as an upper triangular and diagonal 0, avoiding analyzing a variable with itself. Through this quantitative work, the researchers sought to identify those variables that have a greater weight in the literature. In this way, the authors were able to analyze the impact that different approaches to thermal comfort analysis have had.

To conclude this stage, the network diagram of the TCVs was represented using Gephi software, together with the network metrics generated, so that it was possible to carry out a qualitative study by obtaining the most central variables and thus analyze those that were most influential within the network.

### 2.3. Identification of TCV Capture Devices

At this stage, the devices that capture each TCV were identified. A table was drawn up with the variables that could be measured and their corresponding measuring instruments, which was obtained from the classification table of variables previously used in the variable identification stage, in addition to the literature review carried out in the TCV frequency study stage, where the capture devices for TCVs were found based on certain articles.

This section discusses the main problems encountered, the need for standardization, and the guidelines required for thermal comfort evaluations to be carried out as homogeneously as possible.

## 3. Results and Discussion

This section presents the main results obtained from the bibliometric study and the literature review. Characterization of the articles considered in this work and a map of co-occurrence of keywords are presented. Then, the TCVs, their frequency in the literature, and network analysis are presented to determine the most influential ones. Further, the TCVs capture devices or tools and an analysis of the conditions that should be considered when making the measurements.

### 3.1. Characterization of Articles

This section seeks to validate the articles selected for this research. To that end, a characterization of the selected articles is presented, including an annual distribution by year of article publication, distribution of articles by country of publication, and the most influential journals of publication.

#### 3.1.1. Associated Publications Metrics

Figure 3 shows the annual distribution of publications of the selected articles. The 138 articles are distributed according to their year of publication, ranging from 2002 to 2021. Thus, it is possible to observe a trend in the increase in article publications from 2013 onwards. Then, it is visualized that the number of publications does not vary greatly, ranging between 8 and 10 annually; however, they have increased considerably since 2018.

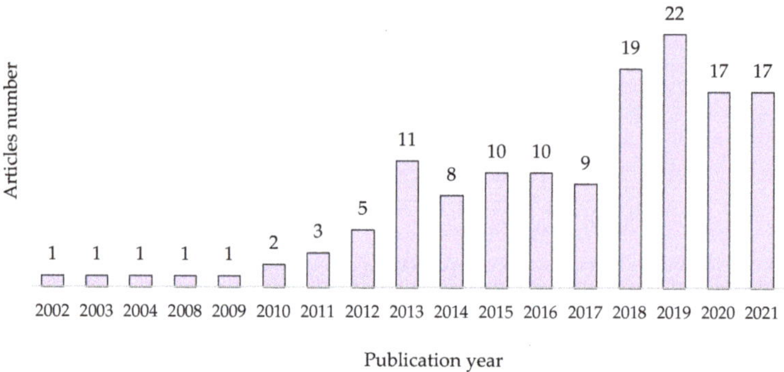


**Figure 3.** Articles' annual distribution.

Even though for 2021, there is an incomplete number of publications due to the date of recovery of the database, there has been a growing increase in research in recent years. Based on what was observed during the review, possible causes may be the need for research on the impact of thermal comfort on people's health and new technologies. The first is given by the climatic conditions that afflict certain areas of extreme climates, and the second by the development of automatic learning models that can optimize energy resources without transgressing the thermal comfort of occupants. It should be noted that one of the keywords used for this research was management, to focus on the operative phase of infrastructure so that within the research are monitoring systems that focus on creating ventilation and heating systems with new technologies that develop continuous learning models which deliver tools that optimize energy resources, which opens a new area of research. On the other hand, the increase in sustainable structures and energy efficiency is a topic that has gathered moment through the years, hence the increase in this research [18].

Concerning the impact of the selected articles in the literature, according to the Web of Science databases, 81% have at least one citation. Of the total, 43% have 20 or more citations.

Figure 4 shows the countries that have researched and published on thermal comfort. A light shade is observed for those countries in which fewer articles have been published, starting from a single publication, and a darker shade for those in which more research has been carried out, reaching a maximum of 22 publications. Thus, the countries with the highest scientific production on this subject are the United States, China, and the United Kingdom, with 22, 20, and 17 publications, respectively.

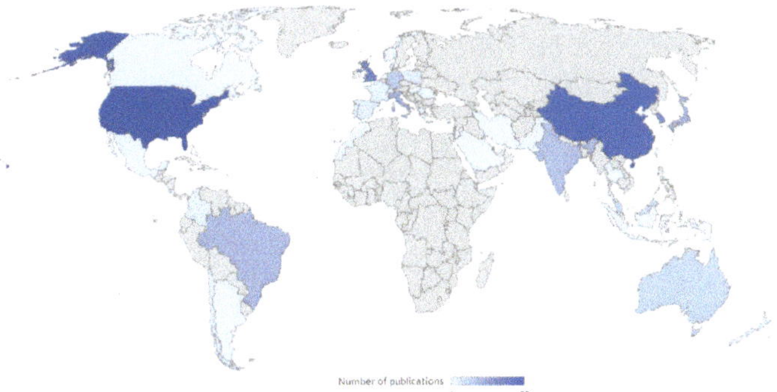


**Figure 4.** Distribution of published articles by country.

It is possible to observe which countries have thermal comfort as a research front with this. Although countries appear as leaders in publications, such as the United States or the United Kingdom, some countries researching thermal comfort in health and sustainability gave their geographical location with complex climatic conditions. Examples of this are China, South Korea, Japan, India, and Brazil; as such, although China presents publications to a greater extent than the other countries, during the review, it was observed that the other countries present a large number of evaluations and field studies to visualize the conditions of occupants in the infrastructures. This was due to the extreme climatic conditions prevailing in each area, which was one of the causes for research on several occasions, as was the case with Brazil and India. One possible cause may be the prevailing climatic conditions of each area. Field studies reflected a greater variability in thermal comfort perceived by subjects based on different climates in naturally ventilated environments. Therefore, to conduct an in-depth investigation to quantify these adaptations, the collection of sufficient data from field measurements becomes a useful avenue of research.

Figure 5 shows the number of articles published by the journals identified in this study. In this way, it is possible to evaluate the impact generated by thermal comfort research in the different disciplines. Of the 138 articles selected, a total of 30 journals were identified, covering research areas focused on engineering and architecture, sustainability, and energy, among other research areas. However, within these categories, it is also possible, to a lesser extent, to find journals that focus on evaluating the impacts of thermal comfort on occupants' health. This shows that the research focuses analyzed in the articles considered in this work cover a broad spectrum of disciplines, leading to a comprehensive review of TCVs. The literature review provides many studies from which conclusions could be generalized, and certain recommendations for future work could be made.

### 3.1.2. Keyword Co-Occurrence Analysis

Figure 6 shows the map of co-occurrence of keywords, where each node with its respective label represents a keyword, and its size reflects the number of occurrences it had in total; the larger it is, the higher its frequency. The centrality reflects the most influential nodes within the network since they are the ones that present a greater amount of co-occurrence with another term within the network [19]. Thus, the central domain of the research is found in the thermal comfort term, being the node with the largest size and centrality within the network. With this, it is concluded that the main research area that unifies the whole network is thermal comfort and the research branches taken for its evaluation. On the other hand, the color of each node represents the grouping that VOSviewer delivers, which is called a cluster [19]. Figure 6 shows a total of four clusters represented by four different colors: the red and blue clusters correlate with energy efficiency that interrelates thermal comfort and energy optimization strategies using machine learning; and the yellow and green clusters focus on operational strategies and behavior perception.

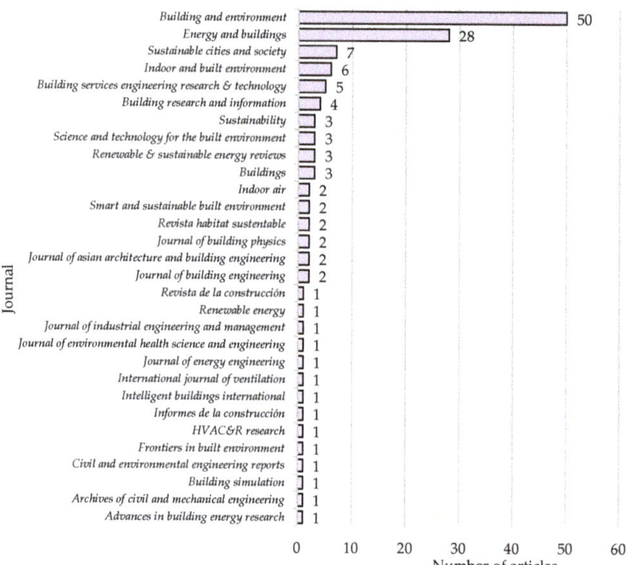

**Figure 5.** Number of articles published by journal.

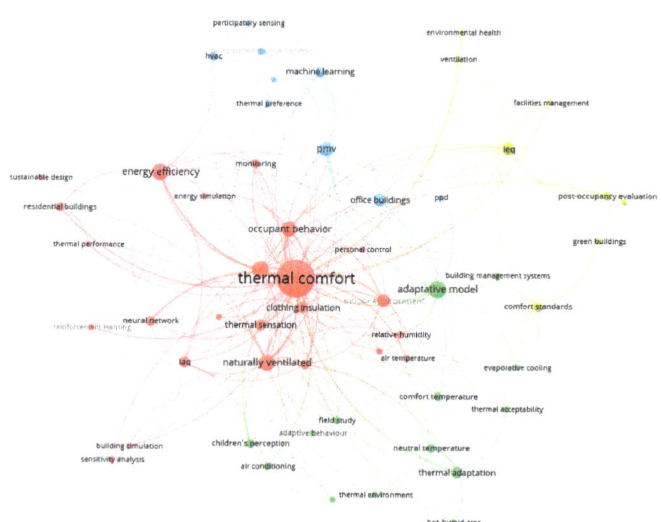

**Figure 6.** Co-occurrence of authors' keywords.

1. The red cluster displays the terms thermal comfort, energy efficiency, naturally ventilated, and occupant behavior as those with the highest number of occurrences given their size. In addition, the energy efficiency term acts as an intermediary between the red and blue clusters. This cluster contains great variability of terms, including keywords of authors attributable to TCVs, such as relative humidity or air temperature, which represents that the articles in this cluster study thermal comfort from different

research fronts, with a focus on one TCV or several variables. On the one hand, the personal attitudes of users represent a key factor to consider when predicting the overall thermal–energetic performance of buildings [20]. It has often been observed that the operational energy used is higher than that predicted in the design, and adaptive behavior driven by occupant comfort is a potential source of this discrepancy [21]. For example, in the field study by Liu J. et al. [22], it can be seen that, although there are central heating systems (radiators), occupants use heating equipment (heaters) to satisfy their thermal preferences. Thus, occupants' comfort requirements regarding thermal conditions and indoor air quality in buildings represent a high energy expenditure. Therefore, in the challenge of reducing the environmental impact, it is important to understand occupants' interactions with the indoor environment to provide comfort conditions in the most efficient way [20,22]. In contrast, in naturally ventilated spaces, this ventilation condition could effectively decrease energy consumption. For example, Park B. and Lee S. [23] found that improving natural ventilation strategies reduced energy consumption by 30% compared to a mechanical ventilation strategy.

2. In the blue group are those terms linked to machine learning models. Here, larger nodes are distinguished as the predicted mean vote (PMV) model, office buildings, machine learning and heating, ventilation, and air conditioning (HVAC) systems. Both PMV and adaptive models have limitations when used to predict occupant comfort in buildings under real conditions. Both models show poor predictive accuracy when applied to a small group of people or individuals because they are designed to predict the average comfort of a large group of people [24]. On the other hand, machine learning is a subfield of artificial intelligence (AI) mainly concerned with programming computers to interpret complex data and evolve in performance with experience [11]. Thus, integrating machine learning algorithms with thermal comfort data analysis could be a very promising direction [25]. Predictions about the thermal preference of individuals can help HVAC systems to provide preventive or corrective control strategies to improve the comfort satisfaction of building occupants [26]. This is evidenced by the growing number of such studies and model development using these techniques. For example, Kim J. et al. [26] developed a predictive model on the thermal preference of individuals based on six machine learning algorithms. Now, regarding the input parameters, different studies may vary for the development of these algorithms. However, most of them used common comfort attributes such as the same environmental and personal factors incorporated in the PMV model [25].

3. The yellow color shows the cluster associated with those terms associated with the evaluations and field studies in the operational stage, evaluating based on standardized guidelines. Thus, the terms with the highest occurrence correspond to indoor environment quality (IEQ), comfort standards, and post-occupancy evaluation. IEQ refers to the quality of indoor spaces concerning the health and wellbeing of users. It is a concept that considers several components of overall indoor comfort: thermal, lighting, air quality, and acoustic [27]. Improving workplace comfort levels positively affects several domains, affecting wellbeing, productivity, energy efficiency, and related economic benefits. However, these areas may conflict each other, so managing a building during the operation phase must take all these into account to balance the weight of each aspect [28]. At present, there are different standards to follow regarding the quality of the environment. There are standards for its evaluation and measurements regarding thermal comforts, such as ASHRAE and ISO7730. On the other hand, if we talk about air quality, $CO_2$ concentration as the main variable can increase in indoor environments due to the combination of human respiration and insufficient ventilation. This is linked to thermal comfort due to the action of doors and windows. Thus, the WELL Building Standard certification recommends a certain threshold of parts per million of $CO_2$ concentration to avoid falling into the "sick building syndrome" [29,30].

4.  In green color, it is possible to observe the terms adaptive model, thermal adaptation, comfortable temperature, and children's perception. Thus, it is possible to observe that the research front of these studies is to evaluate the behavior and perception caused by thermal comfort levels in occupants. According to the adaptive approach, if a user is in a state of discomfort, they will restore their state of wellbeing [31]. Psychological adaptation has been considered the most important factor in explaining discrepancies in observed and predicted thermal sensations and acceptability [22]. People's responses are highly dependent on their thermal experiences [32].

### 3.2. Variables Affecting Thermal Comfort

This section presents the variables selected for the systematic review. They are classified, and a definition is provided to contextualize each one of them. A search is conducted to determine the presence of each in the literature, and then a discussion is carried out based on the results obtained.

### 3.2.1. TCV Classification

Based on the article Development of the ASHRAE Global Thermal Comfort Database II [14], the categories are defined, and variables that affect thermal comfort are described in accordance with analysis in field studies. In addition, variables that were not found in the database but were relevant during the review are added, complementing the initial set of variables. It should be noted that these were classified based on what was visualized within the 138 articles analyzed, giving them the classification that resembled each of them. Thus, Table 1 classifies the TCVs. The classification column is the category to which each variable is assigned, the variable column, and finally, an identifier corresponding to an abbreviation.

**Table 1.** TCV Classification.

| Classification | Variable | Identifier |
|---|---|---|
| Basic identifiers | Structure type | STC |
| | Season | SN |
| | Climate | CLI |
| | Cooling strategy | CSTR |
| Occupant's personal information | Gender | GEN |
| | Age | AGE |
| Occupant characteristics | Metabolic rate | MET |
| | Clothing insulation | CI |
| | Skin temperature | ST |
| | Window/door use | WD |
| Building characteristics | Orientation | OR |
| | Artificial air conditioning | AR |
| | Materials | MAT |
| | Natural ventilation | NV |
| Environmental conditions | Operative temperature | OPT |
| | Dry bulb temperature | DBT |
| | Air temperature | AIRT |
| | Globe temperature | GLT |
| | Mean radiant temperature | MRT |
| | Mean outdoor temperature | MOT |
| | Air velocity | AIRV |
| | Relative humidity | RH |
| | Solar radiation | SR |
| | $CO_2$ concentration | $CO_2$ |

Table 1 shows the five groups into which the TCVs are classified. Within the category of basic identifiers, some variables are attributable to contextual factors specific to each building. Given geographic location, variables such as the predominant season and climate of the area are in this category. Also within this category, we have the TCV type of structure and refrigeration, which are contextual variables to know the building and ventilation conditions. Then, this group provides in a general way the conditions for the evaluation of thermal comfort in a building.

In the category of building characteristics, two variables are given by the cooling strategy (artificial air conditioning and natural ventilation), although there is also a combination called mixed mode. However, for this study, only the first two ones mentioned are considered. There are also variables attributable to the materials of the structure, although it is important to note that this variable is not considered in greater depth in this study since the scope of the study includes a literature review at the operative stage. Therefore, the authors do not delve into construction issues by not focusing on the materials in the design stages; these are only considered in comparisons in operational structures and evaluations, among others. Finally, the structure's orientation is taken as a particular characteristic of the infrastructure since in the same building, different rooms can have different orientations, and different conclusions can be drawn from this variable in the same structure.

For the category of personal information of the users, the TCVs contained are attributable to the occupants who are participants in an evaluation carried out on the premises of the structure to be analyzed or users in the development of a thermal comfort measurement model. Therefore, variables such as gender and age are descriptive and are variables from which conclusions can be drawn.

On the other hand, the category of user characteristics includes those TCVs that are used within the comfort measurement process, either directly, such as metabolic rate and clothing insulation, or physiological variables of the user that are part of the PMV model, or user behavioral variables, such as interaction with the environment through the opening of doors and windows or the temperature of the skin.

Finally, environmental conditions classify the physical TCVs specific to the environment, where variables that can be measured utilizing measuring devices or tools are distinguished. In addition, it should be borne in mind that differences between indoor and outdoor environment must be distinguished due to the model used, either considering static conditions for the PMV model or variations in the environment. For this reason, variables such as temperature, relative humidity, and air velocity are analyzed with an indoor/outdoor denomination.

Table 2 presents the definition of each variable found in the literature review carried out for the classified variables. However, to create Table 2, it was necessary to create an analogy between the variables found since TCVs were related to each other. For example, some variables could be obtained from another, or some variables could be found under different names but alluding to the same concept. In this way, it was possible to find relationships between the variables described throughout this study.

**Table 2.** Definition of variables.

| Id | Variable | Definition |
|:---:|:---:|:---:|
| STC | Structure type | Classroom, multi-family residence, office, others [16]. |
| SN | Season | Spring, summer, autumn, winter [16]. |
| CLI | Climate | Koppen climate classification [16]. |
| CSTR | Cooling strategy | Mechanically ventilated, naturally ventilated, mixed [16]. |
| GEN | Gender | Female, male, indefinite [16]. |
| AGE | Age | Age of participants [16]. |

**Table 2.** *Cont.*

| Id | Variable | Definition |
|---|---|---|
| MET | Metabolic rate | Transformation rate of chemical energy into heat and mechanical work by metabolic activities within an organism [33]. |
| CI | Clothing insulation | Resistance to heat transfer provided by a set of clothing. Heat transfer from the whole body thus also includes uncovered body parts, such as the head and hands [33]. |
| ST | Skin temperature | Surface temperature in radiative equilibrium. It forms the interface between the body and the atmosphere [34]. |
| WD | Window/door use | Status of doors and windows as open or closed [16]. |
| OR | Orientation | It measures which cardinal point a building or space is oriented to the north to south (0–180°) [35]. |
| AR | Artificial air conditioning | The action and effect of air conditioning, i.e., providing an enclosed space with the conditions of temperature, relative humidity, air purity, and sometimes pressure, is necessary for people's wellbeing or the preservation of things [36]. |
| MAT | Materials | Materials of construction of the structure. |
| NV | Natural ventilation | Process of air renewal in the premises by natural means (wind action or thermal draught), the action of which can be favored by the opening of elements of the exterior walls [36]. |
| OPT | Operative temperature | The uniform temperature of an imaginary black enclosure in which a person would exchange the same amount of heat by radiation and convection as in the real non-uniform environment [37,38]. |
| DBT | Dry bulb temperature | The temperature recorded by the standard thermostat with a non-wetted bulb protected from radiant exchange [35]. |
| AIRT | Air temperature | The temperature of the air surrounding the occupant [33]. |
| GLT | Globe temperature | The temperature obtained with a globe thermometer measured radiant heat [35]. |
| MRT | Mean radiant temperature | The uniform temperature of an imaginary enclosure in which the radiant heat transfer of the human body is equal to the radiant heat transfer in the real non-uniform enclosure [36]. |
| MOT | Mean outdoor temperature | Mean monthly outdoor temperature at the time of the field study [16]. |
| AIRV | Air velocity | The velocity of air movement at a point, regardless of direction [33]. |
| RH | Relative humidity | Water vapor concentration at the existing temperature [2]. |
| SR | Solar radiation | The energy emitted by the sun, which propagates in the form of electromagnetic waves. [39]. |
| $CO_2$ | $CO_2$ concentration | $CO_2$ level measured in the air. |

In research focused on thermal comfort, Vellei M. et al. [40] previously carried out a review of variables. It is visualized that, in certain studies, variables such as indoor temperature have usually been measured as dry bulb temperature and globe or operative temperature. However, this is identified by previous studies that have managed to find correlations or minimal variations between some temperatures, so depending on the instruments and the circumstances in which they must perform the measurements, they will opt for one variable or another [41].

One of the relationships is given by a standardization, where mean radiant temperature (MRT) is calculated from the air temperature, velocity, and globe temperature in accordance with the International Organization for Standardization (ISO 7726) [33,34]. On the other hand, for outside temperature, a concept called "running mean temperature", calculated as a ratio based on outside temperature, was used in some thermal comfort evaluation studies. This is supported by the fact that an average of outdoor temperatures weighted in accordance with their distance in the past reflects the thermal experience better than instantaneous monthly average temperature [32,42].

In another aspect, based on the conditions of the field studies, several conditions produce the use of different concepts to refer to indoor environment temperature. For example, Dai J. and Jiang S. [43] express, according to the ANSI/ASHRAE 55 and BS EN15251 standards, the operating temperature based on indoor air temperature, air-velocity-dependent factor, and mean radiant temperature, which in turn are expressed by a relationship with globe temperature. Further, in the study by Kim A. et al. [44], due to an incomplete recording of the globe temperature, they defined the estimated operative temperature through indoor air temperature, ensuring minimal data loss via linear regression.

### 3.2.2. Frequency of Variables in the Literature

Based on the description TCVs in Table 2, a manual review was performed on the selected article base. As previously mentioned, the type of structure for this research corresponds to offices, residences, and educational buildings.

In Figure 7, the TCVs were ordered based on their frequency in the items in descending order. Thus, indoor relative humidity, air temperature, and air velocity are visualized as the TCVs with the highest occurrence within the set of articles. On the other hand, skin temperature and materials are the least frequently occurring TCVs. However, since most of the analyzed documents correspond to field studies, it is observed that many of the variables present in the literature correspond to environmental ones, where they can be presented with another denomination, as discussed in the previous section. Variables such as dry bulb temperature, globe temperature, or operative temperature were not unified and were searched under their denomination. However, it is important to note that they represent the same concept of "operative temperature" as required in the ASHRAE55 standard for measuring thermal comfort [36].

In this context, the field studies follow a standardization under the guidelines of the ISO7730 and ASHRAE55, among other norms or standards using the input variables of the PMV model, which correspond to four environmental variables: air temperature, mean radiant temperature, relative humidity, and air velocity; and two physiological variables: clothing insulation and metabolic rate. On the other hand, most machine learning models also use the PMV model parameters as input variables within their algorithm, which would replicate these TCVs. This would mean that the PMV variables are present in many of the documents analyzed. For this reason, the artificial air conditioning variable also appears, understanding that the environment would be under stable conditions, where there would be no significant variability in the indoor environment.

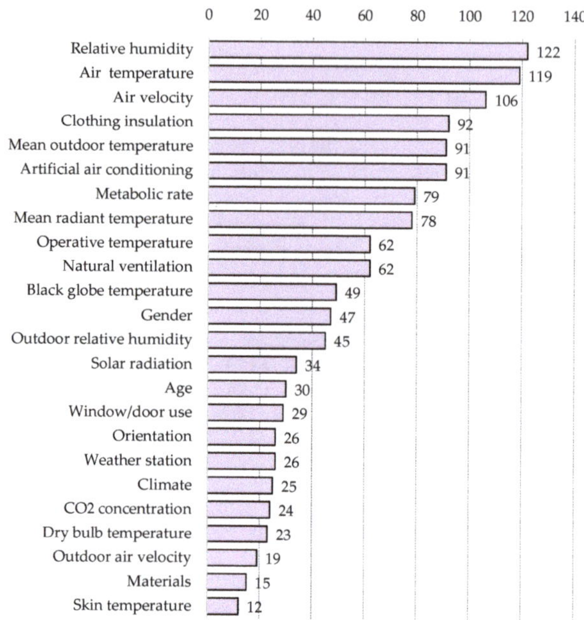

**Figure 7.** Frequency of appearance of variables in reviewed articles.

Likewise, the average outdoor temperature variable appears, representing a variability of the environment and not a static state as in the PMV model. Thus, this variable appears in the scenario of naturally ventilated spaces under the adaptive approach.

On the other hand, it is possible to observe contextual variables such as climate or season, which appear to see their relationship with the thermal comfort perceived by the users. It should be noted that a considerable number of field studies were carried out in extreme climates where the actual requirements of the users were far from the values provided by the standards, which could also be seen in Figure 4, where countries such as Brazil and South Korea presented the highest research productions under these conditions. In addition, taking the climate variable as a measure of perception can significantly affect the thermal sensation. Given the same indoor thermal environment, occupants in warmer climates tend to feel significantly colder than their counterparts in colder climates, suggesting a long-term acclimatization effect [12]. In an office setting, outdoor temperature appears to be the most influential factor in opening windows upon arrival, where users modify their environment through this action, suggesting that users are still influenced by the perception of outdoor conditions [33].

Further, personal information variables such as gender and age appear more frequently, used as contextual variables of the study or to obtain conjectures when concluding about thermal perception [45]. For example, Zhang F. and de Dear R. [17] concluded that gender shows significant main and interaction effects on thermal sensation. Women perceived the same thermal environment to be significantly colder than their male counterparts, and thermal sensitivity was also systematically higher in women. For age, Teli D. et al. [46] detected that children are more sensitive to higher temperatures than adults, with comfortable temperatures about 4 °C lower than PMV predictions. Therefore, they concluded that children have a different thermal perception than adults and that it is necessary to adjust the models to reflect their thermal sensation adequately.

Another variable, i.e., window/door use, represents occupants' interaction with their environment. Indoor conditions such as temperature, humidity, and $CO_2$ can explain the driving forces of window opening behavior [47]. In turn, window operation is related to ventilation, indoor air quality, and energy consumption [21]. In studies such as Jeong B. et al. [48], it was evidenced that thermal stimuli explained window opening and closing behaviors. In the opening case, the outside temperature exceeded 12.7 °C, presumably to achieve a cooling effect. On the other hand, for closing, 90.6% of the openings were closed before the inside temperature reached 4 °C.

Finally, variables such as materials and skin temperature appear to a lesser extent. the former is related to the fact that articles that included the design stage of an infrastructure were not analyzed due to the selection of articles in the operative stage. Therefore, aspects such as thermal envelopes or materials insulation were not analyzed. As for skin temperature, this variable was covered in more specific studies. For example, Faridah F. et al. [49] detected thermal sensations from facial skin temperature using a thermal camera.

Figure 8 shows the network diagram created in Gephi, based on the co-occurrence of two variables in the same text. The nodes are shown with each TCV according to its identifier, and the colors are organized in accordance with the classification in Table 1. Thus, environmental TCVs are colored lilac, occupants' personal information is colored orange, occupants' characteristics are colored light blue, basic identifiers are colored dark green, and building characteristics are colored light green.

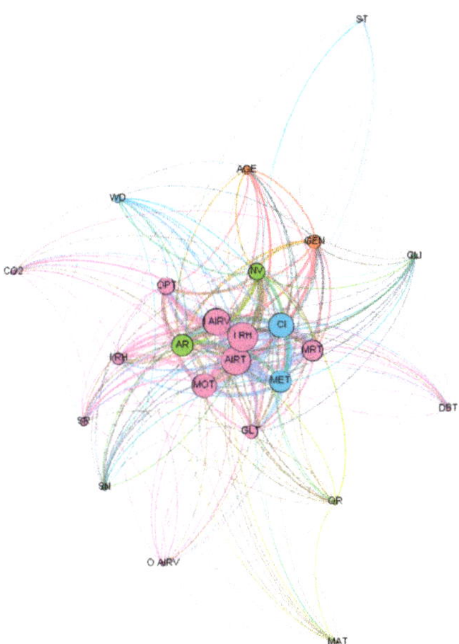


**Figure 8.** TCV network diagram.

The diagram was designed with a ranking in the size of the nodes by degrees with weights. When analyzing by pairs of variables, it was enough that they appeared together in a single article for them to be related. Hence, it was more interesting to see the degree to which they were connected.

On the other hand, the variables that appear at the ends of the network correspond to variables on which research fronts have been set. In this way, we find articles that seek to make new models focused on occupant behavior. Some of these variables are clothing insulation, gender, age of users, use of doors and windows, building orientation, and acclimatization. The above are determined under the premise that the PMV model provided by ISO 7730 and similar standards, although very accurate in neutral and controlled environments, does not always accurately describe occupant perception in a real environment [50].

In the central part of the network, the most influential nodes or TCVs are displayed, where the six variables in the stationary Fanger model are observed. Environmental variables such as mean radiant and air temperature, airspeed, and relative humidity are included, together with the personal variables corresponding to clothing insulation and metabolic rate.

In addition to the Fanger variables, the cooling strategy of the buildings is visualized, where it is possible to see two variables: artificial air conditioning and natural ventilation. To adopt a research line, it is necessary to know the predominant type of ventilation in the building; thus, it will be possible to define the conditions that exist and evaluate thermal comfort from an adaptive or stationary model. This would explain the appearance of the mean outside temperature variable using the first model mentioned.

### 3.3. Capture Devices

Standardizations such as ASHARAE55 [33] provide norms to capture the variables more accurately. Most of the studies carried out in indoor environments are performed according to ISO 7726 [46]. In this literature review, most of the articles analyzed corresponded to field studies and evaluations of thermal comfort in real environments, governed by standards. However, as scenarios and conditions may vary depending on the type of structure, it is necessary to see what factors should be considered when collecting measurements to have greater accuracy.

Within the measurement tools, it is possible to find a great variety of sensors that measure TCV; however, given that there are different types of contextual or user-dependent variables, not all can be measured. Table 3 presents the generalized measurement devices found in the literature, without going into further detail, such as the accuracy of each one. Two large groups are generated here, one focused on sensors that capture environmental variables and the other on measures occupant-related variables. Additionally, environmental variables can be obtained based on data from weather stations, which will depend on the type of study being carried out [51].

**Table 3.** TCV measurement tool.

| Measurement Approach | Tool | Measured Variable |
| --- | --- | --- |
| Occupant-Centered Variables | Questionnaire and survey | Thermal sensation (TSV) |
| | | Thermal preference (TP) |
| | | Thermal acceptability (TA) |
| | | Age, gender, clothing |
| | Skin temperature sensor | Skin temperature |
| | Door and window open sensor | Window/door use |
| Environmental Conditions | CO$_2$ sensor | CO$_2$ concentration |
| | Air Temperature sensor Thermo-hygrometric sensor | Air temperature |
| | Globe thermometer Heat stress meter | Globe temperature |
| | Humidity sensor Thermo hygrometric sensor | Relative humidity |
| | Air velocity sensor Anemometer | Air velocity |
| | Pyranometer | Solar radiation |
| | Thermal camera | Surface temperature |

Instrument accuracy, measurement range, and response time should be indicated in thermal comfort investigations. ISO 7726 and the ASHRAE55 standard [33] require minimum accuracy for measuring certain environmental variables: air velocity, indoor air temperature, globe temperature, and relative humidity. Moreover, measurements must be taken close to the interviewed subjects if the objective is to analyze how people perceive thermal conditions [47].

The instruments must meet the accuracy and range requirements of the standards for monitoring thermal comfort conditions [52]. The deployment strategy of the instruments would be a function of proximity to the people occupying the space, proximity to thermal machines, area of coverage, and height position of the devices [46]. During the review, it was observed that researchers were able to acquire instruments that met the precision requirements of the standards for measuring equipment. According to ISO 7726, the temperature acceptability ranges are from 10 to 40 °C, with instruments with an accuracy of +−5 °C. For air velocity, there is a range of 0.05 to 1 m/s with an instrument accuracy of +−(0.05 + 0.05 Va). For humidity, mentioned as vapor pressure, a measurement range of 0.5 to 3.0 KPa is established with an instrument accuracy of +−0.15 KPa [53]. Now, the capture of these variables is complemented with questionnaires that deliver the thermal preference that users answer, contrasting the real perception they obtain with what is provided in the standard. However, much of the literature shows this realization at a specific time. The vast majority focused on conducting physical surveys that studied the applicability of perceptions in real time. In this way, steps are taken to improve this process of obtaining people's perceptions. Acquiring these data in real time can complement behavioral models and thus help the management of thermal comfort in indoor environments.

They are now countering the perceptions in physical surveys and new technologies that address this digitally. For the former, the applicability to many people and distribution of the devices for measuring environmental variables become difficult. In addition, repeated surveys could lead to attrition in participation. On the other hand, digital channels allow people to give feedback through a daily use channel such as a cell phone and can be applied to thermal comfort management systems [54].

A better understanding of occupants' perception of indoor environment quality helps occupants to be more productive and healthier. At the same time, analyzing occupant behavior patterns through occupancy sensors helps to improve energy savings in the built environment [55]. Thus, it was possible to identify sensors associated with user adaptability behaviors during the review, evidence that these variables are a point of research interest.

## 4. Discussion

### 4.1. Practical Considerations

To achieve a range of acceptance of thermal comfort for a group of people, as many people as possible must be satisfied. Environmental variables such as temperature or humidity are the most visible when discussing thermal comfort. In this context, the three variables found with the greatest presence in the literature were relative humidity, air temperature, and air velocity. These represent the basis for establishing relationships, as seen in Section 3.2.1, wherein the case of not obtaining a variable directly could be obtained from these three.

These variables correspond to the main ones analyzed in the different thermal comfort studies since they represent the basis of the human body's process with the physical environment to perceive a thermal environment. Through a convection process, air temperature determines how much heat the body loses to the air. On the other hand, the relative humidity present in the air regulates the conditions for evaporating sweat, which is one of the mechanisms for cooling the body; the higher the humidity, the more difficult it is for the air to absorb. As for air velocity, the loss of body heat by convection and evaporation processes is influenced by air movement, with those that cool the body the most being called air currents. By establishing ranges for each variable based on certain conditions given in each study, such as climate or season, thermal comfort can be analyzed through different research areas.

The benefits delivered by a better prediction of thermal comfort are:

- Contextual variables and those associated with occupant behavior can contribute to early design stages. For example, recognizing preferences for air-conditioned or naturally ventilated environments based on the climatic context and determining the habit of closing windows, it is possible to improve the architectural design to meet these preferences [28]. Thus, stakeholders such as real estate companies should consider these aspects to improve thermal comfort in operational stages;
- In terms of energy and health, the adaptive actions taken by users in their thermal environments affect energy consumption through the use of heating/cooling systems and generated $CO_2$ levels, respectively. In this way, the search is developed to establish a synergy between the factors of a design that meets people's expectations while complying with a sustainable approach.

Under this context, indoor air quality (IAQ), one of the factors of indoor environment quality (IEQ), impacts ventilation systems associated with controlling the $CO_2$ levels of indoor environments, bringing with it impacts on energy resources. Thus, it becomes necessary to maintain general comfort levels with decreased energy requirements [56]. Considering current pandemic scenarios (COVID-19), controlling indoor air quality takes on great relevance. Reducing viral load through ventilation systems increases energy costs, so it is necessary to consider new ways to provide adequate ventilation, for example, customized ventilation, ensuring synergy between IEQ perception, health, and energy efficiency [57].

One way to integrate these points corresponds to a management system that incorporates a thermal comfort prediction base delivering optimal ranges, complementing this with real-time perceptions that avoid overestimating the thermal comfort acceptance ranges and integrating this into the air conditioning systems to operate only at times when it is required. In this way, a large saving in electrical resources is achieved, which translates into energy efficiency.

*4.2. Limitations*

While the results and conclusions of this research are based on a systematic literature review and aligned with the conclusions of previous research, some limitations of the work need to be indicated:

- Although the impact that each of the variables obtained referring to user behavior can have on thermal comfort is analyzed, there is no emphasis on which causes the greatest impact or which are essential, so the question of which could have the greatest influence at times of gaps in the values of thermal comfort evaluations is not presented;
- Since this review is attributable to three types of infrastructures, i.e., residences, educational establishments, and offices, it is not attributable to other infrastructures such as hospitals, subways, etc. However, many evaluations are focused on these interior spaces since they represent daily scenarios where people spend most of their time. Hence, the conclusions generated from this review manage to cover a large part of potential users;
- This review was carried out for an operational stage, so other variables focused on design, such as material insulation, were not considered. However, the spectrum of variables seen helps to understand users' perceptions and can provide important considerations to be included in the design stages of a building.

## 5. Conclusions

From the literature review, the most influential variables when measuring thermal comfort were obtained. First, through characterization and analysis of the authors' keywords, it was possible to identify the research areas to which the studies were related. Among these, we found mainly indoor environment quality standards, variables focused on human behavior, and machine learning models for a more accurate thermal comfort prediction. Moreover, it was possible to classify these variables and then relate them to the capture devices. Within the four categories, the environmental variables can be measured by devices, which are standardized to obtain greater accuracy in the measurements. It was seen that the important factors to consider are height, data collection intervals, position close to the user, and proximity to other thermal devices.

Regarding a co-occurrence analysis, it was found that the most influential variables are given by the six variables proposed in Fanger's stationary model: air temperature, mean radiant temperature, airspeed, and relative humidity. Further, the physiological variables are given by metabolic rate and clothing insulation. This could be seen both in the frequency graph of the variables and in the TCV network diagram in Figure 8, which is explained by the fact that the input variables for both the standards and the algorithms of the machine learning models were the same as those for the PMV model. Moreover, a high frequency could be seen in the mean outdoor temperature variable linked to the adaptive model. Additionally, the research sector that has focused on the behavior and adaptability of the occupants has given way to new variables that also influence thermal comfort. Those reviewed in this study correspond to gender, age, acclimatization, window/door use, and clothing insulation.

The main contribution of this is giving a wide overview of the most considered variables—in the literature—that influence thermal comfort and, in this way, making pertinent considerations in their management stages. The importance of considering all the variables when evaluating thermal comfort in a building lies in improving its accuracy. While there are standards which provide guidelines and temperature values for which a group of people in the same indoor environment would be satisfied, understanding the individual behaviors and thermal perceptions of each occupant will help to close the gap of differences between the value provided by a standard and actual thermal perception. An individualized approach is necessary to achieve thermal satisfaction in an indoor environment, rather than simply meeting universal design criteria for thermal comfort.

In terms of sustainability and the economy, having the spectrum of variables that influence occupants' behavior to adapt the environment to their thermal preferences contributes to generating a positive impact on the energy savings of buildings. The above, considering the energy used during the operational stage, is greater than the designed stage. Therefore, emphasizing behavioral models will allow a better understanding of occupant comfort, which in early stages could even lead to a better design of the structure, contributing to energy efficiency. Under this same context, new models have been developed for predicting thermal comfort in centralized control systems where thermal comfort ranges are established for the public through algorithms that can optimize thermal comfort with lower energy consumption, making the use of the correct variables to avoid biases in the predictions indispensable.

Variables such as gender and age show differences in thermal perceptions. Therefore, in the context of an educational establishment, where children and adults share the same space, it is necessary to link these variables to the models used to find a neutral temperature that fully satisfies the users. On the other hand, opening doors and windows also indicates users' thermal perceptions. In the early stages, it can contribute to a better design or, in operational stages, help to employ strategies to restore structures.

As for the limitations of this study, given that the scope of the literature review was proposed only for structures such as residences, offices, and educational buildings, the conclusions obtained from the results are not attributable to industries or facilities such as hospitals or a subway. In addition, because the operative stage of a structure was covered, other variables focused on design, such as material insulation, were not considered. However, the spectrum of variables seen helps to understand users' perceptions and can provide important considerations to be included in the design stages of a building.

Finally, a future line of research is opened for a discussion that adapts the current standardization norms of thermal comfort evaluation (ISO7730, ASHARAE55) to new models based on dynamic and behavioral variables in occupants according to a preference scale. For this purpose, it would be interesting to carry out a statistical study based on those variables that generate a greater gap in the thermal comfort evaluations to generate an order of impact.

**Author Contributions:** Conceptualization, R.F.H. and T.M.; methodology, R.F.H. and T.M.; software, T.M.; validation, F.M.-L.R. and E.A.; formal analysis, T.M. writing—original draft preparation, T.M. writing—review and editing, R.F.H., E.A. and F.M.-L.R.; visualization, T.M. and F.M.-L.R.; supervision, R.F.H. and F.M.-L.R. All authors have read and agreed to the published version of the manuscript.

**Funding:** This research was funded (APC) by Pontificia Universidad Católica de Valparaíso. This research was funded by CONICYT grant number CONICYT-PCHA/InternationalDoctorate/2019-72200306 for funding the graduate research of Muñoz-La Rivera.

**Institutional Review Board Statement:** Not applicable.

**Informed Consent Statement:** Not applicable.

**Data Availability Statement:** Not applicable.

**Acknowledgments:** The authors wish to thank the TIMS space (Technology, Innovation, Management, and Innovation) of the School of Civil Engineering of the Pontificia Universidad Católica de Valparaíso (Chile), where part of the research was carried out.

**Conflicts of Interest:** The authors declare no conflict of interest.

## References

1. Lan, X.; Cao, J.; Lv, G.; Zhou, L. Simulation method for indoor airflow based on the Industry Foundation Classes model. *J. Build. Eng.* **2021**, *39*, 102251. [CrossRef]
2. Molina, C.; Veas, L. Evaluación del confort térmico en recintos de 10 edificios públicos de Chile en invierno. *Rev. Constr.* **2012**, *11*, 27–38. [CrossRef]
3. Arballo, B.; Kuchen, E.; Alamino, Y.; Frank, A.A. Evaluación de Modelos de Confort Térmico Para Interiores. 2016. Available online: https://www.researchgate.net/publication/309477141 (accessed on 13 September 2021).
4. Vischer, J.C. The effects of the physical environment on job performance: Towards a theoretical model of workspace stress. *Stress Health* **2007**, *23*, 175–184. [CrossRef]
5. Alzahrani, H.; Arif, M.; Kaushik, A.; Goulding, J.; Heesom, D. Evaluating the effects of thermal comfort on teacher performance using Artificial Neural Network. *Int. J. Build. Pathol. Adapt.* **2018**, *39*, 20–32. [CrossRef]
6. Ekici, C. A Review of Thermal Comfort and Method of Using Fanger's PMV Equation. 2013. Available online: https://www.researchgate.net/publication/289201295 (accessed on 15 September 2021).
7. Volkov, A.A.; Sedov, A.V.; Chelyshkov, P.D. Modelling the thermal comfort of internal building spaces in social buildings. *Procedia Eng.* **2014**, *91*, 362–367. [CrossRef]
8. Zhou, Y.; Su, Y.; Xu, Z.; Wang, X.; Wu, J.; Guan, X. A hybrid physics-based/data-driven model for personalized dynamic thermal comfort in ordinary office environment. *Energy Build.* **2021**, *238*, 110790. [CrossRef]
9. de Dear, R. Thermal comfort in practice. *Indoor Air* **2004**, *14*, 32–39. [CrossRef]
10. Rodríguez, C.M.; Coronado, M.C.; Medina, J.M. Thermal comfort in educational buildings: The Classroom-Comfort-Data method applied to schools in Bogotá, Colombia. *Build. Environ.* **2021**, *194*, 107682. [CrossRef]
11. Ghahramani, A.; Galicia, P.; Lehrer, D.; Varghese, Z.; Wang, Z.; Pandit, Y. Artificial Intelligence for Efficient Thermal Comfort Systems: Requirements, Current Applications and Future Directions. *Front. Built Environ.* **2020**, *6*, 49. [CrossRef]
12. Yu, J.; Cao, G.; Cui, W.; Ouyang, Q.; Zhu, Y. People who live in a cold climate: Thermal adaptation differences based on availability of heating. *Indoor Air* **2013**, *23*, 303–310. [CrossRef]
13. Wang, L.; Kim, J.; Xiong, J.; Yin, H. Optimal clothing insulation in naturally ventilated buildings. *Build. Environ.* **2019**, *154*, 200–210. [CrossRef]
14. Wang, Z.; de Dear, R.; Luo, M.; Lin, B.; He, Y.; Ghahramani, A.; Zhu, Y. Individual difference in thermal comfort: A literature review. *Build. Environ.* **2018**, *138*, 181–193. [CrossRef]
15. Dong, B.; Prakash, V.; Feng, F.; O'Neill, Z. A review of smart building sensing system for better indoor environment control. *Energy Build.* **2019**, *199*, 29–46. [CrossRef]
16. Ličina, V.F.; Cheung, T.; Zhang, H.; De Dear, R.; Parkinson, T.; Arens, E.; Chun, C.; Schiavon, S.; Luo, M.; Brager, G.; et al. Development of the ASHRAE Global Thermal Comfort Database II. *Build. Environ.* **2018**, *142*, 502–512. [CrossRef]
17. Zhang, F.; de Dear, R. Impacts of demographic, contextual and interaction effects on thermal sensation—Evidence from a global database. *Build. Environ.* **2019**, *162*, 106286. [CrossRef]
18. Yoon, Y.R.; Moon, H.J. Performance based thermal comfort control (PTCC) using deep reinforcement learning for space cooling. *Energy Build.* **2019**, *203*, 109420. [CrossRef]

19. van Eck, N.J.; Waltman, L. Software survey: VOSviewer, a computer program for bibliometric mapping. *Scientometrics* **2010**, *84*, 523–538. [CrossRef]
20. Pisello, A.L.; Castaldo, V.L.; Piselli, C.; Fabiani, C.; Cotana, F. How peers' personal attitudes affect indoor microclimate and energy need in an institutional building: Results from a continuous monitoring campaign in summer and winter conditions. *Energy Build.* **2016**, *126*, 485–497. [CrossRef]
21. Rijal, H.B.; Tuohy, P.; Humphreys, M.A.; Nicol, J.F.; Samuel, A.; Clarke, J. Using results from field surveys to predict the effect of open windows on thermal comfort and energy use in buildings. *Energy Build.* **2007**, *39*, 823–836. [CrossRef]
22. Liu, J.; Yao, R.; McCloy, R. An investigation of thermal comfort adaptation behaviour in office buildings in the UK. *Indoor Built Environ.* **2014**, *23*, 675–691. [CrossRef]
23. Park, B.; Lee, S. Investigation of the energy saving efficiency of a natural ventilation strategy in a multistory school building. *Energies* **2020**, *13*, 1746. [CrossRef]
24. Auffenberg, F.; Stein, S.; Rogers, A. A Personalised Thermal Comfort Model using a Bayesian Network. 2015. Available online: https://www.ijcai.org/Proceedings/15/Papers/361.pdf (accessed on 20 September 2021).
25. Zhou, X.; Xu, L.; Zhang, J.; Niu, B.; Luo, M.; Zhou, G.; Zhang, X. Data-driven thermal comfort model via support vector machine algorithms: Insights from ASHRAE RP-884 database. *Energy Build.* **2020**, *211*, 109795. [CrossRef]
26. Kim, J.; Zhou, Y.; Schiavon, S.; Raftery, P.; Brager, G. Personal comfort models: Predicting individuals' thermal preference using occupant heating and cooling behavior and machine learning. *Build. Environ.* **2018**, *129*, 96–106. [CrossRef]
27. Liang, H.H.; Chen, C.P.; Hwang, R.L.; Shih, W.M.; Lo, S.C.; Liao, H.Y. Satisfaction of occupants toward indoor environment quality of certified green office buildings in Taiwan. *Build. Environ.* **2014**, *72*, 232–242. [CrossRef]
28. Devitofrancesco, A.; Belussi, L.; Meroni, I.; Scamoni, F. Development of an Indoor Environmental Quality assessment tool for the rating of offices in real working conditions. *Sustainability* **2019**, *11*, 1645. [CrossRef]
29. Roskams, M.J.; Haynes, B.P. Testing the relationship between objective indoor environment quality and subjective experiences of comfort. *Build. Res. Inf.* **2021**, *49*, 387–398. [CrossRef]
30. Kim, J.; Hong, T.; Lee, M.; Jeong, K. Analyzing the real-time indoor environmental quality factors considering the influence of the building occupants' behaviors and the ventilation. *Build. Environ.* **2019**, *156*, 99–109. [CrossRef]
31. Fabi, V.; Andersen, R.V.; Corgnati, S.; Olesen, B.W. Occupants' window opening behaviour: A literature review of factors influencing occupant behaviour and models. *Build. Environ.* **2012**, *58*, 188–198. [CrossRef]
32. Indraganti, M.; Ooka, R.; Rijal, H.B. Thermal comfort in offices in summer: Findings from a field study under the 'setsuden' conditions in Tokyo, Japan. *Build. Environ.* **2013**, *61*, 114–132. [CrossRef]
33. ANSI/ASHRAE Standard 55. Thermal Environmental Conditions for Human Occupancy. 2004. Available online: https://www. ashrae.org/technical-resources/bookstore/standard-55-thermal-environmental-conditions-for-human-occupancy (accessed on 16 October 2021).
34. Hachem, W.E.L.; Khoury, J.; Harik, R. Combining several thermal indices to generate a unique heat comfort assessment methodology. *J. Ind. Eng. Manag.* **2015**, *8*, 1491–1511. [CrossRef]
35. Gamero-Salinas, J.; Kishnani, N.; Monge-Barrio, A.; López-Fidalgo, J.; Sánchez-Ostiz, A. The influence of building form variables on the environmental performance of semi-outdoor spaces. A study in mid-rise and high-rise buildings of Singapore. *Enegy Build.* **2021**, *230*, 110544. [CrossRef]
36. Cámara Chilena de Refrigeración y Climatización A. G.; División Técnica de Aire Acondicionado y Refrigeración. *Reglamento de Instalaciones Térmicas en los Edificios en Chile (RITCH)*; C.Ch.R.: Santiago, Chile, 2007. Available online: http://cchryc.cl/ reglamento-de-instalaciones-termicas-en-los-edificios-en-chile-ritch/ (accessed on 30 December 2021).
37. Natephra, W.; Motamedi, A.; Yabuki, N.; Fukuda, T. Enriching Building Information Modeling (BIM) with Sensor Data and Thermal Images for Thermal Comfort Analysis. 2017. Available online: https://www.researchgate.net/publication/319122536 (accessed on 20 October 2021).
38. Castilla, M.; Álvarez, J.D.; Ortega, M.G.; Arahal, M.R. Neural network and polynomial approximated thermal comfort models for HVAC systems. *Build. Environ.* **2013**, *59*, 107–115. [CrossRef]
39. Cordero, R.; Caballero, M.; Quiroz, F.; Damiani, A.; Jorquera, J.; Sepúlveda, E.; Rayas, J.; Feron, S. *Radiación Solar en Chile*; CONICYT: Santiago, Chile, 2016.
40. Vellei, M.; Herrera, M.; Fosas, D.; Natarajan, S. The influence of relative humidity on adaptive thermal comfort. *Build. Environ.* **2017**, *124*, 171–185. [CrossRef]
41. Sikram, T.; Ichinose, M.; Sasaki, R. Assessment of Thermal Comfort and Building-Related Symptoms in Air-Conditioned Offices in Tropical Regions: A Case Study in Singapore and Thailand. *Front. Built Environ.* **2020**, *6*, 187. [CrossRef]
42. Zhang, Y.; Chen, H.; Meng, Q. Thermal comfort in buildings with split air-conditioners in hot-humid area of China. *Build. Environ.* **2013**, *64*, 213–224. [CrossRef]
43. Dai, J.; Jiang, S. Passive space design, building environment and thermal comfort: A university building under severe cold climate, China. *Indoor Built Environ.* **2020**, *30*, 1323–1343. [CrossRef]
44. Kim, A.; Wang, S.; Kim, J.E.; Reed, D. Indoor/outdoor environmental parameters and window-opening behavior: A structural equation modeling analysis. *Buildings* **2019**, *9*, 94. [CrossRef]
45. Naspi, F.; Arnesano, M.; Zampetti, L.; Stazi, F.; Revel, G.M.; D'Orazio, M. Experimental study on occupants' interaction with windows and lights in Mediterranean offices during the non-heating season. *Build. Environ.* **2018**, *127*, 221–238. [CrossRef]

46. Teli, D.; Jentsch, M.F.; James, P.A.B. Naturally ventilated classrooms: An assessment of existing comfort models for predicting the thermal sensation and preference of primary school children. *Energy Build.* **2012**, *53*, 166–182. [CrossRef]
47. Park, J.; Choi, C.S. Modeling occupant behavior of the manual control of windows in residential buildings. *Indoor Air* **2019**, *29*, 242–251. [CrossRef]
48. Jeong, B.; Jeong, J.W.; Park, J.S. Occupant behavior regarding the manual control of windows in residential buildings. *Energy Build.* **2016**, *127*, 206–216. [CrossRef]
49. Faridah, F.; Waruwu, M.M.; Wijayanto, T.; Budiarto, R.; Pratama, R.C.; Prayogi, S.E.; Nadiya, N.M.; Yanti, R.J. Feasibility study to detect occupant thermal sensation using a low-cost thermal camera for indoor environments in Indonesia. *Build. Serv. Eng. Res. Technol.* **2021**, *42*, 389–404. [CrossRef]
50. Kosmopoulos, P.; Galanos, D.; Anastaselos, D.; Papadopoulos, A.M. An assessment of the overall comfort sensation in workplaces. *Int. J. Vent.* **2012**, *10*, 311–322. [CrossRef]
51. Kuru, M.; Calis, G. Data acquisition technologies for assessing thermal comfort in the built environment. *Pol. J. Environ. Stud.* **2021**, *30*, 1017–1027. [CrossRef]
52. Johansson, E.; Thorsson, S.; Emmanuel, R.; Krüger, E. Instruments and methods in outdoor thermal comfort studies-The need for standardization. *Urban Clim.* **2014**, *10*, 346–366. [CrossRef]
53. *International Standard ISO-7726*; Ergonomics of the Thermal Environment-Instruments for Measuring Physical Quantities ISO-7726. 1998. Available online: https://www.sis.se/std-615884(accessed on 3 November 2021).
54. Konis, K.; Blessenohl, S.; Kedia, N.; Rahane, V. TrojanSense, a participatory sensing framework for occupant-aware management of thermal comfort in campus buildings. *Build. Environ.* **2020**, *169*, 106588. [CrossRef]
55. Kong, M.; Dong, B.; Zhang, R.; O'Neill, Z. HVAC energy savings, thermal comfort and air quality for occupant-centric control through a side-by-side experimental study. *Applied Energy* **2022**, *306*. [CrossRef]
56. Alfano, F.R.D.; Olesen, B.W.; Palella, B.I.; Riccio, G. Thermal comfort: Design and assessment for energy saving. *Energy Build.* **2014**, *81*, 326–336. [CrossRef]
57. Anand, P.; Cheong, D.; Sekhar, C. A review of occupancy-based building energy and IEQ controls and its future post-COVID. *Sci. Total Environ.* **2022**, *804*, 150249. [CrossRef]